Lecture Notes in Computer Science 7337

Commenced Publication in 1973
Founding and Former Series Editors:
Gerhard Goos, Juris Hartmanis, and Jan van Leeuwen

Gosse Bouma   Ashwin Ittoo
Elisabeth Métais   Hans Wortmann (Eds.)

Natural Language Processing and Information Systems

17th International Conference on Applications
of Natural Language to Information Systems, NLDB 2012
Groningen, The Netherlands, June 26-28, 2012
Proceedings

Springer

Volume Editors

Gosse Bouma
University of Groningen
Information Science Department
Oude Kijk in 't Jatstraat 26, 9712 EK Groningen, The Netherlands
E-mail: g.bouma@rug.nl

Ashwin Ittoo
Hans Wortmann
University of Groningen
Faculty of Economics and Business
Nettelbosje 2, 9747 AE Groningen, The Netherlands
E-mail: {r.a.ittoo, j.c.wortmann}@rug.nl

Elisabeth Métais
CNAM-Laboratoire Cédric
292 rue St. Martin, 75141 Paris Cedex 03, France
E-mail: elisabeth.metais@cnam.fr

ISSN 0302-9743   e-ISSN 1611-3349
ISBN 978-3-642-31177-2   e-ISBN 978-3-642-31178-9
DOI 10.1007/978-3-642-31178-9
Springer Heidelberg Dordrecht London New York

Library of Congress Control Number: 20129396643

CR Subject Classification (1998): I.2.7, H.3, H.2.8, I.5, J.5, I.2.6, J.1

LNCS Sublibrary: SL 3 – Information Systems and Application, incl. Internet/Web
and HCI

Typesetting: Camera-ready by author, data conversion by Scientific Publishing Services, Chennai, India

Printed on acid-free paper

Springer is part of Springer Science+Business Media (www.springer.com)

Preface

These are the proceedings of the 17th International Conference on Applications of Natural Language to Information Systems, also known as NLDB 2012, that was organized in Groningen, The Netherlands, during June 26–28. Since the first NLDB conference in 1995, the main focus of the conference has widened from using natural language processing techniques in the area of databases and information systems to more general applications of NLP that help to make large and complex collections of data and information accessible and manageable.

The rapidly evolving state of the art in NLP and the shifting interest to applications targeting document and data collections available on the Web, including an increasing amount of user-generated content, is reflected in the contributions to this conference. Topics covered are, among others, information retrieval and text classification and clustering, summarization, normalization of user generated content, 'forensic' NLP (addressing plagiarism, cyberbullying, and fake reviews), ontologies and natural language, sentiment analysis, question answering and information extraction, terminology and named entity recognition, and NLP tools development.

For this edition of NLDB, we received over 90 submissions. The Program Committee, consisting of renowned researchers in the area of natural language processing and information systems, did an excellent job in providing detailed and well-motivated reviews. In all, 12 papers were selected as full papers for the conference and a number of other contributions as short papers (24) or posters (16).

We would like to thank all reviewers for their time and effort, and our invited speakers, Philipp Cimiano and Bernhard Thalheim, for accepting our invitation and making the conference attractive by their contributions. Finally, we would like to thank everyone involved in the local organization, especially the Department of Business and ICT and the Department of Information Science of the University of Groningen, the *Groningen Congres Bureau*, for helping us with all practical issues, *Het Kasteel*, for hosting the conference, and *Gezamenlijk Gastheerschap Groningen*, for sponsoring the welcome reception.

June 2012

Gosse Bouma
Ashwin Ittoo
Elisabeth Métais
Hans Wortmann

Organization

Conference Chairs

Hans Wortmann   University of Groningen, The Netherlands
Elisabeth Métais   CEDRIC/CNAM, France

Program Chair

Gosse Bouma   University of Groningen, The Netherlands

Local Organization

Valerio Basile   University of Groningen, The Netherlands
Gosse Bouma   University of Groningen, The Netherlands
Ashwin Ittoo   University of Groningen, The Netherlands
Laura Maruster   University of Groningen, The Netherlands
Hans Wortmann   University of Groningen, The Netherlands

Program Committee

Jacky Akoka   CNAM, France
Frederic Andres   National Institute of Informatics, Japan
Akhilesh Bajaj   University of Tulsa, USA
Tim Baldwin   University of Melbourne, Australia
Herman Balsters   University of Groningen, The Netherlands
Johan Bos   University of Groningen, The Netherlands
Antal van den Bosch   Radboud University, The Netherlands
Bert de Brock   University of Groningen, The Netherlands
Paul Buitelaar   DERI, Ireland
Samira Si-Said Cherfi   CNAM, France
Philipp Cimiano   Universität Bielefeld, Germany
Roger Chiang   University of Cincinnati, USA
Isabelle Comyn-Wattiau   ESSEC, France
Alfredo Cuzzocrea   ICAR-CNR and University of Calabria, Italy
Walter Daelemans   University of Antwerp, Belgium
Stefan Evert   University of Osnabrück, Germany
Dan Flickinger   Stanford University, USA
Alexander Gelbukh   Mexican Academy of Science, Mexico
Jon Atle Gulla   NTNU, Norway
Karin Harbusch   Koblenz University, Germany
Arjan van Hessen   University of Twente, The Netherlands
Dirk Heylen   University of Twente, The Netherlands

Erhard Hinrichs	Tübingen University, Germanny
Helmut Horacek	Saarland University, Germany
Paul Johannesson	Stockholm University, Sweden
Epaminondas Kapetanios	University of Westminster, UK
Sophia Katrenko	Utrecht University, The Netherlands
Zoubida Kedad	Université de Versailles, France
Christian Kop	University of Klagenfurt, Austria
Valia Kordoni	Saarland University, Germany
Leila Kosseim	Concordia University, Canada
Georgia Koutrika	Stanford University, USA
Zornitsa Kozareva	University of Southern California, USA
Nadira Lammari	CNAM, France
Dominque Laurent	Synapse, France
Jochen Leidner	Thomson Reuters, USA
Piroska Lendvai	Hungarian Academy of Sciences, Hungary
Johannes Leveling	Dublin City University, Ireland
Deryle Lonsdale	Brigham Young Uinversity, USA
Rob Malouf	San Diego State University, USA
Heinrich C. Mayr	University of Klagenfurt, Austria
Farid Meziane	Salford University, UK
Luisa Mich	University of Trento, Italy
Andres Montoyo	Universidad de Alicante, Spain
Rafael Muñoz	Universidad de Alicante, Spain
John Nerbonne	University of Groningen, The Netherlands
Guenter Neumann	DFKI, Germany
Gertjan van Noord	University of Groningen, The Netherlands
Jan Odijk	Utrecht University, The Netherlands
Stephan Oepen	Oslo University, Norway
Manuel Palomar Sanz	Universidad de Alicante, Spain
Pit Pichappan	Al Imam University, Saudi Arabia
Lonneke van der Plas	Université de Genève, Switzerland
Gabor Proszeky	Morphologic, Hungary
Mike Rosner	University of Malta, Malta
Fabio Rinaldi	University of Zurich, Switzerland
German Rigau	University of the Basque Country, Spain
Patrick Saint-Dizier	Université Paul Sabatier, France
Max Silberztein	Université de Franche-Comté, France
Ielka van der Sluis	University of Groningen, The Netherlands
Veda Storey	Georgia State University, USA
Vijayan Sugumaran	Oakland University Rochester, USA
Bernhard Thalheim	Kiel University, Germany
Michael Thelwall	University of Wolverhampton, UK
Krishnaprasad Thirunarayan	Wright State University, USA
Jörg Tiedemann	Uppsala University, Sweden

Table of Contents

Short Papers

Posters

Syntax, Semantics and Pragmatics of Conceptual Modelling

Bernhard Thalheim

Computer Science Institute,
Christian-Albrechts-University Kiel
thalheim@is.informatik.uni-kiel.de

Abstract. Models, modelling languages, modelling frameworks and their background have dominated conceptual modelling research and information systems engineering for last four decades. Conceptual models are mediators between the application world and the implementation or system world. Currently conceptual modelling is rather a craft and at the best an art. We target on a science and culture of conceptual modelling. Models are governed by their purpose. They are used by a community of practice and have a function within application cases. Language-based models use a language as a carrier. Therefore, semiotics of models must be systematically developed. This paper thus concentrates on the linguistic foundation of conceptual modelling.

1 Introductory Notions for Conceptual Modelling

Conceptual modelling is a widely applied practice in Computer Science and has led to a large body of knowledge on constructs that might be used for modelling and on methods that might be useful for modelling. It is commonly accepted that database application development is based on conceptual modelling. It is however surprising that only very few publications have been published on a *theory of conceptual modelling*. We continue our research [2, 3, 15–18] and aim in a theory of linguistic foundations for modelling within this paper.

1.1 Goals, Purposes and Deployment Functions of Conceptual Models

Purpose is often defined via intention and mixed with function. *Goal* (or intention or target or aim) is a ternary relation between a current state, envisioned states, and people (community of practice). Typical - sometimes rather abstract - intentions are perception support, explanation and demonstration, preparation to an activity, optimisation, hypothesis verification, construction, control, and substitution.

The *purpose* is a binary relation between intentions and means or instruments for realisation of the intention. The main mean we use is the language. Semiotics is widely intentionally used for modelling; however without paying attention to it.

G. Bouma et al. (Eds.): NLDB 2012, LNCS 7337, pp. 1–10, 2012.

The *deployment function* of a model relates the model purpose to a practice or application cases or application 'game' similar to Wittgenstein's language game (we call it better *deployment case* and is characterised by answering the classical W-questions: how, when, for which/what or why, at what/which (business use case), etc. We add to purpose: application, conventions, custom, exertion, habit, handling, deployment, service, usage, use, and way of using. The model has a role and plays its behaviour within this application game.

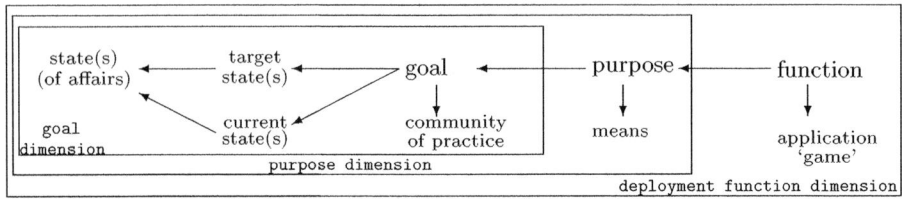

Fig. 1. Distinction of Intention, Purpose,and Deployment Function of a Model

1.2   Abstraction and Conceptual Modelling

Abstraction is one of the most overloaded conceptions in Computer Science and at the same time one of the most under-specified. Abstraction means development of general concepts by abstracting common properties of specific concepts. Using the approach in [14] we develop three dimensions of abstraction:

Structural abstraction is used for highlighting essential, necessary, and general structural elements of the origin. Structural abstraction has three main constituents: *combining structural abstraction* (often called) *classifying structural abstraction* (often called meta(-meta(-meta)) abstraction) combines things of interest into collections that contain these things as elements; *generalising structural abstraction* (often called pattern or templates).

Context abstraction "factors out" repeating, shared or local patterns of things and functionality from individual things. Context abstraction assumes that the surroundings of a things under consideration are commonly assumed by a community of practice or within a culture and focuses on the concept, turning away attention from its surroundings such as the environment and setting. Models use ambiguities, ellipses, metaphors and commonly assumed conceptions.

Behavioural abstraction is used for concentrating of essential and general behavioural elements. We may distinguish between *combining, classifying,* and *generalising behavioural abstraction.* Aspect separation, encapsulation, and modularisation are specific techniques.

The opposite of abstracting is detailing. *Refinement* is a specific kind of faithful detailing. Refinement uses the principle of modularisation and information hiding. Developers typically use conceptual models or languages for representing and conceptualising abstractions. Classical forms of abstraction are generalisation, isolation, and idealisation.

We thus may concentrate on three main tasks for abstraction within a community of practice:

- *Choose the right scope* within the application area in dependence on goals.
- *Choose the right focus* at the right level and in the right granularity .
- *Choose the right observation* with the right behaviour and right properties.

2   The Notion of the Model

It is common misbelief (e.g., [1] or more generally almost all Computer Science textbooks) that there is no definition of the concept of the model. We consider this claim as the *big misunderstanding* of the science and art of modelling.

A model is simply a material or virtual *artifact* (1) which is called model within a community of practice (2) based on a judgement (3) [7] of appropriateness for representation of other artifacts (things in reality, systems, ...) and serving a *purpose* (4) within this community. We distill thus criteria for artifacts to become a model. We can use on two approaches: abstract properties or we criteria for artifacts.

2.1   Stachowiak, Aristoteles, Galilei and Mahr Properties of Models

Models are often defined through abstract properties they must satisfy [15, 18].

(1) *Mapping* property: Each model has an origin and is based on a mapping from the origin to the artifact.

(2) *Truncation* property: The model lacks some of the ascriptions made to the original and thus functions as an Aristolean model by abstraction of irrelevant.

(3) *Pragmatic* property: The model use is only justified for particular model users, tools of investigation, and period of time.

(4) *Amplification* property: Models use specific extensions which are not observed for the original,

(5) *Distortion* property: Models are developed for improving the physical world or for inclusion of visions of better reality, e.g. for construction via transformation or in Galilean models.

(6) *Idealisation* property: Modelling abstracts from reality by scoping the model to the ideal state of affairs.

(7) *Carrier* property: Models use languages and are thus restricted by the expressive power of these languages.

(8) *Added value* property: Models provide a value or benefit based on their utility, capability and quality characteristics.

(9) *Purpose* property: Models and conceptual models are governed by the purpose. The model preserves the purpose.

The first three properties have been introduced by Stachowiaks [12]. The fourth and fifth property have been introduced by Steinmüller [13]. The seventh property is discussed by Mahr [10]. The sixth, eight and nine properties [18] are however equally if not the most important ones.

2.2 Criteria for Appropriateness of an Artifact to Become a Model

The separation into goal, purpose and deployment function for models provides three main appropriateness criteria:

(1) The *adequacy* of a model defines its *potential* for the goals. Adequacy is given by the similarity of the model with its origin in dependence on its goal, the regularity for the application (within a well-founded system that uses rules for derivation of conclusions), the fruitfulness (or capacity) for goals, and the simplicity of the model through the reduction to the essential and relevant properties in dependence on the goal.

(2) A model is *fit* for its purpose if it is *usable* for the purpose, *suitable* within the given context and for the prescribed purposes, *robust* against small changes in the parameters, *accurate* to the level of precision that is necessary for the purpose, and *compliant* with the funding concepts, application context and meta-model. The model must be *testable* and, if false, it can be disconfirmed by a finite set of observations (finitely testable) and by any of superset of these observation (irrevocably testable).

(3) The *usefulness* for deploying is given by *effectiveness* for complete and accurate satisfaction of the goal, *understandability* for purposeful deployment of the model by users, *learnability* of the model within the deployment stories, *reliability* and a high *degree of precision* of the the model, and *efficiency* of the model for the function of the model within the application. *testability*

Additional criteria are *generality* of the model beside its direct goals and intentions and the extend of *coverage* in the real world for other goals.

2.3 The Hidden Background of Models

The structure and function of a model are based on a correspondence relation between the reality or the augmented and the model. The model can be constructed incrementally. It represents a number of facets of the origin (topology/geometry, state, interaction, causal). The *model pragmatism* is however hidden. It consists of at least three background dimensions:

Founding concepts: A model uses the cultural background within the application area and within a community of practice. Base conceptions (scope, expressions, concept space organisation, quantification/measurement), a namespace/ontology/carrier, a number of definitions (state, intrinsic, object, interaction descriptors and depictors), and a language as cargo [10] characterise these founding concepts.

Application context: The application domain binds the model to some common understanding that is not explicitly defined in the model. Each model has an empirical scope of the model, has specific application-domain driven correspondences, and must satisfy a number of laws and regulations.

Meta-model: Each model has a basement, is restricted by paradigms and theories, has a status in the application; context, displays elements on certain abstraction level and granularity, and uses a scale. It is also prone for paradigmatic evolution within the epistemological profile of community of practice.

These dimensions are partially known in didactics (of modelling) [5]. The three background dimensions drive however the model deployment and development.

3   Language-Backed Modelling

3.1   Language Selection Matters

Languages may however also restrict modelling. This restriction may either be compensated by over-development of language components or by multi-models. The relational database modelling language uses integrity constraint as compensation component for the inadequate expressibility of the language. The Sapir-Whorf hypothesis [19] results in the following principle:

Principle of linguistic relativity: Actors skilled in a language may not have a (deep) understanding of some concepts of other languages. This restriction leads to problematic or inadequate models or limits the representation of things and is not well understood.

The principle of linguistic relativity is not well understood. In [15] we demonstrated via a crossroad example that Petri nets are often not the right tool for representation of behaviour. A similar observation on UML is made by Krogstie [9].

3.2   The Cognitive Insufficiency of the Entity-Relationship Modelling Language

Lakoff introduces six basic schemata of cognitive semantics without stating that this list of schemata is complete.

- The container schema define the distinction between in and out. They have an interior, a boundary and an exterior.
- The part-whole schema define an internal structuring and uses whole, part and configuration as construction units.
- The link schema connects thing of interest. It uses various kinds of links for associating or un-associating things.
- The center-periphery schema is based on some notion of a center. Peripherical elements are not as important than those in the center.
- The source-path-goal schema uses source (or starting point), destination, path, and direction. It allows also to discuss main and side tracks.
- Typical ordering schemata are the up-down, front-back and the linear ordering schema. They use spatial and temporal associations.

We call a modelling language *cognition-complete* if these six schemata can be represented.

The classical ER modelling language suffers from a number of restrictions. It uses the container and the link schemata. It allows to mimic the part-whole schema via special links (called IsA). This work-around is however badly misunderstood. In order to become cognition-complete integrity constraints must be used. Their cognitive complexity is however beyond surveyability of humans. A typical flaw of the classical ER model is the use of monster types that integrate stabile - almost not changing - properties and transient - often changing - properties. Objects are then taken as a whole. Unary relationship types easily resolve this problem if higher-order types are permitted.

Extended ER modelling languages are however also not cognitive complete. The center-periphery schema can only be emulated. The source-path-goal schema can be represented by higher-order relationship types. The part-whole schema is supported by the specialisation via unary relationship types and by generalisation via cluster types. Ordering schemata can be defined using the order types and bulk types.

4 Syntax of Conceptual Models: Structuring and 'Functioning'

Syntax of models is build on deictic context-based rules [20] for construction of complex expressions using a domain-dependent vocabulary and governed by a set of meta-rules for construction (styles, pattern, abstraction).

4.1 Morphology of Conceptual Models and the Form of Elements

Morphology is the science of word form structure. The part of syntax is completely neglected in conceptual modelling. It is however equally important. Elements of a modelling language can be classified according to their categories and roles within a model and according to their specific expression within a model. Expression might similarly ruled by inflection, deviation, and composition. Therefore, techniques like lemmatisation (reduction of words to their base form) and characterisation by the (morpho-syntactic) role within a model.

Based on [11] we distinguish five morphological features: full or partial specification, layering within a model, integrity constraints, cyclic or acyclic structuring, complete set of schemata for cognitive semantics, open or closed context, and kind of data types.

Syntax for models is context-dependent. Constructs are bound by an implicit construction semantics [14] that is an integral component of any language. Models are governed by syntactic rules or explicit and implicit social norms. They are constructed with implicit styles and architectures.

4.2 The Lexicography and the Namespace of Models

Lexicography has developed a number of principles for coding and structuring lexical elements based on the lexicon on the application domain. Most ontological research in Computer Science does not got beyond lexicography and uses a topical annotation of model elements while hoping that every stakeholder has the same interpretation for words such as 'name', 'description', 'identifier' etc. If we consider however more complex entries such as 'address' then we detect that such kind of annotation does not work even for one language. The situation becomes far worse if we consider different languages, cultures, or application domains. Then the nightmare of "integration" becomes a challenge.

Models typically use a general and an application-dependent namespace. Moreover, the model is a product of a community of practice with its needs, its common-speak, its specific functions of words, its specific phrases and abbreviations, and its specific vocabulary.

5   Semantics of Conceptual Models

5.1   Kinds of Semantics

Semantics is the study of meaning, i.e. how meaning is constructed, interpreted, clarified, obscured, illustrated, simplified, negotiated, contradicted and paraphrased. It has been treated differently in the scientific community, e.g., in the area of knowledge bases and by database users.

- The scientific community prefers the treatment of 'always valid' semantics based on the mathematical logic. A constraint is valid if this is the case in any correct database.
- Database modellers often use a 'strong' semantics for several classes of constraints. Cardinality constraints are based on the requirement that databases exist for both cases, for the minimal and for the maximal case.
- Database mining is based on a 'may be valid' semantics. A constraint is considered to be a candidate for a valid formula.
- Users usually use a weak 'in most cases valid' semantics. They consider a constraint to be valid if this is the usual case.
- Different groups of users use an 'epistemic' semantics. For each of the group its set of constraints is valid in their data. Different sets of constraints can even contradict.

Semantics is currently one of the most overused notions in modern computer science literature. Its understanding spans from synonyms for structuring or synonyms for structuring on the basis of words to precise defined semantics. This partial misuse results in a mismatch of languages, in neglecting formal foundations, and in brute-force definitions of the meaning of syntactic constructions.

Semantics of models uses also commonsense, intended and acceptable meanings, various kinds of quantifications, (logical) entailment, deduction, induction, and abduction.

5.2   The Lexicology and the Namespace of Models

Ontologies are becoming very popular in Computer Science research. Philosophy developed now a rather restrained usage of ontologies. Lexicology [4] - as a part of philology - or semasiology is based on semantic relations in the vocabulary of a language. Lexicology of models studies elements of models and their meaning, relations between these elements, sub-models and the namespace in the application domain. Classical linguistic relations such as homonym, antonym, paronym, synonym, polysemy, hyponym, etc. are used for stereotyped semantics in the namespace.

Models combine two different kinds of meaning in the namespace: referential meaning establishes an interdependence between elements and the origin ('what'); functional meaning is based on the function of an element in the model ('how'). The referential meaning is well investigated and uses the triangle between element, concept and referent. The functional meaning relates elements in a model to the model context, to the application context, and to the function of this element within the model. It thus complements the referential meaning. Additionally model lexicology use the intext (within the model), the general, the part-of-model, and the differential (homonym-separating) meaning. Further, we need to handle the change of meaning for legacy models.

6   Pragmatics of Conceptual Models

6.1   General Pragmatics of Modelling in a Community of Practice

Pragmatics for modelling is the study how languages are used for intended deployment functions in dependence on the purposes and goals within a community of practice. Functions, purposes and goals are ruling the structure and function of the model. We distinguish the descriptive-explanatory and persuasive-normative functions of a model. Models are used for (1) acting (2) within a community, especially the modeller and have (3) different truth or more generally quality [8]. We may distinguish between *far-side* and *near-side* pragmatics separating the 'why' from the 'what' side of a model. General pragmatics allows to describe the overall intentions within the community and strategies for intention discovery.

We may distinguish between the development and deployment modes. The first mode starts with abstraction and mapping and then turns to the representation. The second mode inverses the first mode. Based on this distinction we infer a number of basic principles of modelling pragmatics similar to Hausser [6]: model surface compositionality (methodological principle), model presentation order's strict linearity relative to space (empirical principle), model interpretation and production analysed as cognitive processes (ontological principle), reference modelled in terms of matching an model's meaning with context (functional principle).

Models must be methodologically valid, support subjective deductive (paradeductive) inference with an open world interpretation, allow context-dependent reasoning (implicature), provide means for collaborative interaction, weaken connectives and quantifications, and integrate deductive, inductive, abductive and paradeductive reasoning.

6.2   Visualisation or the 'Phonetics' of Conceptual Models

Phonology is the science of language sounds. *Phonetics* investigates the articulatory, acoustic, and auditive process of speech. It is not traditionally considered not considered to be a part of a grammar. But it is equally important for practical language deployment. Phonology of models is concerned with the ways in which intentions can be conveyed using conventional and non-conventional resources. The modeller uses reference for transferring the specific point of view that has been used for modelling.

Visualisation is the 'phonetics' in modelling. It is based on three principles:

Principles of visual communication are based on three constituents: *Vision, cognition,* and *processing and memorizing characteristics.* We may use specific visual features such as contrast, visual analogies, presentation dramaturgy, reading direction, visual closeness, symmetric presentation and space and movement.

Principles of visual cognition refer to *ordering, effect* delivery, and *visualisation.* We base those on *model organisation, model economy, skills of users,* and *standards.*

Principles of visual design are based on *optical vicinity , similarity , closeness, symmetry, conciseness, reading direction.*

These principles help to organise the model in a way that correspond to human perception.

7   Summary

Models are artifacts that can be specified within a $(W^4+W^{17}H)$-frame that is based on the classical rhetorical frame introduced by Hermagoras of Temnos[1]. Models are primarily characterised by W^4: wherefore (purpose), whereof (origin), wherewith (carrier, e.g., language), and worthiness ((surplus) value). Secondary characterisation $W^{17}H$ is given by:

- user or stakeholder characteristics: by whom, to whom, whichever;
- characteristics imposed by the application domain: wherein, where, for what, wherefrom, whence, what;
- purpose characteristics characterising the solution: how, why, whereto, when, for which reason; and
- additional context characteristics: whereat, whereabout, whither, when.

Modelling is the art, the systematics and the technology of model (re)development and model application. It uses model activities and techniques. This paper is going to be extended by more specific aspects of the modelling art in the context of semiotics and linguistics.

---

[1] The rhetor Hermagoras of Temnos, as quoted in pseudo-Augustine's De Rhetorica defined seven "circumstances" as the loci of an issue: Quis, quid, quando, ubi, cur, quem ad modum, quibus adminiculis(W^7: Who, what, when, where, why, in what way, by what means). See also Cicero, Thomas Aquinas, and Qunitillian's loci argumentorum as a frame without quesrioning. The Zachman frame uses an over-simplification of this frame.

References

1. Agassi, J.: Why there is no theory of models. In: Niiniluoto, I., Herfel, W.E., Krajewsky, W., Wojcicki, R. (eds.) Theories and Models in Scientific Processes, Amsterdam, Atlanta, pp. 17–26 (1995)
2. Dahanayake, A., Thalheim, B.: Towards a Framework for Emergent Modeling. In: Trujillo, J., Dobbie, G., Kangassalo, H., Hartmann, S., Kirchberg, M., Rossi, M., Reinhartz-Berger, I., Zimányi, E., Frasincar, F. (eds.) ER 2010. LNCS, vol. 6413, pp. 128–137. Springer, Heidelberg (2010)
3. Dahanayake, A., Thalheim, B.: Enriching Conceptual Modelling Practices through Design Science. In: Halpin, T., Nurcan, S., Krogstie, J., Soffer, P., Proper, E., Schmidt, R., Bider, I. (eds.) BPMDS 2011 and EMMSAD 2011. LNBIP, vol. 81, pp. 497–510. Springer, Heidelberg (2011)
4. Ginsburg, R.S., Khidekei, S.S., Knyazeva, G.Y., Sankin, A.A.: A course in modern English lexicology, 2nd edn. Vysschaja Schkola, Moscow (1979) (in Russian)
5. Halloun, I.A.: Modeling Theory in Science Education. Springer, Berlin (2006)
6. Hausser, R.: Foundations of computational linguistics - human-computer communication in natural language, 2nd edn. Springer (2001)
7. Kaschek, R.: Konzeptionelle Modellierung. PhD thesis, University Klagenfurt, Habilitationsschrift (2003)
8. Korta, K., Perry, J.: Critical Pragmatics. Cambridge University Press, Cambridge (2011)
9. Krogstie, J.: Quality of uml. In: Encyclopedia of Information Science and Technology (IV), pp. 2387–2391 (2005)
10. Mahr, B.: Information science and the logic of models. Software and System Modeling 8(3), 365–383 (2009)
11. Ritchey, T.: Outline for a morphology of modelling methods - Contribution to a general theory of modelling
12. Stachowiak, H.: Modell. In: Seiffert, H., Radnitzky, G. (eds.) Handlexikon zur Wissenschaftstheorie, pp. 219–222. Deutscher Taschenbuch Verlag GmbH & Co. KG, München (1992)
13. Steinmüller, W.: Informationstechnologie und Gesellschaft: Einführung in die Angewandte Informatik. Wissenschaftliche Buchgesellschaft, Darmstadt (1993)
14. Thalheim, B.: Entity-relationship modeling – Foundations of database technology. Springer, Berlin (2000)
15. Thalheim, B.: Towards a theory of conceptual modelling. Journal of Universal Computer Science 16(20), 3102–3137 (2010), http://www.jucs.org/jucs_16_20/towards_a_theory_of
16. Thalheim, B.: The art of conceptual modelling. In: Proc. EJC 2011, Tallinn, pp. 203–222 (2011)
17. Thalheim, B.: The Science of Conceptual Modelling. In: Hameurlain, A., Liddle, S.W., Schewe, K.-D., Zhou, X. (eds.) DEXA 2011, Part I. LNCS, vol. 6860, pp. 12–26. Springer, Heidelberg (2011)
18. Thalheim, B.: The theory of conceptual models, the theory of conceptual modelling and foundations of conceptual modelling. In: The Handbook of Conceptual Modeling: Its Usage and Its Challenges, ch.12, pp. 543–578. Springer, Berlin (2011)
19. Whorf, B.L.: Lost generation theories of mind, language, and religion. Popular Culture Association, University Microfilms International, Ann Arbor, Mich. (1980)
20. Wojcicki, R.: Wyklady z logiki z elementami teorii wiedzy. Wyd. Naukowe SCHOLAR (2003) (in polish)

Multi-dimensional Analysis
of Political Documents

Heiner Stuckenschmidt and Cäcilia Zirn

KR & KM Research Group,
University of Mannheim, Germany
{heiner,caecilia}@informatik.uni-mannheim.de

Abstract. Automatic content analysis is more and more becoming
an accepted research method in social science. In political science researchers are using party manifestos and transcripts of political speeches
to analyze the positions of different actors. Existing approaches are limited to a single dimension, in particular, they cannot distinguish between
the positions with respect to a specific topic. In this paper, we propose
a method for analyzing and comparing documents according to a set of
predefined topics that is based on an extension of Latent Dirichlet Allocation for inducing knowledge about relevant topics. We validate the
method by showing that it can reliably guess which member of a coalition was assigned a certain ministry based on a comparison of the parties'
election manifestos with the coalition contract.

Keywords: Topic Models, Political Science.

1 Motivation

Data Analysis has a longstanding tradition in social science as a main driver
of empirical research. Traditionally, research has focused on survey data as a
main foundation. Recently, automatic text analysis has been discovered as a
promising alternative to traditional survey based analysis, especially in the political sciences [6], where policy positions that have been identified automatically
based on text can for example be used as input for simulations of party competition behavior [7]. The approach to text analysis adopted by researchers in this
area is still strongly influenced by statistical methods used to interpret survey
data [2]. A typical application is to place parties on a left-right scale based on
the content of their party manifestos [11]. While it has been shown that existing
methods can be very useful for analyzing and comparing party positions over
time, existing methods are limited to a single dimension, typically the left-right
scale. This means that positions of a party on various topics are reduced to a single number indicating an overall party position independent of a specific policy
area. In this paper, we argue that there is a need for new analysis methods that
are able to discriminate between positions on different policy areas and treat
them independently. We propose a new approach on multidimensional analysis
of party positions with respect to different policy areas. Often, we are interested

G. Bouma et al. (Eds.): NLDB 2012, LNCS 7337, pp. 11–22, 2012.

in the position of a party *with respect to a certain topic* rather than an overall position. Existing methods are only able to answer questions of that kind if the input are texts talking exclusively about the topic under consideration (e.g. [9]). In contrast, there is a good reason why party manifestos have been the primary subject of attempts to identify party positions [13], as they are independent of personal opinions and opportunistic statements that influence for instance political speeches. This means that on the one hand manifestos are an important reference point for various comparisons and party position analyses, but on the other hand are hard to analyze with existing approaches as they cover a large variety of topics and the respective party's position towards this topic. We conclude that there is a need for methods that allow for position analysis based on multi-topic documents that takes these different topics into account.

In this paper, we address the problems of current one-dimensional analyses of political positions by proposing a content analysis method based on topic models that identifies topics put forward by parties in connection with a certain policy area. We use a variant of topic models that allows the inclusion of seed words for characterizing the respective policy areas. This approach has a number of advantages over conventional topic models where topics are solely formed based on the analysis of a corpus:

- It allows to define certain policy areas that the topics in the model are supposed to represent.
- This in turn makes it possible to compare party interests in a certain policy area defined by a set of seed words.
- The use of seed words provides the flexibility to adapt the areas to be analyzed to the given question, e.g. policy areas that are of interest in a regional election will not necessarily be of interest in the context of a federal election and vice versa.

The approach is based on an existing extension of Latent Dirichlet Allocation with the possibility to include knowledge in terms of seed words into the model [1]. We extended the implementation of the logicLDA approach to incorporate external knowledge sources as seed words. As a main contribution of this work we show that external knowledge can be used to fit topic models to certain topical areas as a basis for a comparison of different documents with respect to these areas. We test this by showing that the method can be used to predict the distribution of ministries between the parties of a winning coalition based on the distance of the positions extracted from their manifestos to the positions in the coalition agreement. We explain the rationale of this experiment in more detail later on. We also show that although of course the result of the analysis depends on the choice of the seed words, the general principle works independently from a specific set of keywords.

The paper is organized as follows. In section 2 we present our multidimensional content analysis method that uses an extension of Latent Dirichlet Allocation for generating a topic model according to a predefined set of policy areas. Section 3 describes the experiments we conducted to validate the method by describing the rational of the experiment as well as the data sources used and the experimental setting. The results of the experiments are presented in section 4. We conclude with a discussion of the results and the implications for computer-aided content analysis in the social sciences.

2   Multi-dimensional Analysis

The goal of this work is the creation of a method for analyzing the position a certain document takes towards a topical area. The method follows a number of assumptions that have to be explicated before discussing the method itself. First of all, we assume that there is a well defined set of topic (or policy) areas and that the document(s) to be analyzed actually contains information related to these topic areas. The second fundamental assumption is that topic areas and specific positions can be described in terms of words associated with the respective topical area. This does not only allow us to characterize a topical area in terms of a number of seed words, it also justifies the use of topic models as an adequate statistical tool for carrying out the analysis. Finally, we assume that the distance between topic descriptions in terms of distributions over words is an indicator for the actual distance between the positions of the authors of the documents analyzed, in our case the parties stating their political programme. Based on these assumptions, we have designed the following method for analyzing (political) positions based on documents such as party manifestos.

2.1   Data Preparation

Data preparation is an important step for any content analysis as the quality of the raw data has significant influence on the quality of the analysis. For our method, we need to carry out two basic preprocessing steps: the first one is the creation of the corpus to be analyzed, the second one is to determine the vocabulary that should be the basis for the creation and the comparison of the topics.

As topic models rely on co-occurrence statistics of words within a corpus, the data preparation step has to generate a corpus of documents with meaningful co-occurrences. Typically, topic models assume that a corpus consists of a set of documents rather than a single document. However, the data we want to analyze is single document only, e.g. a party manifesto. As a solution to create a sufficiently large corpus, we split this single document into several parts, which are considered as separate documents. While this can of course be done manually by reading the document and dividing it in a thematically coherent way, we aim at automating the analysis as far as possible to be able to carry out large scale analyses with limited manpower. Please note that the documents analyzed by topic models are allowed to cover various topics and are not limited to a single one. In our approach, we use TextTiling [5], which is a popular method for automatically cutting texts into topically coherent subparts using lexical cohesion as a main criteria. TextTiling determines thematic blocks in a document in three steps. First, the document is segmented into individual tokens (roughly words) that can be compared. Further, the method splits the document into sequences of tokens with equal length called token sentences. In the second step the cosine similarity between adjacent token-sentences is determined and plotted into a graph. In the final step, thematic boundaries between token sentences are determined based on changes in the similarity. We chose this segmentation method because of its underlying assumption that text segments always contain a number of parallel information threads ([5], end of page 3) is very close to the underlying assumption of Latent Dirichlet Allocation, that a single document always addresses a number of different topics to a certain extent which is given by the Dirichlet distribution. A positive side effect of the TextTiling method is that it is domain independent and does not require external parameters to be set.

Another decision that has to be taken when preparing the data is which types of words should be taken into account when building the statistical model. Of course, all words occurring in a document can in principle be used, however, this often leads to rather meaningless topics that contain a lot of words that do not actually carry a meaning. A rather natural restriction is to only use words of a certain type. For this purpose, we determine word types in our documents using a state of the art part-of-speech tagger [12] and filter the documents based on word types. For the purpose of our experiments it turned out that using nouns only works best, as they are best suited to describe a topic. For some questions it might also be useful to include adjectives to identify how certain words are perceived by the respective party (e.g. 'unfair' vs. 'effective' tax system) or verbs to get an idea of planned actions ('raise' vs. 'lower' taxes). Regardless of the chosen word types it can make sense to exclude infrequent words or stop words from the analysis. Stop words are function words that appear with high frequency in all kinds of text and are therefore useless for content analysis. As we restrict our vocabulary to nouns only, we do not have to care about stop words. Addressing the issue of very infrequent words, we only take into account terms that occur at least twice in the corpus.

2.2   Topic Creation

The actual creation of the topic model consists of two steps. In the first step the topic structure is determined by setting the number of topics, selecting seed information for each topic and linking the seed information to the vocabulary created in the preparation step. In the second step, a topic model is generated using corpus statistics and the seed information using the logicLDA system.

The logicLDA system [1] extends Latent Dirichlet Allocation [3] with the possibility to include first order knowledge. It extends the standard generative model with a filter for possible groundings of the first order model. Adopting the notation of [1] we define $\mathbf{w} = w_1, \cdots, w_N$ to be a set of word tokens from a given vocabulary in a corpus and $\mathbf{d} = d_1, \cdots, d_N$ the document indices of the word tokens ($d_k = l$ means that token k occurs in document l) and $\mathbf{z} = z_1, \cdots, z_N$ a topic assignment for word token. Each topic $t = 1 \cdots T$ is represented by a multinomial distribution ϕ_t and each document $j = 1 \cdots D$ as a multinomial distribution θ_j over the vocabulary with Dirichlet priors β and α respectively. The generative model can then be described as $P(\mathbf{w}, \mathbf{z}, \phi, \theta | \alpha, \beta, \mathbf{d}) \propto$

$$\left(\prod_{t \in T} p(\phi_t | \beta) \right) \left(\prod_{j \in D} p(\theta_j | \alpha) \right) \left(\prod_{i \in N} \phi_{z_i}(w_i) \cdot \theta_{d_i}(z_i) \right)$$

where $\phi_{z_i}(w_i)$ is the w_i-th argument in the vector representing ϕ_{d_i}, and $\theta_{d_i}(z_i)$ is the z_i-th element in a vector representing θ_{d_i}. In order to identify topics in a given corpus, we have to estimate the multinomial distributions ϕ_t from a given corpus $(\mathbf{w}, \mathbf{d})$. This estimation can be done using standard methods from statistical inference. A method often used in this context is Gibbs sampling [4].

The logicLDA approach now extends this model with a set of weighted first-order clauses $\mathbf{c} = (\lambda_1, \psi_1), \cdots, (\lambda_L, \psi_L)$, where each ψ_l is a clause and λ_l the associated weight describing background information about the corpus and the associated topics. The connection to the generative model is now created via possible groundings of the clauses. Let $\mathbf{o}$ be external observations influencing the validity of clauses in $]$ and $G(\psi_l)$ the set of all possible groundings of ψ_l. For each grounding $g \in G(\psi_l)$ Andrzejewski et al introduce an indicator function

$$\mathbb{I}_g(\mathbf{z}, \mathbf{w}, \mathbf{d}, \mathbf{o}) = \begin{cases} 1, \text{ g is true under } (\mathbf{z}, \mathbf{w}, \mathbf{d}, \mathbf{o}) \\ 0, \text{otherwise} \end{cases}$$

Using this indicator function, the conditional probability that a word token occurring in a certain context given a corpus, a first order knowledge base and possibly some external observations $P(\mathbf{z}, \phi, \theta | \alpha, \beta, \mathbf{w}, \mathbf{d}, \mathbf{o},])$ is proportional to

$$\exp \left(\sum_{l \in L} \sum_{g \in G(\psi_l)} \lambda_l \mathbb{I}_g(\mathbf{z}, \mathbf{w}, \mathbf{d}, \mathbf{o}) \right) \times$$

$$\left(\prod_{t \in T} p(\phi_t | \beta) \right) \left(\prod_{j \in D} p(\theta_j | \alpha) \right) \left(\prod_{i \in N} \phi_{z_i}(w_i) \cdot \theta_{d_i}(z_i) \right)$$

Using this formulation, the most likely joint distribution $(\mathbf{w}, \psi, \theta)$ can be estimated using Maximum a Posteriori inference, i.e. by solving the following optimization problem.

$$\underset{\mathbf{w}, \psi, \theta}{\operatorname{argmax}} \sum_{l \in L} \sum_{g \in G(\psi_l)} \lambda_l \mathbb{I}_g(\mathbf{z}, \mathbf{w}, \mathbf{d}, \mathbf{o}) + \sum_{t \in T} \log p(\phi_t | \beta) +$$

$$\sum_{j \in D} \log p(\theta_j | \alpha) + \sum_{i \in N} \log \phi_{z_i}(w_i) \cdot \theta_{d_i}(z_i)$$

This problem is non-convex and cannot be solved using standard methods. However, Andrzejewski et al propose an approximate method called mirror descent for solving the problem and determining the most likely distribution. For details on this method, we refer the reader to [1].

We use this extension of Latent Dirichlet Allocation to predefine topics in terms of first order clauses that state background knowledge about the nature of the topics of interest. In particular Andrzejewski et al define special predicates modeling the assignment of word tokens to documents

- $Z(i, t)$ is true iff the hidden topic $z_i = t$.
- $W(i, v)$ is true iff word $w_i = v$.
- $D(i, j)$ is true iff $d_i = j$.

Using these predicates, different forms of knowledge can be induced in the model. We primarily use them to link the topics to be created with seed words taken from external sources. For this purpose, we introduce a new predicate $SEED(w, t)$ that is true if a word w is a seed word for topic t. The general impact of seed words on the topic model is then described by the following knowledge base:

$$\bigwedge_{i=1}^{N} W(i, w) \wedge SEED(w, t) \implies Z(i, t)$$

Based on this general definition, we can now introduce additional rules for defining the SEED predicate, thereby defining what kind of words act as seed words for a certain topic. The most straightforward way is to simply enumerate seed words, however, we can also use external resources and force all words occurring in a certain document (e.g. Wikipedia articles) to be seed words for a specific topic.

3   Experiments

We test the method described above in a number of experiments in the context of political science research. The purpose of these experiments is to test the ability of the proposed method to determine positions on particular topics stated in documents rather than to answer an actual research question in political science. In the following, we first provide a more detailed justification and the rational for the experiments carried out. Afterwards the data sources and the detailed experimental design are described.

3.1   Setting

As mentioned in the introduction, the goal of this work is to develop a content analysis method that is able to determine the (relative) position with respect to a certain topic as stated in a document. As we have explained in the last section, we do this by creating a topic model whose topics are partially predefined by the use of seed words to make them comparable. We claim that the distribution of words in a topic of the resulting model represents the position expressed in the document. In particular, we claim that the distance between the topic multinomials generated from different documents represent the distance of the positions stated in the two documents.

In the experiments, we test this hypothesis in an indirect way, analyzing party manifestos and coalition contracts. In particular, we determine the distances between the parties' positions stated in their manifestos and the coalition contract, and compare those distances among the two parties participating in the coalition. The underlying assumption is that the party that was to get control over the respective ministry has a stronger influence on the position stated in the coalition agreement on the topics represented by that ministry. Therefore, we can assume that the position on a topic stated in the coalition agreement is more similar to the position stated in the manifesto of the party that was assigned the ministry.

Now instead of directly comparing party manifestos, we first generate topic models from the party manifestos and the coalition agreement based on a classification of policy areas provided by Seher and Pappi [10]. They originally used the scheme for comparing positions across different federal states. We then measure the distance between the topic stated by the parties involved in the coalition and the same topic expressed in the coalition agreement. We consider our method to work as planned if out method is able to 'guess' the party that is in control of a certain ministry based on the positions generated from the party manifestos and the coalition contract with a certain level of confidence.

3.2   Data Sources

We base our experiments on Data from the last three German national elections (2002, 2005 and 2009). In all three cases the coalition was formed by two parties. Interestingly, the combination was always a different one: SPD and Greens in 2002, CDU/CSU and SPD in 2005 and CDU/CSU and FDP in 2009. While the 2005 election resulted in a grand coalition with CDU and SPD as (almost) equal partners, the Greens in 2002 were a junior partner as was the FDP in 2009. This variety of combinations is quite useful for our purposes as it allows us to test whether our method works independently from a particular party or coalition.

We use plain text versions of party manifestos provided by the Mannheim Centre for European Social Research (MZES)[1]. As it turned out that in some cases using only the manifesto from a single election does not provide sufficient data to obtain meaningful statistics, we supplemented the election manifestos with the general programs of the respective parties[2] that we retrieved from the web and semi-automatically converted to plain text format. Finally, we used plain text versions of the coalition agreements provided by Sven-Oliver Proksch from the MZES.

In [10] Seher and Pappi investigate the topics addressed by German Parties on the level of federal states. For their analysis they use the following set of policy areas, each characterized by a set of portfolios whose descriptions can be used as seed information.

---

[1] http://www.mzes.uni-mannheim.de/projekte/polidoc_net/
   index_new.php?view=home

[2] The general programs originate from the following years: FDP: 1997; SPD: 1998; Greens: 2002; CDU: 1994/2007 respectively.

1. Social Affairs and Labour Market ('Arbeit und Soziales')
2. Development and Reconstruction ('Aufbau, Wiederaufbau')
3. Building ('Bau')
4. Culture and Education ('Kultus')
5. National and European Affairs ('Bund und Europa')
6. Post War Effects ('Kriegsfolgen')
7. Agriculture ('Landwirtschaft')
8. Finance ('Finanzen')
9. Justice ('Justiz')
10. Internal Affairs ('Inneres')
11. Environment and Regional Planning ('Umwelt und Landesplanung')
12. Economics and Transport ('Wirtschaft und Verkehr')
13. Special Topics ('Sonderaufgaben')
14. Chancellery ('Staatskanzlei')
15. Security and Foreign Affairs ('Aussen- und Sicherheitspolitik')

It turned out that not all of the policy areas made sense in our setting. In particular, Development and Reconstruction as well as Building were not reflected in our data at all. The same holds for Post War Effects, Social Affairs and Chancellery. We therefore excluded these areas from the investigating leaving us with a set of nine policy areas that were used to compare documents.

3.3   Experimental Design

In the course of our experiments, we first transformed all documents into plain text format. We manually removed indices and tables of contents. We appended the general program of a party to its party manifesto in order to extend the data. For each election, we applied the TextTiling Method to the extended manifestos of the two parties under consideration and to the coalition contract, obtaining three sets of documents. In the next step, we ran a POS-tagger on all documents and filtered for nouns, resulting in corpora whose documents consist of nouns only. For each corpus, we then generated the vocabulary which consists of all nouns that appear at least twice in the corpus. At this point, the intersection between the vocabulary and the seed information is computed and parameter files for the logicLDA system are generated fixing the number of topics, the α and β values and the seed information. The system is then run using standard settings, but setting the number of words to be considered in a topic to 100. The resulting information is stored in a vector representation and the similarity of the vectors is computed using the Stanford OpenNLP API.

4   Results

In the following, we present the results of our experiments. In particular, we compare the outcome of the application of our method to the actual assignment of ministries to the coalition parties. We tested different measures for comparing the topic models, in particular Skew divergence, Jensen-Shannon Divergence and

Cosine Similarity[8]. It turned out that using the cosine similarity consistently outperforms the use of the divergence measures. We therefore only present the results based on using cosine similarity and predicting the party whose topic is more similar to the topic created from the coalition contract to be in charge of the respective ministry. As it turns out, our method very rarely makes a wrong prediction some of which can even be explained by the specifics of the topics and the coalition. We present the results for each election individually as the parties involved and the ministries finally created differ from each election making it impossible to aggregate results in a meaningful way.

German National Election 2002. In table 1 we can see that our method correctly predicted most of the ministries. In the case of Security and Foreign affairs the result is ambiguous as the responsibility for this area is split between two ministries, namely Foreign Affairs and Defense. The method made a mistake on the area of Economics and Transport, this mistake can be explained, however, by the high relevance of environmental issues which is traditionally a green topic for the Transport area. Another mistake was made on the Social Affairs and Labour Market where the method predicted the Greens to be in charge, whereas the ministry was taken by the SPD. Overall, we can see that the method was able to correctly predict six out of eight unambiguous areas (75%).

Table 1. Result of the Analysis of the German national elections 2002

Policy Area	SPD	Greens	Ministry	correct
Social Affairs and Labour Market	0.67	**0.77**	SPD	no
Culture and Education	**0.69**	0.56	SPD	yes
Agriculture	0.10	**0.59**	Greens	yes
Finance	**0.17**	0.04	SPD	yes
Justice	**0.38**	0.21	SPD	yes
Internal Affairs	**0.45**	0.43	SPD	yes
Environment and Regional Planning	0.53	**0.59**	Greens	yes
Economics and Transport	0.45	**0.63**	SPD	no
Security and Foreign Affairs	0.08	**0.14**	Greens/SPD*	yes/no

*foreign affairs:Greens, Defence: SPD

German National Election 2005. For the 2005 election, we obtain a similar picture as shown in table 2. In the new configuration of this election, we have two ambiguous cases. In addition to Security and Foreign Affairs, we have a second area that is split between different ministries that are occupied by different parties, namely Economics and Transport which is represented in the Ministries of Economics and Technology occupied by the CDU and the Ministry of Transport which was given to the SPD. Out of the remaining seven unambiguous cases our method predicted six correctly (86%) only making a mistake on the Ministry of Justice. It is interesting to see that the values for the ambiguous cases are very close to each other indicating an almost identical influence of the parties in the respective topics.

Table 2. Result of the Analysis of the German national elections 2005

Policy Area	CDU	SPD	Ministry	correct
Social Affairs and Labour Market	0.39	**0.44**	SPD	yes
Culture and Education	**0.71**	0.58	CDU	yes
Agriculture	**0.07**	0.05	CDU	yes
Finance	0.06	**0.22**	SPD	yes
Justice	**0.46**	0.03	SPD	no
Internal Affairs	**0.56**	0.49	CDU	yes
Environment and Regional Planning	0.18	**0.26**	SPD	yes
Economics and Transport	**0.67**	0.66	CDU/SPD*	yes/no
Security and Foreign Affairs	**0.03**	0.01	SPD/CDU**	yes/no

*Economics: CDU, Transport: SPD
**foreign affairs:SPD, Defence:CDU

German National Election 2009. The best result was obtained on the 2009 election as we show in table 3. Here all unambiguous cases were correctly predicted by our method.

Table 3. Result of the Analysis of the German national elections 2009

Policy Area	CDU	FDP	Ministry	correct
Social Affairs and Labour Market	**0.62**	0.37	CDU	yes
Culture and Education	**0.81**	0.72	CDU	yes
Agriculture	**0.52**	0.16	CDU	yes
Finance	**0.23**	0.04	CDU	yes
Justice	0.07	**0.32**	FDP	yes
Internal Affairs	**0.41**	0.30	CDU	yes
Environment and Regional Planning	**0.22**	0.20	CDU	yes
Economics and Transport	**0.79**	0.72	FDP/CDU*	yes/no
Security and Foreign Affairs	0.08	**0.55**	FDP/CDU**	yes/no

*Economics: FDP, Transport: CDU
**Foreign Affairs: FDP, Defence: CDU

5   Conclusions

The work presented in this paper shows that it is possible to use topic modeling as a basis for multi-dimensional analysis of political documents pushing the limits of automatic content analysis in the social sciences. Our experiments show that we can relatively reliably determine topics related to predefined policy areas and compare individual topics across documents. As we have mentioned before, these results do not have a direct value for research in political science. Yet they provide a proof of concept that we can build upon for addressing questions in the area of party competition which currently cannot be addressed without the need for manual coding. There are a number of issues that have to be investigated in more detail in future work before we can apply this method to open research questions in political science.

The most central problem is the choice of the right policy areas and seed information for a given question. As mentioned before, we used a coding scheme introduced by Seher and Pappi for policy analysis on the regional level. This already led to some problems when applying it to the national level and we had to exclude some areas not relevant in the context of national elections. On the European level, again different policy areas with different scope are relevant[3]. While for example Competition is a central area on the European level, it almost does not play any role on the regional level. In a similar way, policy areas as well as their focus change over time. While in the early phase of German politics post-war issues like Compensation and Denazification was a major issue, later periods were dominated by the cold war, these topics are not relevant today any more. Similarly new areas like environmental protection have emerged and gained importance. This means that the determination of relevant topics and seed information is a scientific problem that requires expertise in political science. From a technical point of view, the extraction of relevant seed information from existing knowledge resources is a problem that needs to be addressed.

In this work, we have investigated a basic setting in which only nouns are used to model topics in various documents. As already briefly mentioned, we can imagine situations where restricting ourselves to nouns is not enough. This way, we can analyze where a certain document puts a thematic emphasis inside a policy area, but we can neither determine the sentiment towards individual aspects of a topic nor proposed actions lines. In order to be able to extend the approach in these directions, we will have to include other word types. Such an extension, however, makes it harder to compare topics due to a higher degree of heterogeneity we can expect in the different vocabularies. We also have restricted our attention to the generation of topics and the distance between topics. Most related work in the political sciences, however, focuses on the projection of party positions on different scales (i.e. left-right or liberal-conservative). In future work we will investigate the projection of the generated models on a multidimensional scale. This will support researchers to carry out well established scale-based analysis while taking the different topic areas into account. Such an approach would solve the problems of one-dimensional analysis outlined in the motivation.

References

1. Andrzejewski, D., Zhu, X., Craven, M., Recht, B.: A framework for incorporating general domain knowledge into latent dirichlet allocation using first-order logic. In: Proceedings of the 22nd International Joint Conference on Artificial Intelligence, IJCAI 2011 (2011)
2. Benoit, K., Mikhaylov, S., Laver, M.: Treating words as data with error: Uncertainty in text statements of policy positions. American Journal of Political Science 53(2), 495–513 (2009)

---

[3] Compare http://europa.eu/pol/index_en.htm

3. Blei, D.M., Ng, A.Y., Jordan, M.I.: Latent dirichlet allocation. Journal of Machine Learning Research (JMLR) 3, 993–1022 (2003)
4. Casella, G., George, E.I.: Explaining the gibbs sampler. The American Statistician 46(3), 167–174 (1992)
5. Hearst, M.: Texttiling: Segmenting text into multi-paragraph subtopic passages. Computational Linguistics 23(1), 33–64 (1997)
6. Laver, M., Garry, J.: Estimating policy positions from political texts. American Journal of Political Science 44(3), 619–634 (2000)
7. Laver, M., Sergenti, E.: Party Competition: An Agent-Based Model. Princeton University Press (2011)
8. Lee, L.: On the effectiveness of the skew divergence for statistical language analysis. In: Artificial Intelligence and Statistics, pp. 65–72 (2001)
9. Pappi, F.U., Seher, N.M., Kurella, A.-S.: Das politikangebot deutscher parteien in den bundestagswahlen seit 1976 im dimensionsweisen vergleich: Gesamtskala und politikfeldspezifische skalen. Working Paper 142, Mannheimer Zentrum für Europäische Sozialforschung, MZES (2011)
10. Seher, N.M., Pappi, F.U.: Politikfeldspezifische positionen der landesverbände der deutschen parteien. Working Paper 139, Mannheimer Zentrum für Europäische Sozialforschung, MZES (2011)
11. Slapin, J.B., Proksch, S.-O.: A scaling model for estimating time-series policy positions from texts. American Journal of Political Science 52(3), 705–722 (2008)
12. Toutanova, K., Klein, D., Manning, C., Singer, Y.: Feature-rich part-of-speech tagging with a cyclic dependency network. In: Proceedings of HLT-NAACL 2003, pp. 252–259 (2003)
13. Volkens, A., Lacewell, O., Lehmann, P., Regel, S., Schultze, H., Werner, A.: The Manifesto Data Collection. Manifesto Project (MRG/CMP/MARPOR), Wissenschaftszentrum Berlin für Sozialforschung, WZB (2011)

Fake Reviews: The Malicious Perspective

Theodoros Lappas

Boston University, Computer Science Dept.
tlappas@cs.bu.edu

Abstract. Product reviews have been the focus of numerous research efforts. In particular, the problem of identifying fake reviews has recently attracted significant interest. Writing fake reviews is a form of attack, performed to purposefully harm or boost an item's reputation. The effective identification of such reviews is a fundamental problem that affects the performance of virtually every application based on review corpora. While recent work has explored different aspects of the problem, no effort has been done to view the problem from the attacker's perspective. In this work, we perform an analysis that emulates an actual attack on a real review corpus. We discuss different attack strategies, as well as the various contributing factors that determine the attack's *impact*. These factors determine, among others, the *authenticity* of fake review, evaluated based on its linguistic features and its ability to blend in with the rest of the corpus. Our analysis and experimental evaluation provide interesting findings on the nature of fake reviews and the vulnerability of online review-corpora.

1 Introduction

Item reviews are nowadays found in abundance in a plethora of websites on the World Wide Web. The reviewed items include a wide range of different products (e.g. electronics, books, movies) and service providers (e.g. restaurants, hotels). Users write reviews to rate a particular item and express their opinion on the item's various attributes. The accumulated volume of reviews on the item then serves as a valuable source of information for a potential customer. Not surprisingly, it has been shown that reviews have a tremendous effect on the popularity of an item [5, 8, 21, 23].

Before the establishment of the web and the modern e-commerce model, reviews were authored by eponymous experts, typically considered authorities in their respective fields. In today's online world, however, anyone can author and publish an anonymous review on a major review-hosting site such as *Amazon* or *Yelp*. The option of anonymity, combined with the major impact of reviews on the users' purchase decisions, make review corpora a prime target for malicious attacks. An attack on the review corpus of an item includes the injection of fake reviews in order to harm or boost the item's reputation. A fake review does not provide an accurate representation of the item's quality and can thus be misleading to users. In addition, false information can have a negative effect on any application that is defined in the context of the review corpus (e.g. review search [15], selection [24],summarization [27]).

The detection of fake reviews is a well-recognized problem that has attracted significant interest from the research community [10, 11, 16–18, 25]. While previous work

G. Bouma et al. (Eds.): NLDB 2012, LNCS 7337, pp. 23–34, 2012.

has made advances in the identification of fake reviews, no effort has been made to approach the problem from the point of view of the attacker. In this paper, we present an analysis of attack scenarios on review corpora. As we demonstrate in our work, the design and implementation of such an attack is a non-trivial task. In fact, the attacker needs to address a number of relevant questions, such as:

– *What information should I include in a fake review?*
– *How can I maximize the impact of the attack on the reputation of the target item?*
– *How can I make a fake review to appear as authentic as possible?*

Contribution: Our work is the first to formalize the parameters that need to be considered by the aspiring attacker. By unraveling the attack-process, our findings can be used toward the immunization of review corpora and the creation of enhanced methods for the detection of fake reviews. Our corpus-sensitivity measure can be used by a review-hosting website in order to properly allocate its limited resources for fraud detection. Further, a merchant (e.g. a restaurant owner) can be more vigilant given the knowledge that the corpus of reviews on their item is highly sensitive to a potential attack.

2 Writing Fake Reviews

A successful attack needs to take into consideration the factors that are most influential to its success. In our work, we identify and examine the following two factors:

1. **Authenticity:** In order for a fake review to be convincing, and thus less likely to be detected, it needs to be as authentic-looking as possible.
2. **Impact:** The injected fake review needs to be written in a way that maximizes the (positive or negative) impact on the target's reputation.

In the following sections, we formalize and discuss each of these two factors in detail.

2.1 Authenticity

First, we identify the three factors that affect the authenticity of a fake review:

1. **Stealth** measures the ability of the review to blend in with the corpus.
2. **Coherence** evaluates whether the assigned rating is in accordance with the opinions expressed in the review's text.
3. **Readability** depends on the structural features of a review's text and captures how easy it is to parse.

Next, we discuss and motivate these factors via the use of appropriate examples.

[Stealth]: Consider a restaurant evaluated on four attributes: *food quality, service quality, parking options* and *price*. For these four attributes, we assume the following opinions are observed in the review corpus:

- food quality: 200 positive, 5 negative
- service quality: 100 positive, 90 negative
- parking options: 5 positive, 1 negative
- price: 200 positive, 0 negative

Now consider the following fake review, assigning the minimum 1-star rating (out of 5) to this restaurant.

Review 1: *I'm never going to this restaurant again. First of all, finding a place to park around this place is impossible. It took us forever. Things only got worse after we actually got there. Our server was rude and entirely unprofessional. He pretty much ignored our requests and made us feel unwanted. On top of that, the food was terrible and seriously overpriced. One of the worst eating experiences of my life.*

This review will obviously have a high impact, if injected to the corpus. Observe that, in addition to assigning the minimum possible rating, the review also expresses a negative opinion on *all* the attributes. However, when one considers the distribution of opinions in the existing corpus, the review stands out as suspicious. In particular, the reviewer expresses (among others) highly negative opinions on the food and the price. This contradicts the very strong positive consensus that is observed in the corpus on these particular attributes, and hints that the reviewer is malicious or at least biased. Even if a review-management system does not eliminate this review, it can consider decreasing its visibility, in order to avoid presenting such questionable information to its users.

The above example clearly illustrates the following: in order to avoid having the fake review blacklisted or demoted, the attacker needs to consider the distribution of opinions in the existing corpus. The stronger the consensus on a given attribute, the more suspicious it appears if the fake review contradicts it. Further, the more positive and less negative opinions there are on an attribute, the stronger the positive consensus on an that attribute. Similarly for a negative consensus. To capture this intuition, we formalize the *stealth* of a fake review r in the context of a review-corpus R as follows:

$$stealth(r, R) = \frac{\sum_{\alpha \in \mathcal{A}_r} |\{r' : r' \in R, r[\alpha] = r'[\alpha]\}|}{\sum_{\alpha \in \mathcal{A}_r} P(\alpha, R) + N(\alpha, R)} \tag{1}$$

where $P(\alpha, R)$ is the number of positive opinions on attribute α in R, and $N(\alpha, R)$ is the respective number of negative opinions. Also, $\mathcal{A}_r$ is the set of attributes evaluated in r, and $r[\alpha] \in \{-1, +1\}$ returns the polarity of the opinion expressed in review r on attribute α. A value of +1 (-1) is returned for a positive (negative) polarity. The numerator represents the number of reviews that agree with r on attribute α, for every attribute $\alpha \in \mathcal{A}_r$. The denominator is equal to the sum of all the opinions expressed on all the attributes in $\mathcal{A}_r$. Obviously, the lower the fraction, the more suspicious the review appears.

[Coherence]: While the above formalization captures the concept of stealth, it disregards the review's *rating*. In the context of the previous example, a savvy attacker can easily write a stealthy review, such as the following:

Review 1: *Great food and very reasonably priced. That aside, I was not happy with our server. She was impolite and always in a hurry. Parking was also a bit of a pain, it took us a while to find a spot.*

After making sure that the opinions expressed in the review are in accordance with the rest of the corpus (see the distribution in the previous example), the attacker can then assign a very low rating (e.g. 1 star) in order to have a major impact on any rating-based ranking function. However, such a review would not be *coherent*, since the opinions expressed in the text are mixed and cannot justify such a low rating. This example motivates the concept of *coherence* for reviews. Intuitively, opinions expressed in a review should be in accordance with not only the other reviews in the corpus, but also with its own rating. For simplicity, we organize ratings into 3 mutually-exclusive groups: positive (+1), negative (-1) and neutral (0). For example, for the 5-star scale, the respective groups could be {4,5}, {1,2} and {3}. We then formalize *coherence* as follows:

$$coh(r) = \begin{cases} P(r)/|r| & rating(r) = +1 \\ N(r)/|r| & rating(r) = -1 \\ \frac{(|r|-|P(r)-N(r)|)}{|r|} & rating(r) = 0 \end{cases} \tag{2}$$

where $P(r)$ and $N(r)$ return the number positive and negative opinions expressed in r, respectively. If the assigned rating is positive (negative), then coherence is the probability that a randomly chosen opinion expressed in the review is positive (negative). For neutral ratings (zero), the ideal scenario occurs when the review contains an equal number of positive and negative opinions. In other words, in the ideal scenario, every negative opinion is balanced by a positive one. Thus, for a neutral rating, coherence is the fraction of matched opinions.

[Readability]: The final factor that contributes to a review's authenticity is its *readability*. This is a well-studied concept that has motivated a significant body of relevant work [7]. In our methodology, we apply the well-known Flesh-Reading Ease (FRE) formula [6] for the evaluation of readability. Our choice is motivated by the popularity of this measure (incorporated in systems like Microsoft Office Word, WordPerfect and WordPro), as well as its successful application in the domain of reviews in relevant work [14, 19]. FRE is formally defined as follows:

$$FRE(r) = 206.835 - 1.015 \times \frac{words(r)}{sents(r)} - 84.6 \times \frac{syllables(r)}{words(r)} \tag{3}$$

where *words(r), sents(r)* and *syllables(r)* return the number of words, sentences and syllables in r, respectively. The Flesch Reading Ease yields numbers from 0 to 100, expressing the range from *very difficult* to *very easy*.

2.2 Impact

An attack on a review corpus aims at manipulating the reputation of the target-item. Toward an appropriate formalization of *impact*, we need a review-based measure that objectively captures the *reputation* concept. On a high level, we assume the existence of a reputation-evaluation function $f(\cdot)$ which, given the set of reviews on an item, assigns a score that captures the item's popularity, as encoded in the review corpus. Most modern review-hosting sites employ the average rating, observed over all the reviews in the corpus. We include this definition in our analysis, formalized as:

$$f_1(R) = \frac{\sum_{r \in R} rating(r)}{|R|}. \tag{4}$$

where R is the corpus of reviews for a given item and $rating(r)$ returns the rating for a given review r. The rating can be defined in the popular 5-star scale or on any other ordinal scale. Despite its popularity, the average rating often fails to deliver an accurate evaluation of the item's quality, since it treats every item as a single unary entity. In practice, however, reviewers evaluate an item based on the quality of its numerous *attributes*. For example, a digital camera can be evaluated based on its battery life, the picture-quality, the design, the price etc. The individual opinions that focus on particular attributes can be much more informative than the assigned star rating. Consider the following 3-star review on an Italian restaurant, taken from a major review-hosting site:

Example 1. This place has a lot of personality. The ambiance is great and the greeters were terrific despite the fact that we had a huge party coming in. They seated us at a table that was comfortable even though we were packed in pretty tightly. We started off with bread and some garlic oil concoction that one of my friends had ordered. It was delicious. I then ordered the Lasagna which was meatless or so I was told and I must say was quite mouthwatering. The sauce was tasty and the pasta itself was very good. I know the rest of my party was very pleased with their meal so I think I will definitely come back if not for just the bread and the oil dipping thingy. Overall, good experience, nothing over the top but nothing subpar either. My only complaint would be the service which i thought was slow and they didn't refill my drink once even though I had asked for it.

If one were to ignore the text of the review and focus only on the rating, the 3 stars assigned by the reviewer imply an impression of a mediocre restaurant. However, the text reveals attributes worthy of praises like "delicious", "mouthwatering", "tasty" and "very pleased". In fact, the only complaint of the reviewer was related to the service. For a potential customer who is mostly interested in the quality of the food, this review would be considered extremely favorable.

Motivated by the above discussion, we consider an *attribute-based* evaluation function. Thankfully, a long line of related research has provided us with methods for extracting opinions from reviews [4, 20]. Specifically, given a review r, we can identify the attributes of the item that the reviewer comments on, as well as the polarity (positive/negative) of each expressed opinion. We can thus define the following reputation function for an item, given its corpus of reviews R and the set of its attributes $\mathcal{A}$:

$$f_2(R) = \frac{\sum_{\alpha \in \mathcal{A}} (P(\alpha, R) - N(\alpha, R))}{|R|} \tag{5}$$

Here, $\mathcal{A}$ is the set of the item's attributes. $P(\alpha, R)(N(\alpha, R))$ is the number of positive (negative)opinions expressed in the entire corpus R on attribute α. The fraction captures the average difference between positive and negative opinions per review.

At this point, we have defined two alternative functions that evaluate the reputation of an item, as captured in its review corpus (see Equations 4 and 5). Let $f(\cdot)$ represent either of these functions. Then, we define the *impact* of a fake review r in the context of a review-corpus R as follows:

$$impact(r, R) = f(R \cup \{r\}) - f(R) \tag{6}$$

The assigned impact value can be positive or negative, depending on whether the fake review was injected to boost or harm the item's reputation. Given our formal definition of impact and authenticity, we can now formalize the *sensitivity* of a review corpus to attacks. Formally, given a corpus R, a stealth threshold λ and a coherence threshold μ, the sensitivity of R to attacks that aim to *boost* the item's reputation is defined as:

$$sensitivity^{(+)}(R, \mu, \lambda) =$$

$$\max_{r}\{impact(r, R) : stealth(r, R) \geq \lambda, coh(r) \geq \mu, \} \tag{7}$$

To compute the sensitivity to attacks that aim to *harm* the item's reputation, it is sufficient to find the review with the *minimum* possible impact, under the given constraints. In our experiments, we use this measure to evaluate the vulnerability of real datasets. Observe that we do not consider readability, since it can be optimized independently.

3   Attack Strategies

Next, we formalize two attack strategies that consider authenticity and impact. For the remainder of this section, we assume attacks that want to *boost* the reputation of a given item. The analysis for attacks that attempt to *hurt* the item trivially follows.

First, we formalize the first attack strategy, which we refer to as the *single-item attack*:

Problem 1. **[Single-Item Attack]:** Given the corpus of reviews R on an item and a reputation function $f(\cdot)$, we want to inject a fake review r^* into R so that $stealth(r^*, R) \geq \lambda$, $coherence(r) \geq \mu$, $readability(r) \geq \nu$ and:

$$impact(r^*, R) = \max_{r}\{impact(r, R)\} \tag{8}$$

If the attacker is interested in short-term results or is confident in the inability of the review management system to identify fake reviews, λ can be set to a low value. Otherwise, a higher value for λ is appropriate. The parameters μ and ν are similarly tuned to determine coherence and readability. In our experiments, we show how these parameters can be learned from real review corpora.

By focusing exclusively on the review corpus of the target, single-item attacks overlook a valuable resource: the reviews on the target's *competitors*. Consider the case when the target item I competes with a competitor I' for the same market share. A single-item attack would try to increase the market share of I by improving its reputation. However, the same can be accomplished by attacking the corpus of I'. By damaging the competitor's reputation, the attack will make it less appealing to customers, who are then likely to turn to I. Formally, we define this attack strategy as follows:

Problem 2. **[Competitor Attack]:** Let $\mathcal{C}$ be the set of immediate competitors for the target item I, and let $\mathcal{R}_\mathcal{C}$ be the respective collection of review corpora for the items in $\mathcal{C}$. Then, given a reputation function $f(\cdot)$, we want to inject a fake review r^* into a corpus $R^* \in \mathcal{R}_\mathcal{C}$, so that $stealth(r^*, R^*) \geq \lambda$, $coherence(r) \geq \mu$, $readability(r) \geq \nu$ and

$$impact(r^*, R^*) = \min_{r, R \in \mathcal{R}_\mathcal{C}}\{impact(r, R)\} \tag{9}$$

We observe that a solution for Problem 1 can be easily extended to optimally solve Problem 2 (i.e. by identifying the review with the highest negative impact for each competitor in $\mathcal{C}$). Thus, we focus on solving Problem 1. We observe that readability is completely independent of impact, and can thus be optimized separately. Thus, we want to find the review with the maximum impact, among those that respect the thresholds λ, μ and ν. As we show next, a single-item attack can be intuitively formed as a 0-1 integer linear programming problem. First, we consider the maximization of the review's impact for each of the two reputation functions considered in our work.

For the rating-based reputation function $f_1(\cdot)$, we want the injected fake review to maximize the average rating over the entire corpus. This trivially translates into submitting the review with the highest possible rating, as allowed by the constraints. For the attribute-based reputation function $f_2(\cdot)$, maximizing $impact(r, R)$ translates into maximizing $f_2(R \cup \{r\})$. Thus, we want to maximize:

$$\propto f_2(R \cup \{r\}) = \sum_{\alpha \in \mathcal{A}} P(\alpha, R) + x_\alpha - N(\alpha, R) - y_\alpha \tag{10}$$

where:

$$y_\alpha + x_\alpha \leq 1, y_\alpha, x_\alpha \in \{0, 1\} \tag{11}$$

x_α is set to 1 when the review expresses a positive opinion on α. Similarly, $y_\alpha = 1$ for a negative opinion. The $y_\alpha + x_\alpha \leq 1$ constraint ensures that at most one of the two variables can be set to 1. Now, based on Eq. 1, the stealth constraint can be written as:

$$\sum_{\alpha \in \mathcal{A}} x_\alpha \times [P(\alpha, R) - \lambda(P(\alpha, R) + N(\alpha, R))] + \tag{12}$$
$$y_\alpha \times [N(\alpha, R) - \lambda(P(\alpha, R) + N(\alpha, R))] \geq 0$$

where x_α and y_α are the variables introduced in Equation 10 above.

Finally, we need to incorporate the coherence threshold μ. Considering Eq. 2, we need to separately consider the assignment of a positive, negative or neutral rating:

positive: $\sum_{\alpha \in \mathcal{A}} x_\alpha / (\sum_{\alpha \in \mathcal{A}} x_\alpha + y_\alpha) \geq \mu$ (13)   negative: $\sum_{\alpha \in \mathcal{A}} y_\alpha / (\sum_{\alpha \in \mathcal{A}} x_\alpha + y_\alpha) \geq \mu$ (14)

neutral: $1 - \left| \sum_{\alpha \in \mathcal{A}} x_\alpha - \sum_{\alpha \in \mathcal{A}} y_\alpha \right| / \left(\sum_{\alpha \in \mathcal{A}} x_\alpha + y_\alpha \right) \geq \mu$ (15)   (13)

If multiple ratings satisfy the coherence threshold μ, then we simply choose the rating that leads to the highest impact.

For both reputation functions, standard techniques for binary integer programming can be used to create the reviews with the highest impact. In order to obtain the solution to Problem 1 for the reputation function $f_2(\cdot)$, we need to solve the optimization problem that maximizes Equation 10, satisfies both constraints given by Equations 11 and 12, and also satisfies *at least one* of three inequalities defined in Equations 13, 14 and 15. For the reputation function $f_1(\cdot)$, we only need to check whether there is an assignment of values for the variables $x_\alpha, y_\alpha, \forall \alpha \in \mathcal{A}$ that satisfies the inequality system consisting of Equations 11, 12 and 13. If no such a solution exists, then we know that there is no positive-rating review that respects the given thresholds. Then, the only option is to lower the thresholds. If a solution exists, we simply assign the maximum possible rating that falls within the "positive" group (e.g. 4 or 5 in the 5-star scale). For this function, we observe that an adversary can bypass the stealth and coherence thresholds by submitting a review that does not include any opinions (just irrelevant or generic text). This can be handled by an additional constraint, asking for at least M opinions to be expressed in a valid review. We explore this direction in our experiments.

4   Experiments

In this section, we present the experiments that we conducted to evaluate our methodology. In our evaluation, we use the openly available data introduced by Jindal and [17]. The dataset consists of more than 5.8 million reviews from Amazon.com. For our evaluation, we focus on the domains of MP3 players and Digital Cameras. These domains were chosen based on the plethora of available reviews. Other than that, our study does not benefit from this choice. In particular, we extracted the reviews for 666 different MP3 players and for 589 cameras. For each review, we retrieve the star rating and the actual text. We refer to these datasets as MP3 and CAM, respectively. The method by Ding et al. [4] was used for opinion extraction. The authors show that their method is superior to previous state-of-the art techniques. Nonetheless, our methodology is compatible with any other alternative method for this task [9, 22].

4.1   Authenticity

In this experiment, we study the observed values of the three components of authenticity: stealth, coherence and readability. This evaluation provides insight on the tuning of the respective thresholds (λ, μ, ν). First, we compute the average value for each measure, over all items in each of the two datasets. We then present the distribution of the three measures across different value intervals in Figure 1. As shown in the figure, the vast majority of the reviews exhibited high readability values (60-80). For comparison, consider that the readability of the popular *Reader's Digest* is about 65, while *Time*

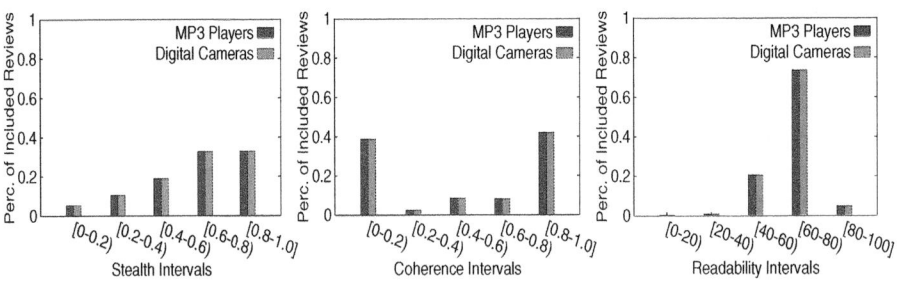

Fig. 1. Distribution of stealth, coherence and readability in real review corpora

magazine scores at about 52 [1]. For coherence, two major clusters emerged from the analysis: about 40% of the reviews achieved high values, indicating that their ratings matched the opinions expressed in their text. However, another 40% of the reviews exhibited very low values (0-0.2). This cluster can serve as the starting point for a review-detection algorithm. However, its large size also hints at the ambiguity of the 5-star scale, which has been pointed out by others in the past [2]. Further, the fact that the ratings assigned by users are often discordant with the expressed opinions, motivates the use of more sophisticated reputation functions than the standard star-scale. With respect to stealth, about 65% of the reviews achieved a value greater than 0.6, while the rest are those that generally contradict the opinions expressed by the majority of the reviewers.

4.2 Corpus Sensitivity

In this experiment we use our sensitivity measure (Eq. 7) to evaluate the vulnerability of real review corpora to fake reviews.

[Attribute-Based Reputation]: We begin our analysis with the attribute-based reputation function (Eq. 5). For the first experiment, the coherence threshold is set to $\mu = 0.8$ (in accordance with the findings of the previous experiment). Then, for different values of the stealth threshold λ, we compute the sensitivity of each corpus to positive (POS) and negative (NEG) fake reviews. The results are shown in Figure 2, where we show the average sensitivity over all items in each dataset. As can be seen in the figure,

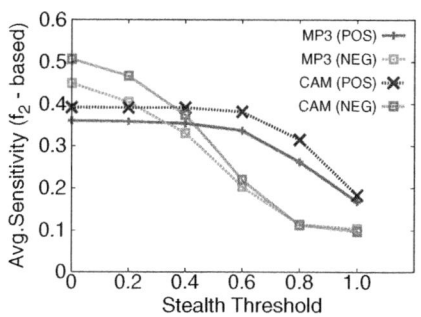
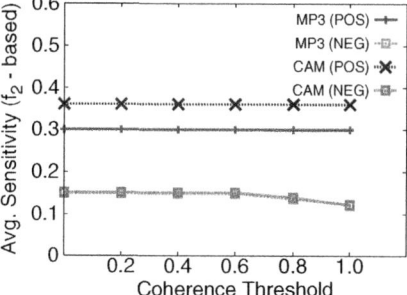

Fig. 2. Sensitivity Vs. Stealth   **Fig. 3.** Sensitivity Vs. Coherence

the maximum sensitivity to fake positive reviews is observed for $\lambda \leq 0.6$. For higher values the impact declines, as it becomes harder to included positive opinions while respecting the stealth threshold. The respective results on the sensitivity to negative fake reviews are very interesting, with the impact curve declining a lot faster. This can be explained by the bias toward positive reviews that ails most review corpora [12]. In other words, the plethora of positive opinions makes it harder for a negative review to maintain stealth. This finding indicates that, for an attacker who wants to hurt an item's reputation, a competitor-attack that tries to boost the reputation of its competitors can be more effective to an attack that tries to directly hurt the item's reputation.

To complete our evaluation of sensitivity, we fix the stealth threshold to $\lambda = 0.6$ and tune the coherence threshold μ. The results are shown in Figure 3. The figure illustrates that the coherence threshold has little or no effect on the sensitivity of the corpus. In other words, it is typically feasible for an adversary to create fake (positive or negative) review that abides by even the strictest of thresholds.

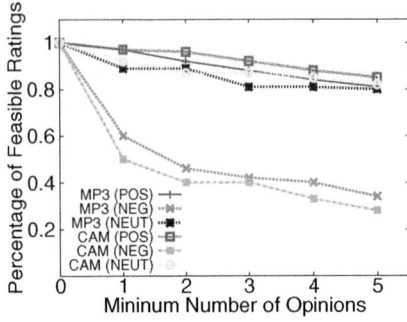

	Single-Item	Competitor
Success Perc.	71%	91%
#Reviews (Avg.)	35.8	15.8
#Reviews (Stdev.)	36.1	12.8

Fig. 4. Rating feasibility Vs. Min Number of Opinions in a fake review (M)

Fig. 5. Results for the two Attack Strategies

[Rating-Based Reputation]: As discussed in our analysis on the rating-based function, it is sufficient to check which ratings are feasible under the stealth and coherence thresholds. However, an adversary can bypass the stealth and coherence thresholds by submitting a review that does not include any opinions. As mentioned in our analysis, this can be handled by asking for at least M opinions to be expressed in a valid review. Next, we explore the effect of this parameter on the sensitivity of the corpus. As before, the stealth and coherence threshold as set to $\lambda = 0.6$ and $\mu = 0.8$, respectively. In Figure 4, we report the percentage of corpora for which a positive, neutral and negative rating was feasible, for different values of M (i.e. the minimum number of required opinions).

As can be seen by the figure, M has a minor effect on the feasibility of positive and neutral reviews. On the other hand, the feasibility of negative ratings clearly decreases as M rises. This can be attributed to the bias toward positive evaluations in real review corpora, which also emerged for the attribute-based function. For a negative rating to be possible as more opinions are required, the coherence threshold requires the majority of these opinions to be negative. However, the positive bias makes it harder for a review with numerous negative opinions to respect the stealth threshold.

4.3   Comparing Attack Strategies

Finally, we compare the two attack strategies that we formalized in our work (*single-item* and *competitor*). For this experiment, we use all the corpora from MP3. For every possible pair, we identify the item with the lowest reputation, based on the attribute-based function $f_2()$ (the findings for the rating-based function were similar and are omitted for lack of space). We refer to this item as the *client*, while the other item of the pair is referred to as the *victim*. Our goal is to inject enough fake reviews for the *client* to achieve an equal or higher score than the *victim*. Using the *single-item attack*, we repeatedly inject the fake review with the highest positive impact until the goal is reached. In addition to this option, the *competitor attack*, allows us to inject reviews with the highest possible negative impact to the *victim*'s corpus. For this strategy, we greedily choose the option that brings the scores of the two items closer. We record the number of fake reviews required for each strategy. If more than a 100 reviews are required to reach the goal, the attack is considered a failure. The results are shown in Figure 5. The first observation is that the competitor-attack reaches the goal for 90% of the pairs, cleary outperforming the single-item strategy. For those successful pairs, the two methods required about the same number of fake reviews on average. Finally, the high standard variation illustrates that sensitivity can vary across review corpora, a finding that is in accordance with our own previous evaluation.

5   Related Work

A significant amount of work has recently been devoted to the identification of fake reviews [10, 11, 17]. In addition to the textual and linguistic features of reviews, relevant papers have explored different aspects of the topic, including review-ratings [3, 26], group spamming [18], and atypical (suspicious) reviews [13, 15]. Atypical reviews are an interesting concept that we also consider in our own work. Another relevant line of work explores the behavior of reviewers [16, 25]. By tracking a user's reviewing and rating patterns, these methods aim at finding suspicious behavior that hints to malice or bias against particular brands or products. Our contribution is complementary to that made by these papers: while this line of work examines the problem from the defensive point of view (trying to identify and eliminate spam or malicious reviews) our work is the first to explore the opposite perspective, examining the creation of fake reviews from the attacker's side. Finally, our work has ties to the extensive literature on opinion mining and sentiment analysis [20]. In particular, we employ the method by Ding et al. [4] for the extraction of opinions from reviews. Nonetheless, our framework is compatible with any alternative technique for opinion extraction [9, 22].

6   Conclusion

In this paper we presented the first study of fake reviews from the perspective of the attacker. We formalized the factors that determine the success of an attack and explored different attack strategies. Our analysis and experimental findings on real datasets provide valuable insight on the domain of fake reviews, and can be used as the basis for the immunization of review corpora and the improvement of fraud-detection techniques.

References

1. http://en.wikipedia.org/wiki/flesch_kincaid_readability_test
2. http://www.wellesleywinepress.com/2011/09/
   on-5-star-system-and-100-point-scale.html
3. Dellarocas, C.: Immunizing online reputation reporting systems against unfair ratings and discriminatory behavior. In: EC 2000, pp. 150–157. ACM, New York (2000)

4. Ding, X., Liu, B., Yu, P.S.: A holistic lexicon-based approach to opinion mining. In: WSDM 2008, pp. 231–240. ACM, New York (2008)
5. Duan, W., Gu, B., Whinston, A.B.: Do Online Reviews Matter? - An Empirical Investigation of Panel Data. Social Science Research Network Working Paper Series (November 2004)
6. Flesch, R.: A new readability yardstick. J. Appl. Psychol. 32, 221–224 (1948)
7. Flesch, R.: The Art of Readable Writing. Harper and Row (1949)
8. Hankin, L.: The effects of user reviews on online purchasing behavior across multiple product categories. PhD thesis (2007)
9. Hu, M., Liu, B.: Mining and summarizing customer reviews. In: KDD 2004, pp. 168–177. ACM, New York (2004)
10. Jindal, N., Liu, B.: Analyzing and detecting review spam. In: ICDM, pp. 547–552 (2007)
11. Jindal, N., Liu, B.: Review spam detection. In: WWW 2007, pp. 1189–1190. ACM, New York (2007)
12. Jindal, N., Liu, B.: Opinion spam and analysis. In: WSDM 2008, pp. 219–230. ACM, New York (2008)
13. Jindal, N., Liu, B., Lim, E.-P.: Finding unusual review patterns using unexpected rules. In: CIKM 2010, pp. 1549–1552. ACM, New York (2010)
14. Korfiatis, N., Rodríguez, D., Sicilia, M.-Á.: The Impact of Readability on the Usefulness of Online Product Reviews: A Case Study on an Online Bookstore. In: Lytras, M.D., Damiani, E., Tennyson, R.D. (eds.) WSKS 2008. LNCS (LNAI), vol. 5288, pp. 423–432. Springer, Heidelberg (2008)
15. Lappas, T., Gunopulos, D.: Efficient Confident Search in Large Review Corpora. In: Balcázar, J.L., Bonchi, F., Gionis, A., Sebag, M. (eds.) ECML PKDD 2010, Part II. LNCS, vol. 6322, pp. 195–210. Springer, Heidelberg (2010)
16. Lim, E.-P., Nguyen, V.-A., Jindal, N., Liu, B., Lauw, H.W.: Detecting product review spammers using rating behaviors. In: CIKM 2010, pp. 939–948. ACM, New York (2010)
17. Mackiewicz, J.: Reviewer motivations, bias, and credibility in online reviews. In: Kelsey, S., Amant, K.S. (eds.) Handbook of Research on Computer Mediated Communication, pp. 252–266. IGI Global (2008)
18. Mukherjee, A., Liu, B., Wang, J., Glance, N.S., Jindal, N.: Detecting group review spam. In: WWW (Companion Volume), pp. 93–94 (2011)
19. O'Mahony, M.P., Smyth, B.: Using readability tests to predict helpful product reviews. In: RIAO 2010, Paris, France, pp. 164–167 (2010)
20. Pang, B., Lee, L.: Opinion mining and sentiment analysis. Found. Trends Inf. Retr. 2, 1–135 (2008)
21. Park, D.-H., Lee, J., Han, I.: The effect of on-line consumer reviews on consumer purchasing intention: The moderating role of involvement. Int. J. Electron. Commerce 11(4), 125–148 (2007)
22. Popescu, A.-M., Etzioni, O.: Extracting product features and opinions from reviews. In: HLT 2005, pp. 339–346. Association for Computational Linguistics, Stroudsburg (2005)
23. Reinstein, D.A., Snyder, C.M.: The influence of expert reviews on consumer demand for experience goods: A case study of movie critics. Journal of Industrial Economics 53(1), 27–51 (2005)
24. Tsaparas, P., Ntoulas, A., Terzi, E.: Selecting a comprehensive set of reviews. In: KDD, pp. 168–176 (2011)
25. Wang, G., Xie, S., Liu, B., Yu, P.S.: Review graph based online store review spammer detection. In: IEEE International Conference on Data Mining, pp. 1242–1247 (2011)
26. Wu, G., Greene, D., Smyth, B., Cunningham, P.: Distortion as a validation criterion in the identification of suspicious reviews. Technical report (2010)
27. Zhuang, L., Jing, F., Zhu, X.-Y.: Movie review mining and summarization. In: CIKM 2006, pp. 43–50. ACM, New York (2006)

Polarity Preference of Verbs: What Could Verbs Reveal about the Polarity of Their Objects?

Manfred Klenner and Stefanos Petrakis

Institute for Computational Linguistics, University of Zurich, Switzerland
{klenner,petrakis}@cl.uzh.ch
http://www.cl.uzh.ch/

Abstract. The current endeavour focuses on the notion of positive versus negative polarity preference of verbs for their direct objects. This preference has to be distinguished from a verb's own prior polarity - for the same verb, these two properties might even be inverse. Polarity preferences of verbs are extracted on the basis of a large and dependency-parsed corpus by means of statistical measures. We observed verbs with a relatively clear positive or negative polarity preference, as well as cases of verbs where positive and negative polarity preference is balanced (we call these bipolar-preference verbs). Given clear-cut polarity preferences of a verb, nouns, whose polarity is yet unknown, can now be classified. We reached a lower bound of 81% precision in our experiments, whereas the upper bound goes up to 92%.

Keywords: sentiment analysis, polarity inference, data mining.

1 Introduction

There are two major classes of polarity lexicons: those that tag polarity at the word level (mostly done semi-automatically), e.g. the Subjectivity Lexicon by Wilson et al. (described in [1]), and those that determine the polarity on the basis of lexical resources, e.g. the SentiWordNet [2]. Accordingly, positive, negative and neutral polarity is attached to word *senses*. Fortunately, quite a number of words do have a clear-cut prior polarity that is invariant over the word's senses. A carefully designed word-level prior-polarity lexicon thus is a valuable resource for sentiment analysis.

The work reported here strives to semi-automatically augment the publicly available word-level prior-polarity lexicon for German by Clematide and Klenner [3]. We base our approach on a large text corpus for German, the DeWaC corpus [4], and try to automatically identify new polar nouns and add them to the lexicon after manual inspection.

The focal point of the proposed method is the verb-object relationship. Verbs play a crucial role in clause-level sentiment determination. Some verbs (e.g. to love) not only have a prior polarity, but they also seem to have a specific polarity 'disposition' (or 'preference') as far as the polarity on their objects is concerned (e.g. I love books). In the case of a human subject, some authors, namely

G. Bouma et al. (Eds.): NLDB 2012, LNCS 7337, pp. 35–46, 2012.

Wiebe et al. in [5], speak of a positive or negative attitude of the opinion holder, the subject, towards the opinion target, the object. Our goal was to automatically identify verbs with such a polarity preference for their objects based on corpus material. Starting from a prior-polarity lexicon for German (see [3]) and the noun polarities defined there, all verbs that have a polar object (a single noun) have been collected. Although there are clear-cut cases, where the prior polarities of all nouns of a verb have a single orientation, most of the time, both polarities, positive and negative, occur with some frequency. Moreover, there are cases, where both polarities occur with (almost) the same frequency. Those verbs, which we name bipolar-preference verbs, do not impose any restriction on the polarity of their objects. In order to operationalize the choice of whether a verb actually has a clear preference or not, we first use a statistical test. Then, after having obtained a list of polarity preference verbs, we measured how well these verbs could actually predict the polarity of (unseen) nouns occurring as their objects. We did this in a held-out setting, where we used nouns from the prior-polarity lexicon. But instead of letting one verb predict the polarity of these unseen nouns, we rather used a Naive Bayes assumption to let all verbs that co-occur with a noun vote for its polarity. It turned out that our approach was able to reproduce from 81% and up to 92% of the true noun polarities.

The paper is organized as follows. First, we introduce the resources being used, then we discuss our approach to separate bipolar-preference from single-preference verbs. We then show how the classification of prior polarities of nouns can be achieved on the basis of the polar verbs and a Naive Bayes model. We finally discuss related work and conclude.

2   Resources

Our experiments were based on a freely available German polarity lexicon comprising 8'400 entries, 3'400 nouns, 1'600 verbs and 3'800 adjectives (see [3]). The word polarities for nouns and verbs were manually tagged, whereas 1000 of the 3800 adjectives were semi-automatically derived.

Our primary data source was the DeWaC corpus (see [4]), a large collection of web-retrieved documents in the German language. We worked with a subset of the DeWaC corpus consisting of 20 Million sentences. We fed this dateset to a dependency parser developed by Sennrich et al. [6], in order to obtain dependency information relating verbs with their objects. We subsequently labeled this automatically parsed dataset with polarity information originating from the lexicon. From this preprocessed collection of sentences we proceeded to extract verbs and their polar objects by applying the following restrictions:

- based on the available dependency information, select all verbs and their objects
- allowed objects are noun phrases consisting of exactly one noun and an optional preceding determiner (an article in most cases)
- the core noun of such a noun phrase must be labeled as positive or negative

Based on these restrictions we produced 22'155 triples, where each triple comprises a verb, a noun being the verb's object and a count of the observed co-occurences of the given verb-object pair. The total number of noun types is 1'781, they co-occur with 1'776 verbs. Fig. 1 gives a couple of examples[1].

verb	noun	translation(en)	freq
abbauen	Stress	stress	62
abbauen	Übergewicht	overweight	18
abbauen	Unsicherheit	unease	5
auslösen	Ekel	disgust	1
auslösen	Empörung	indignation	45
auslösen	Enthusiasmus	enthusiasm	1

Fig. 1. Input Data

3   Verb's Preference Determination

As mentioned above, we extracted triples consisting of verbs, their objects and the triple's observed frequency. From this set of triples, a list of verbs can be further generated with new frequency counts: the number of positive nouns and the number of negative nouns they have as their direct objects. We distinguish three such frequency patterns:

1. verbs, where one polarity is clearly prevailing
2. verbs, where the frequency of a single polarity orientation is higher than the frequency of the opposite polarity, but where the frequency counts do not indicate a clear-cut preference
3. verbs, where - on average - both polarities occur with the same frequency

In all these cases, neutral NPs are to be expected as well, but we can not calculate their frequency as the polarity lexicon we use contains no entries with neutral polarity. With verbs of type (1), these neutral NPs are likely to adopt the polarity preference (cf. Fig. 2, example 1). It is not entirely clear, how verbs of type (2) and type (3) do influence and can define these neutral NPs and if they do it at all. (cf. Fig. 2, example 2 and 3). Moreover, verbs with a clear preference sometimes occur with a NP that has a prior polarity that conflicts with the verbs polarity preference. For instance, the sentence *"He approves the war"*, where *"approve"* has a preference for positive polarity (one normally approves of something positive, which indicates that a neutral object or an object of unknown polarity should be regarded as something positive) and *"war"* is clearly negative. Such conflicts are to be expected given a large text corpus. There are cases where a real conflict was intended by the writer/speaker, but noise can also occur, e.g. stemming from preprocessing (e.g. parsing).

---

[1] *"abbauen"*: *"to reduce"*, *"auslösen"*: *"to trigger"*.

1. She loves books/skating.
2. He expresses interest/contempt.
3. She feels happy/angry.

Fig. 2. Polar and Bipolar Verbs

The presence of neutral nouns cannot tell anything about a verb's preference. So we do not have to care about our inability to classify them as neutral. Also, verbs with a balanced frequency pattern (same number of positive and negative nouns) seem to be less interesting, since one cannot predict anything for a noun occurring as an object of such verbs (we call these verbs bipolar). This is not to say that they are of completely no interest, since some of them might be shifters - turning the polarity from positive to negative and vice versa (e.g. *"lose control"* = negative, *"lose fear"*= positive).

3.1   Hypothesis Testing

As previously described, we are interested in those verbs that could help us predict the polarity of unseen nouns. It poses, however, the question of how to operationalize distinguishing between verbs that exhibit a clear-cut polarity preference in regards to their direct objects and those verbs that tend to be 'indifferent' towards the polarity of their direct objects.

$$\chi^2_{(k-1)} = \sum_{j=1}^{k} \frac{(f_j - e_j)^2}{e_j}$$

where:
e_j = expectation
f_j = frequency
k = number of independent trials

Fig. 3. Chi Square

The chi-square test (see. Fig. 3) used as a hypothesis test allows for evaluation of a postulated prior probability of events in the light of frequency counts from a sample. Based on the expected frequencies, the difference to the real frequencies is used to derive the so-called p-value, which is a value taken from the chi-square distribution. If the likelihood of the p-value and any value greater than it is below a significance threshold, the so-called null hypothesis is rejected and the alternative hypothesis is adopted. The null hypothesis in our case could either be that a) a verb has a clear polarity preference or b) that a verb is a bipolar-preference verb. In either case, we first have to define what we mean by these distinctions. We define it like this:

1. A verb has a clear-cut preference, it is polar, if the probability of one orientation is 1 and the probability of the other orientation is 0.
2. A verb is bipolar, if the probability for both polarities is 0.5

Definition (1) does not lead to a valid null hypothesis, since on one hand the expected frequencies of the null class, namely 0, would lead to a division by zero (one of the denominators) and on the other hand they would not be in line with the requirement that $e_j > 5$. So we had to continue with definition (2). We thus set up the test hypothesis shown in Fig. 4. H_0 states that a verb is bipolar, H_1

$$h_0 : p(verb = POS) = p(verb = NEG)$$

$$h_1 : p(verb = POS) \neq p(verb = NEG)$$

Fig. 4. Hypothesis testing

that is not bipolar. But does being non-bipolar actually implies that the verb is polar, i.e. has a clear polarity preference? Clearly, this test cannot tell us much about the strength of such a preference. Thus, we use conditional probabilities to quantify the strength parameter. Although H_0 seems to be rather strict, defining bipolarity by an equal probability, the chi-square distribution allows for some deviation from such a strict requirement.

For every verb v, we set the null hypothesis to be that v is bipolar. We then took the verb's positive and negative noun distribution ($ndist$) and determined a p-value (one degree of freedom). If the p-value ($pval$) was less or equal to 0.05 we rejected H_0 and adopted H_1, meaning that verb was categorized as polar, i.e.:

$$pval(v, ndist) = chi^2_{(1)} =< 0.05 \rightarrow polar(v)$$

Since we do not have a gold standard of preclassified polar and bipolar verbs, we decided to test our approach by finding out how well as being polar classified verbs predict the prior polarity of unseen nouns. This is described in the next section. We here give some examples of polar verbs derived by this criterion (see Fig. 5). From the total of 1776 verbs, 420 were classified as polar, while 1356 were classified as bipolar. Each row in Fig. 5 shows a polar verb, the conditional probability that a noun occurring with the verb is negative and positive, respectively and the p-value (on the 0.05 significance level). There are clear cases, e.g. *"tilgen"* (to extinguish) with a polarity of 1 for negative polarity and 0 for positive polarity. But also weaker candidates are among the entries, e.g. *"kompensieren"* (to compensate), where the probability of a negative noun is 0.133 and 0.866 for a positive noun. Before we turn to the question how to evaluate this list, we give the formulas to determine the conditional probabilities from Fig. 5. We use the relative frequency of a verb and the nouns that are its objects. Fig. 6 gives the estimation for the conditional probability of a positive orientation POS for a noun given the verb. The formula for a negative orientation is defined accordingly.

verb	translation(en)	$P(n{=}POS|v)$	$P(n{=}NEG|v)$	p-value
kompensieren	to compensate	0.133	0.866	0.004509
hinnehmen	here: to bear	0.125	0.875	0.0027
zusagen	to promise	0.875	0.125	0.03389
eindämmen	to stem	0.111	0.888	0.00096
gewährleisten	to guarantee	0.9444	0.0555	0.00016
beheben	to mend	0.076	0.923	0.002282
erbitten	to ask for	1	0	0.00031
hassen	to hate	0.1	0.9	0.00034
befriedigen	to satisfy	1	0	0.0081
vorgaukeln	to simulate	1	0	0.00091
tilgen	to extinguish	0	1	0.01431
zusichern	to assure	1	0	0.000532

Fig. 5. Polarity preference of verbs

$$P(POS|verb) = \frac{\#pos\_nouns}{\#pos\_nouns + \#neg\_nouns}$$

Fig. 6. Conditional Probability Estimation

4   Classification of Prior Noun Polarities

Having classified verbs according to their positive or negative polarity preference is crucial for the polarity classification of unseen nouns (i.e. nouns that are not in the polarity lexicon). For instance, if the noun *"stability"* is unknown, but occurs in a phrase like *"it guarantees the stability"*, according to the polarity preference of *"to guarantee"* it should be classified as POS, since $P(POS|guarantee) > P(NEG|guarantee) = 0.9444 > 0.0555$. However, with *"to promise stability"* this is not so clear, since $P(NEG|promise)$ is 0.125. Moreover, *"stability"* might occur with a verb that has a negative polarity preference, namely *"to bear"* or *"to compensate"*. That is, the polarity of an unseen noun should be the result of a voting among all verbs that have the noun as their direct object. We can operationalize this under a Naive Bayes Assumption, see Fig. 7. The probability of a noun's polarity given its co-occurrence as an object of a number of verbs, v_i, is approximated by making an independence assumption and computing the product of the probabilities for the probability of a polarity orientation given

$$P(n_j = POS|v_1, ..v_n) = \prod_i P(POS|v_i)$$

Fig. 7. Naive Bayes Assumption

$$pol(n_j) = \begin{cases} POS & if \quad P(n_j = POS|v_1,..v_n) > P(n_j = NEG|v_1,..v_n) \\ NEG & if \quad P(n_j = POS|v_1,..v_n) < P(n_j = NEG|v_1,..v_n) \\ undef \; else \end{cases}$$

Fig. 8. Noun Polarity

that the noun occurs with a single verb. Fig. 8 gives the full definition of the polarity determination for an unseen noun. Strictly spoken, we can only infer the absence of the inverse polarity of the polarity that wins. If *"stability"* was classified as being positive, then, if we want to be precise, we could only say that it is not negative. The reason is, it could be positive *or* neutral. As we mentioned earlier, nouns co-occurring with a polarity preference verb inherit the preference if they are neutral. But this does not mean they necessarily have this as a prior polarity. For instance, if one tells us that he prefers coke (over beer), we would not classify coke as positive in general, but as being positive only for the person (group or institution) that approved to such a statement. However, this is true only as long as we consider a single statement. But if a noun co-occurs many times with different polarity preference words and possibly different opinion holders, we start becoming more confident in deducing a noun polarity that is based on common sense, rather than on personal preference. In the way our approach works we aim to derive a prior polarity of nouns.

5   Evaluation

How can we evaluate our approach? One way is to have a closer look at the upper bound for precision. A simple way to measure this was to use the prior-polarity lexicon and the preclassified nouns therein and let the system try to recover their real orientation. This is described in the next section. A second way for evaluation was to test the learned verb preferences on nouns with unknown prior polarity, and check the predictions of our method against human evaluators. Both of these ways are described in the following two subsections.

5.1   Experiments Based on a Polarity Lexicon

The preference polarity of a verb is determined in our approach based on the prior polarities of the nouns it has as objects in a large corpus. We have sampled over 22'000 triples, where each triple specifies the frequency a single noun co-occurs with the given verb as its direct object. For instance, the verb *"abbauen"* (here *"to reduce"*) has 65 times the negative noun *"Stress"* (stress) as a direct object. But *"Stress"* might occur with other verbs as well, actually is occurs with 91 different verbs with a total frequency of 312. In order to measure how well our approach reproduces the prior polarities of the nouns co-occuring with the verbs, we adopted a held-out scenario. We removed a single noun from our triples (i.e. every triple that counts the frequency of the noun given a single verb) and subsequently ran our algorithm (i.e. trained the system) and classified the heldout noun with the learned statistical model. Fig. 9 shows the core algorithm.

```
forall heldout ∈ nouns
    ho_verbs = all verbs that have heldout as an object
    forall hov ∈ ho_verbs
        npos(hov) = number of nouns types with positive prior polarity (not counting heldout)
        nneg(hov) = number of nouns types with negative prior polarity (not counting heldout)
        pval(hov) = chi(1) based on npos(hov), nneg(hov) and p=q=0.5
        if pval(hov) < 0.05 do (i.e. the verb is polar)
            P(POS|hov) = npos(hov) /(npos(hov) +nneg(hov))
            P(NEG|hov) = nneg(hov) /(npos(hov) +nneg(hov))
            pred_pol(heldout)= POS if P(POS|hov) > P(NEG|hov)
            pred_pol(heldout)= NEG if P(POS|hov) < P(NEG|hov)
            pred_pol(heldout)= no_prediction if P(POS|hov) == P(NEG|hov)
            hit+= if pred_pol(heldout) == true_polarity_of(heldout)
```

Fig. 9. Core Algorithm

A verb is polar, if it passes the threshold set by chi-square. There are however other parameters that influence the precision of the algorithm:

1. minfvn: the number of different verbs the heldout noun occurs with
2. minfn: the frequency of the heldout noun
3. minfv: the frequency of the verb, the heldout noun occurs with

It turned out that the precision ranges from 81.97% up to 92.20 % by varying these parameters. Fig. 10 gives the details of various runs with different settings for minfvn, minfn and minfv. By introducing these thresholds, nouns that do not occur in a context that fulfills the parameter settings are left unclassified,

minfvn	minfv	minfn	prec	#hit	#all
1	1	1	81.97	1210	1476
1	5	2	83.32	1228	1475
2	5	2	85.52	768	898
3	5	2	85.54	497	581
4	5	2	86.68	358	413
4	10	5	87.62	347	396
4	10	10	87.62	347	396
5	10	5	88.53	224	253
5	20	5	91.06	214	235
5	**25**	**5**	**93.34**	**217**	**235**
5	30	5	89.16	210	235
6	10	5	92.43	159	173
6	25	5	91.66	154	168

Fig. 10. Varying Thresholds

i.e. no prediction is made and, thus, the total number of classifications drops the more restrictive the setting is (see column # all). For instance, the last line is the most restrictive setting, here only 168 out of the total number of 1781 nouns are classified. 312 nouns are not classified in any setting, they occur only with bipolar verbs. Please note that we can control this situation, that is, we can actually decide to run for high precision by simply setting the parameters accordingly.

The best result in terms of precision is achieved with minfvn=5, minfv=25 and minfn=5. Here, 235 nouns are classified, 217 were classified correctly (93.34 %), This is the upper bound precision our approach reaches given the corpus at hand. We deliberately have not measured recall, we wanted to fix the parameter setting that gives us the best precision if applied to unseen examples (see the next section) For illustration purposes, we give some examples of misclassified nouns in Fig. 11. Please note that among these conflicts are also cases where

verb v	English	predicted	true
Kompliment	compliment	NEG	POS
Schauder	shiver	POS	NEG
Entwarnung	all-clear	NEG	POS
Begehrlichkeit	greediness	NEG	POS
Spannung	suspense	NEG	POS
Unterwerfung	repression	POS	NEG
Offenheit	frankness	NEG	POS

Fig. 11. Wrong Predictions

the polarity value of a word according to the prior-polarity lexicon seems to be wrong. For example, *"Begehrlichkeit"* (roughly, greediness) is positive according to the lexicon, but negative according to the algorithm's prediction. We actually would argue that the prediction is right, and the lexicon entry wrong. We have not systematically explored the possibility to automatically detect such false entries.

It is an interesting question where the other conflicts arise from. Noise might be a reason, e.g. the wrong parse trees (i.e. wrong objects). Another and quite to be expected factor is word sense ambiguity. Take *"Spannung"*: it is positive if suspense is meant, and negative if it is used in the sense of strain (Anspannung); it is even neutral in the (electrical) sense of voltage. The mix-up of these different polarities for the word might explain the misclassification. If might even be the case that in our corpus actually *"Spannung"* is mostly used negative in the sense of strain. So another function of our approach could be to detect words that are highly ambiguous and should better not be part of a prior-polarity lexicon. Conflicting polarities could also indicate, if the corpus is domain-specific, that the polarity of a word in that domain is different from the *"normal"* one. Further work is needed to explore these possibilities. Finally, there is also the possibility of real conflicts in the data (e.g. someone approves/likes something negative), especially given low frequency cases, where no majority vote can compensate for extreme attitudes towards an object.

5.2 The Case of Unseen Nouns

In order to evaluate our approach not only on heldout nouns that have a known prior positive and negative polarity, but to also see how well it works in the presence of neutral nouns, we extracted 200 unseen nouns that have no entry in the polarity lexicon. We took nouns that occur with a high frequency with the polar verbs our approach have identified. We then ran the system with the parameter setting that proved best (see Fig. 10) and produced a list of 200 polarity classifications of unseen nouns. Fig. 12 shows some examples. Two an-

noun	pred	true
Aufenthaltserlaubnis	POS	POS
Urlaub	POS	POS
Bereitschaft	POS	POS
Hindernis	NEG	NEG
Ausbildungsplatz	POS	POS
Geburtstag	NEG	POS
Wertung	NEG	NEUT
Tisch	POS	NEUT
Kommentar	NEG	NEUT
Rad	NEG	NEUT

Fig. 12. Nouns' Prior-Polarity Predictions

notators have then evaluated this list. The labels were (1) agree, (2) disagree, (3) neutral. That is, if the noun actually was neutral, no disagreement was (and should not be) measured, since the nouns appeared as objects of verbs with a polarity preference (the annotators did not see the verb or the whole sentence). So in the context of their verbs and sentences, the labeling as polar must not necessarily be wrong (the neutral noun inherits the preference by definition). For instance, *"Rad"* (bicycle) was, according to Fig. 12, labelled as negative - so it occurred with negative verbs mostly (e.g. *"I hate bicycles"*). In that context, *"bicycle"* is negative. We simply have not checked this in the present investigation. So neutrals cannot be evaluated properly by just looking at our result list. The annotators agreed on 24 cases (86%) and disagreed on 4 cases, so we found 28 polar nouns, the rest, 144, were neutral. What was somewhat disappointing is the fact, that only few nouns with a prior polarity turned up - most nouns were neutral.

6   Related Work

It is not the first time that a corpus-driven approach has been used for predicting the semantic orientation of a specific word class. Such approaches have in general been widely applied in the field of Sentiment Analysis over the last decade. Exploiting syntactic relations between specific word-classes and patterns of co-occurrences has proven quite effective. The word classes that mostly receive the focus of such approaches are the two usual suspects: adjectives as treated by Clematide and Klenner in [3] and by Hatzivassiloglou and McKeown in [7] and nouns as treated by Rillof et al. in [8]. Especially the latter attempt comes quite close to the idea behind our method. It is however a bootstrapping method while we are equipped with a complete polarity lexicon.

The role that verbs play in sentiment analysis, especially their semantic role, is a very important point and has been well defined in the work of Chesley et al. in [9] and also by Neviarouskaya et al. in [10]. Our method is not incompatible with these ideas, but we choose to neglect any semantic information in favor of simplicity. The semantics of verbs are extremely important, especially in compositional, subsentential frameworks. We choose an engineering approach trying to see how far minimal linguistic knowledge can lead in a prediction task.

7   Conclusion and Future Work

We introduced a purely corpus-based method to both, detecting verbs that have a polarity preference towards their objects and - based on that - a method to derive the prior polarity of nouns co-occurring with these verbs. The polarity preference of a verb allows e.g. to deduce the attitude that an opinion holder has towards the opinion target. So our approach fits well into existing lines of research. Moreover, it demonstrates how to learn the polarity preference of verbs instead of manually assigning it. The advantage is that these polarity preferences are empirically licensed. A statistical measure was used to operationalize the distinction between such *polar* verbs and verbs without a clear-cut tendency, which we called *bipolar* verbs. As our experiments with heldout data showed, the polarity preference of verbs can be reliably learned.

The polarity preference expressed in a single statement based on the basis of a single polar verb does not justify the classification of an opinion (noun) object as bearing a prior, positive or negative polarity. However, if a single noun occurs with different polar verbs produced by different opinion holders in different statements, then we no longer talk about personal preferences but enter the realm of common sense consensus. However, as discussed, noise and ambiguity at the word level interfere with this tendency, a fully automatic identification of nouns with a prior polarity is therefore quite challenging.

We have discussed various applications for our approach, e.g. one could use it to carry out domain adaption (detecting words that have a domain-specific polarity) or to check the entries of a polar lexicon on the basis of a large corpus for false entries. It might be better to remove highly ambiguous words from the lexicon - our method could help to identify such words.

References

1. Wilson, T., Wiebe, J., Hoffmann, P.: Recognizing Contextual Polarity in Phrase-Level Sentiment Analysis. In: Proceedings of the Conference on Human Language Technology and Empirical Methods in Natural Language Processing, pp. 347–354 (2005)
2. Esuli, A., Sebastiani, F.: Sentiwordnet: A publicly available lexical resource for opinion mining. In: Proceedings of the Fifth Conference on Language Resources and Evaluation, pp. 417–422 (2006)
3. Clematide, S., Klenner, M.: Evaluation and extension of a polarity lexicon for German. In: Proceedings of the First Workshop on Computational Approaches to Subjectivity and Sentiment Analysis, pp. 7–13 (2010)
4. Baroni, M., Bernardini, S., Ferraresi, A., Zanchetta, E.: The WaCky Wide Web: A collection of very large linguistically processed Web-crawled corpora. Language Resources and Evaluation, 209–226 (2009)
5. Wiebe, J., Wilson, T., Cardie, C.: Annotating expressions of opinions and emotions in language. Language Resources and Evaluation 39(2), 165–210 (2005)
6. Sennrich, R., Schneider, G., Volk, M., Warin, M.: A new hybrid dependency parser for German. In: Proceedings of the German Society for Computational Linguistics and Language Technology, pp. 115–124 (2009)
7. Hatzivassiloglou, V., McKeown, K.R.: Predicting the semantic orientation of adjectives. In: Proceedings of the Eighth Conference on European Chapter of the Association for Computational Linguistics, pp. 174–181 (1997)
8. Riloff, E., Wiebe, J., Wilson, T.: Learning subjective nouns using extraction pattern bootstrapping. In: Proceedings of the Seventh Conference on Natural Language Learning at HLT-NAACL 2003, vol. (4), pp. 25–32 (2003)
9. Chesley, P., Vincent, B., Xu, L., Srihari, R.K.: Using Verbs and Adjectives to Automatically Classify Blog Sentiment. Training 580(263), 233 (2006)
10. Neviarouskaya, A., Prendinger, H., Ishizuka, M.: Semantically distinct verb classes involved in sentiment analysis. In: Proceedings of the International Conference on Applied Computing, pp. 27–34 (2009)

Labeling Queries for a People Search Engine

Antje Schlaf, Amit Kirschenbaum, Robert Remus, and Thomas Efer

Abteilung Automatische Sprachverarbeitung,
Institut für Informatik, Universität Leipzig
PF 100920, 04009 Leipzig, Germany
{antje.schlaf,amit,rremus,efer}@informatik.uni-leipzig.de

Abstract. We present methods for labeling queries for a specialized search engine: a people search engine. Thereby, we propose several methods of different complexity from simple probabilistic ones to Conditional Random Fields. All methods are then evaluated on a manually annotated corpus of queries submitted to a people search engine. Additionally, we analyze this corpus with respect to typical search patterns and their distribution.

Keywords: query labeling, people search, conditional random fields.

1   Introduction

Understanding users' queries is a key concept in search engine performance since they encode users' intent. Classifying a query correctly may assist search engines in this task to e.g., process the query in a way customized to the query's class, which would then retrieve more relevant results for the user. This, among other factors, lead to increased development of specialized search engines, which focus on indexing content of particular topics, genres, media types etc. The domain of people search has gained attention, in recent years, both as a research topic and in commercial tools. This can be observed given the number of events dedicated to it, such as the Expert search in TREC Enterprise Track[1] and WePS[2] workshops, and the different search engines dedicated to this purpose e.g., pipl[3], zoominfo[4], or Yasni[5].

In the domain of people search, users' intent often means searching for details about a specific person, also known as *person profiling*, i.e. describe that person based on the given query terms. Alternatively, users are interested in finding people who are skilled in a given field, or for a given task, i.e. *expert recommendation*.

Labeling query terms with semantic classes is a way to achieve this goal, since it assists the search engine in categorizing the query. Query labeling is a

---

[1] http://www.ins.cwi.nl/projects/trec-ent/wiki/index.php/Main_Page
[2] http://nlp.uned.es/weps/
[3] www.pipl.com
[4] www.zoominfo.com
[5] www.yasni.com

G. Bouma et al. (Eds.): NLDB 2012, LNCS 7337, pp. 47–57, 2012.

challenging task due to ambiguity of query terms and the very limited context, which does not contribute to decreasing this ambiguity. This is true also in the case of people search where a query term can be e.g., a first name, a last name, a name of a location etc.

This paper presents several methods for labeling people search engine queries. The methods vary in complexity from simple probabilistic ones to Conditional Random Fields (CRFs), which are used to create labeling models based on the assumption that there is a sequential dependency among adjacent labels for query tokens.

2   Related Work

Although searching the web to fulfill our information needs is almost omnipresent in our everyday life, comparatively little work has been published on the structure and analysis of web-search queries, especially on queries to specialized search engines, like product search engines, or in our case, people search engines. We know that web-search queries contain on average 2.35 [14] to 2.4 terms [15]. Barr et al. [3] investigate the linguistic structure of English-language web-search queries and "find the majority of query terms are proper nouns, and the majority of queries are noun-phrases". Thus, they conclude, web-search queries fundamentally differ from language usually observed in standard English-language corpora.

Due to their rather limited context, web-search queries also often tend to be ambiguous. Allan and Raghavan [1] approach these ambiguities by asking the user for feedback regarding a query term's context in question. Paşca [12] uses web-search queries to automatically extract all sorts of named entities from them. Jansen and Booth [6] classify AOL search engine queries using a three-level hierarchy of user intent and by topic. Given manually annotated queries, they analyze topics' predominant user intent.

Possibly closest to our work are those of Li et al. [8] and Manshadi and Li [9]. The method in [8] assigns a pre-defined category to each query term using supervised and semi-supervised CRFs. Their training data consists of a small amount of manually labeled queries and large amounts of a priori unlabeled queries, in which some query terms are then automatically labeled by leveraging additional resources [17]. Li et al. [8] reach up to almost 80% accuracy on query-level and up to 89% accuracy on term-level using semi-supervised CRFs, significantly outperforming supervised CRFs by about 1% to 12%, depending on the amount of manually labeled training data used. Although Li et al. [8] state that "[...] despite the fact that user queries are commonly viewed as bags-of-words, we found that the field ordering in product search queries does display statistically significant patterns that can help sequential labeling." [9] uses a multi-set approach, i.e. a bag-of-words, which, according to them, leads to the main drawback: the complete ignorance of query term ordering. Still, using a combination of a probabilistic context-free grammar to parse web-search queries and a discriminative model on top to re-rank best parses, they outperform their previous best results, even for very small training sets. All their experiments focus on product search queries, whereas we focus on a maybe even more specialized query domain: people search.

3   Query Labeling

Our task is to find a label pattern by tagging each of the tokens in the query with a suitable label. We have used several methods to predict label patterns, which are described below.

3.1   KBP – Labeling Using Knowledge Base Probabilities

We used a knowledge base featuring tokens and their possible labels with corresponding relative weights (cf. Section 4.1). The weights were normalized for each token to create a conditional probability distribution for labeling individual tokens

$$p(l|t) = \frac{w}{\sum\limits_{w'} w'} \qquad (1)$$

where t is a token, l is a possible label for t with respective weight w, and w' are weights corresponding to other possible labels for t. Each query is then represented as a bag of tokens, and for each token the label with the corresponding maximal weight is selected.

3.2   TPP – Labeling by Training-set Pattern Probabilities

The queries which serve as training data were used to construct a distribution over possible label patterns, based on query length. That is, for queries containing l tokens, a distribution of label sequences was created based on their relative frequencies: the probability for a pattern s of length l would be

$$p(s) = \frac{freq(s)}{\sum\limits_{\{s'|\ |s'|=l\}} freq(s')} \qquad (2)$$

where s' is some pattern of length l. A query of length l would then be assigned with the most probable label pattern, i.e., sequence of labels, for this length.

3.3   KBC-TPP – Labeling by Knowledge Base Candidates
Combined with Training-set Pattern Probabilities

We used the knowledge base mentioned earlier, and the distribution over label patterns to assign labels for query tokens: For a each token t_i in a query

$q = (t_1, \ldots, t_n)$, a set of all possible labels were retrieved as candidates from the knowledge base: $L_{t_i} = \{l_{ij}\}$. The label sets corresponding to the tokens were used to construct a Cartesian product of all possible n-tuples of labels

$$L_{t_1} \times \ldots \times L_{t_n} = \{(l_{1m}, \ldots, l_{nk}) | l_{1m} \in L_{t_1}, \ldots l_{nk} \in L_{t_n}\} \tag{3}$$

The selected n-tuple is then the one with the highest probability according to the label pattern distribution constructed from the training data. We introduced also a wild-card label which was used in case the set of possible labels of some token in the query was empty. In this case we seek to match a pattern of same length without considering that label of that token, and assign its label according to the best match.

3.4 KBP-TPP – Labeling by Knowledge Base Probabilities Combined with Training-set Pattern Probabilities

This method can be viewed as a weighted version of the method described above. The probability for each possible n-tuple of labels is further multiplied by the weights of individual token labels for the given tokens, extracted from the knowledge base. The relation between a token and its label in a proposed pattern is thus taken into account, a fact which should better reflect the association between the query and its assigned labels.

3.5 CRFs – Conditional Random Fields

Additionally to the methods described above, we address the problem of query tagging as a sequence labeling task, i.e. mapping a sequence of observations $x = (x_1, \ldots, x_T)$ to a sequence of labels $y = (y_1, \ldots, y_T)$, where the value of each y_t is selected from a given set of categories. For this task we employ Conditional Random Fields [7,16]. CRF models maximizes the conditional probability $P(y|x)$ of a sequence of labels given the sequence of observations. In particular, linear-chain CRFs model the probability:

$$p(y|x; \Lambda) = \frac{1}{Z(x, \Lambda)} exp(\sum_{t=1}^{T} \sum_{k=1}^{K} \lambda_k f_k(y_t, y_{t-1}, x, t)) \tag{4}$$

$Z(x, \Lambda)$ is a normalization factor over possible labels of x, which transforms the exponent form into a probability distribution.

$$Z(x, \Lambda) = \sum_{y} exp(\sum_{t=1}^{T} \sum_{k=1}^{K} \lambda_k f_k(y_t, y_{t-1}, x, t)) \tag{5}$$

$\Lambda = \{\lambda_k\}$ are the model parameter weights corresponding to the set of feature functions $F = \{f_k(y, y', x, t)\}$ where every f_k may depend on the current label

y_t on the previous label y_{t-1} and on the observation sequence x. A query is then represented by a vector of feature functions based on its tokens. The values for parameters Λ are estimated from the training data by an objective function maximizing the penalized log-likelihood:

$$L(\Lambda) = \sum_{i=1}^{N} \log p(y^{(i)}|x^{(i)}; \Lambda) - R(\Lambda) \qquad (6)$$

Where $D = \{(x^{(i)}, y^{(i)})\}_{i=1}^{N}$ is a set of training examples consisting of sequences of observations and their corresponding label sequences. The second term is a regularization function to reduce over-fitting of the parameters to the training data (e.g., ℓ_2 regularization).

4   Experiments

In order to assess the performance of the proposed methods in a query labeling task, we applied them to a corpus of manually annotated queries to a people search engine. We used the tokens predefined by the manual annotation, so the labeling is not affected by potential wrong tokenization. This separates the actual labeling problem from the recognition of multi-word units, which will not be addressed here. First, we describe our resources, i.e. the knowledge base and the corpus of queries. Then we describe the features used by our CRFs and the optimization methods employed to train their models. Finally, we perform a model selection for the CRFs and compare it with the other methods.

4.1   Knowledge Base

The knowledge base contains instances and their categories relevant to people search, as shown in Table 1. This list with instances and probability-weighted category label assignments is based on knowledge acquired by Yasni. This knowledge was partly inferred by a method described by Schlaf and Remus [13]. We access the knowledge in a case insensitive manner but did not apply any further preprocessing.

4.2   Data Sets

Our data set was obtained from real user queries randomly sampled from Yasni's search engine, which focuses on German users, and annotated by one expert annotator from Yasni. Since real user queries might contain data which allows for conclusions about a person's identity, we unfortunately cannot publish our query corpus due to data privacy reasons and applicable law. Therefore, the following example, as presented in Table 2, lists some artificially created but typical queries and their manual annotation marked by a pipe symbol.

Table 1. Categories of the knowledge base and their respective number of instances

Category	Number of instances
company	208,708
lastname	122,827
firstname	106,061
location	31,464
skill	14,235
profession	5,680
badword	2,517
title	1,115
business	568
functional	147
media	26

Table 2. Typical queries and their annotations

```
Max|firstname Mustermann|lastname
Max|firstname Thomas|firstname Mustermann|lastname
Sabine|firstname Thomas|lastname Berlin|location
Meier|lastname Max|firstname Siemens|company
Deutsche Post|company
```

The only pre-processing performed on queries was the removal of potential quotation marks. In particular, we did not correct any spelling errors, since it has only little effect on actual retrieval quality [10].

The queries are split into a training corpus, consisting of 7,004 queries and a test corpus, consisting of 2,000 queries. The training corpus was analyzed regarding the query-based distribution of patterns and pattern lengths, and the token-based distribution of assigned categories. Analyzing the distribution of pattern lengths in the training corpus, as presented in Table 3, reveals an average length of 2.04 tokens. The majority of queries (85.57%) have a length of two tokens. Queries with length greater than three rarely occur (1.26%). Thus, people search engine queries provide even less context than queries to general purpose search engines [14,15].

Table 3. Distribution of pattern lengths

Pattern length	1	2	3	4	5	6	7	8
Amount	443	5,993	480	65	16	4	2	1
Percentage	6.35	85.57	6.85	0.93	0.23	0.06	0.03	0.01

The categories used for labeling as well as their distribution over the manually labeled queries are shown in Table 4. Most of the assigned labels (92.8%) are "firstname" and "lastname". Inspecting the 10 most frequent patterns in Table 5

Table 4. Distribution of assigned categories in the training corpus

Category	Amount	(%)
firstname	6,669	46.78
lastname	6,559	46.01
term	347	2.43
location	195	1.37
company	137	0.96
badword	118	0.83
functional	56	0.39
business	45	0.32
title	44	0.31
media	39	0.27
profession	34	0.24
skill	7	0.05

Table 5. Distribution of patterns in training corpus – 10 most frequent patterns

Pattern	Amount	(%)
firstname lastname	5680	81.11
firstname firstname lastname	270	3.86
lastname	150	2.14
lastname firstname	107	1.53
term	85	1.21
company	84	1.20
firstname lastname lastname	49	0.70
location	46	0.66
term term	31	0.44
firstname	26	0.37

reveals that the pattern "firstname lastname" is by far the most frequent pattern (81.11%), which concurs with our intuition regarding the use of a people search engine.

4.3   CRF Setup

We have experimented with different types of optimization algorithms as well as with different sets of features. The query labeling method was implemented using MALLET [11], a Java based toolkit for machine learning. We used optimization algorithms which were part of MALLET, namely L-BFGS [5] with ℓ_1 and ℓ_2 regularization functions, and Stochastic Gradient Descent [4]. Our feature set consisted of different local and contextual features, some of which relate to our methods before (cf. Section 3):

1. The surface form of the token.
2. Possible labels of the token when applying the labeling method KBP.
3. Possible labels of the token when applying the labeling method TPP.
4. Possible labels of the token when applying the labeling method KBP-TPP.
5. Suffix of length 3 for each token.
6. Prefix of length 3 for each token.

We used features derived from neighboring tokens in various configurations. The following list describes the various settings of using these values. Numbers in parentheses encode the position of the neighbor relative to the current token. If there is more than one position in parentheses, then values from the respective tokens were taken in conjunction to create a new feature.

1. No conjunction – features taken only from the current token.
2. (-1),(1)
3. (-2),(-1),(1),(2)
4. (-1,1)

To retrieve the best setting, each optimization algorithm was evaluated with all possible subsets of features combined with each conjunction type. The surface of the token was always selected as feature.

4.4   Results

Model Selection. To retrieve the best labeling method, based on token-wise micro-average precision, we applied the methods, as described above, to the training corpus using 10-fold cross validation. For the purpose of the evaluation we use only the predicted labeling with the highest probability and compare it to the actual one, though in general the methods return a ranked list of labeling options for each query.

With increasing complexity of the methods, they reach a higher precision. It appears that there were no significant differences in the performance of the different optimization methods we examined in the CRF model selection, with comparable feature set used. The best performance, however, was achieved by the Stochastic Gradient Descent optimization algorithm with the features 1, 2, and 4 and the conjunction type 1. The performance of each method, based on 10-fold cross validation and token-wise micro-average precision, is presented in Table 6.

Table 6. Model selection results

Labeling method	Precision
KBP	0.7303
TPP	0.8951
KBC-TPP	0.8960
KBP-TPP	0.9181
CRF	0.9289

Testing the Methods and Evaluating the Models. Selecting the best performing setting for CRF, we applied all labeling methods to the test corpus using the whole training corpus as training data. The results are presented in Table 7. As can be observed, all methods show better performance than in the training scenario (cf. Table 6). Notable is that the KBP-TPP method now even outperforms the more sophisticated CRF method. That effect cannot be simply explained with an overfitting model selection, since the 10-fold cross validation secures the method's ability for generalization. So the answer has to lie in differences between the training and test sets used, such as the following: Patterns of length 2 account for 86.85% of all queries in the test set in contrast to 85.57% in the training data (cf. Table 3). The most frequent pattern (firstname lastname) matches 83.55% of the test queries instead of 81.11% at training time; see Table 5. The five most frequent patterns have the same rank in both sets but represent 92.90% of the queries in the testing set and only 89.83% in the training set.

All in all the testing analysis exposes a slight short-term vulnerability of the CRF based method towards variations in the input data's statistical properties. That aspect could also be interpreted as "latent knowledge" that could be beneficial in the long run – all assuming that the statistic variations eventually middle out and e.g. longer queries appear again in the "expected" frequency later on in the constant stream of new queries. As an attempt of explanation, one could imagine some queries of length 7 in the training data that provide pattern infomation, that the CRF based method will incorporate in its general model while the KBP-TPP only saves it for queries of that exact length. If no query of length 7 appears in the test set, the non-CRF methods will not regard the corresponding trained data while the CRF model is still somewhat influenced.

Table 7. Testing results

Labeling method	Precision
KBP	0.7396
TPP	0.9243
KBC-TPP	0.9233
KBP-TPP	0.9470
CRF	0.9320

5   Conclusions and Future Work

We made two main contributions: First, we extensively analyzed queries to a people search engine regarding the categories most commonly searched for and the structure of the queries, i.e. the predominant patterns and the distribution of pattern lengths. Second, we successfully applied several methods with different complexity to the task of labeling queries to a people search engine. Thereby, we carried out an extensive model selection for the CRF method using different feature sets, and different optimization methods. Although the chosen CRF model reached very high precision on a held-out test set, it was outperformed by a small margin by the method KBP-TPP. We firmly believe this is a positive outcome, as the latter solution is both simpler and faster to train and apply than the former and thus, more elegant.

While multi-word unit recognition can potentially improve the overall labeling quality, multi-word unit terms exhibited in our training data set were actually quite rare (around 1%). In addition, this task requires highly precise methods for comparatively short queries, which is out of the scope of our current investigations, but is nevertheless interesting to explore.

Despite our focus on a German-language people search engine, the developed methods should be generic enough to be useful for other cultural and linguistic areas, which is to be proven in further research.

Since a knowledge base like described above could also be extracted from external resources like Wikipedia[6] or DBpedia [2], we would like to evaluate our methods with different knowledge bases and compare the results.

Acknowledgements. This research was funded by Sächsische AufbauBank (SAB) and European Regional Development Fund (EFRE). We gratefully acknowledge the annotation efforts of Yasni and thank them for providing us data and insights into their work.

References

1. Allan, J., Raghavan, H.: Using Part-of-speech Patterns to Reduce Query Ambiguity. In: Proceedings of the 25th Annual International ACM SIGIR Conference on Research and Development in Information Retrieval, pp. 307–314 (2002)
2. Auer, S., Bizer, C., Kobilarov, G., Lehmann, J., Cyganiak, R., Ives, Z.G.: DBpedia: A Nucleus for a Web of Open Data. In: Aberer, K., Choi, K.-S., Noy, N., Allemang, D., Lee, K.-I., Nixon, L.J.B., Golbeck, J., Mika, P., Maynard, D., Mizoguchi, R., Schreiber, G., Cudré-Mauroux, P. (eds.) ASWC 2007 and ISWC 2007. LNCS, vol. 4825, pp. 722–735. Springer, Heidelberg (2007)
3. Barr, C., Jones, R., Regelson, M.: The Linguistic Structure of English Web-search Queries. In: Proceedings of the Conference on Empirical Methods in Natural Language Processing (EMNLP), pp. 1021–1030 (2008)
4. Bottou, L.: Stochastic Learning. In: Bousquet, O., von Luxburg, U., Rätsch, G. (eds.) Machine Learning 2003. LNCS (LNAI), vol. 3176, pp. 146–168. Springer, Heidelberg (2004)
5. Byrd, R., Lu, P., Nocedal, J., Zhu, C.: A Limited Memory Algorithm for Bound Constrained Optimization. SIAM Journal on Scientific Computing 16(5), 1190–1208 (1995)
6. Jansen, B., Booth, D.: Classifying Web Queries by Topic and User Intent. In: Proceedings of the 28th International Conference on Human Factors in Computing Systems (CHI), pp. 4285–4290 (2010)
7. Lafferty, J., McCallum, A., Pereira, F.C.: Conditional Random Fields: Probabilistic Models for Segmenting and Labeling Sequence Data. In: Proceedings of the 18th International Conference on Machine Learning (ICML), pp. 282–289 (2001)
8. Li, X., Wang, Y.Y., Acero, A.: Extracting Structured Information from User Queries with Semi-supervised Conditional Random Fields. In: Proceedings of the 32nd International ACM SIGIR Conference on Research and Development in Information Retrieval, pp. 572–579 (2009)
9. Manshadi, M., Li, X.: Semantic Tagging of Web Search Queries. In: Proceedings of the Joint Conference of the 47th Annual Meeting of the ACL and the 4th International Joint Conference on Natural Language Processing (ACL/IJNLP), pp. 861–869 (2009)

---

[6] `www.wikipedia.org`

10. Martins, B., Silva, M.J.: Spelling Correction for Search Engine Queries. In: Vicedo, J.L., Martínez-Barco, P., Muñoz, R., Saiz Noeda, M. (eds.) EsTAL 2004. LNCS (LNAI), vol. 3230, pp. 372–383. Springer, Heidelberg (2004)
11. McCallum, A.K.: Mallet: A machine learning for language toolkit (2002), http://mallet.cs.umass.edu
12. Paşca, M.: Weakly-supervised Discovery of Named Entities using Web Search Queries. In: Proceedings of the 16th Conference on Information and Knowledge Management (CIKM), pp. 683–690 (2007)
13. Schlaf, A., Remus, R.: Learning Categories and their Instances by Contextual Features. In: Proceedings of the 8th International Conference on Language Resources and Evaluation, LREC (2012)
14. Silverstein, C., Henzinger, M., Marais, H., Moricz, M.: Analysis of a very large AltaVista Query Log. Tech. rep., Technical Report 1998-014, Digital SRC (1998)
15. Spink, A., Wolfram, D., Jansen, M., Saracevic, T.: Searching the Web: The Public and their Queries. Journal of the American Society for Information Science and Technology 52(3), 226–234 (2001)
16. Sutton, C., McCallum, A.: An Introduction to Conditional Random Fields for Relational Learning, ch. 4, pp. 93–128. MIT Press (2006)
17. Wang, Y., Hoffmann, R., Li, X., Szymanski, J.: Semi-supervised Learning of Semantic Classes for Query Understanding: from the Web and for the Web. In: Proceeding of the 18th ACM Conference on Information and Knowledge Management (CIKM), pp. 37–46 (2009)

Litmus: Generation of Test Cases from Functional Requirements in Natural Language

Anurag Dwarakanath and Shubhashis Sengupta

Accenture Technology Labs.
Bangalore, India
{anurag.dwarakanath,shubhashis.sengupta}@accenture.com

Abstract. Generating Test Cases from natural language requirements pose a formidable challenge as requirements often do not follow a defined structure. In this paper, we present a tool to generate Test Cases from a functional requirement document. No restriction on the structure of the sentence is imposed. The tool works on each requirement sentence and generates one or more Test Cases through a five step process – 1) The sentence is analyzed through a syntactic parser to identify whether it is testable; 2) A compound or complex testable sentence is split into individual simple sentences; 3) Test Intents are generated from each simple sentence (Test Intents map to the aspects on which the requirement is to be tested); 4) The Test Intents are grouped and sequenced in temporal order to generate Positive Test Cases. A Positive Test Case verifies the affirmative action of the system; 5) Wherever applicable, Boundary Value Analysis and other techniques are used generate Negative Test Cases. Negative Test Cases verifies the behavior of the system in exception conditions. The automated generation of the Test Cases has been implemented in a tool called Litmus. We provide experimental results of our tool on actual requirement documents across domains and discuss the advantages and shortcomings of our approach.

Keywords: Functional Testing, Test Case Generation, NLP, Link Grammar.

1 Introduction

Majority of the software projects have requirements written in Natural Language [1], [10]. Surveys show 79% [7] of all requirements are documented in free-flow natural language and only 7% [1] use formal specifications. Although the use of natural language for requirement description is known to be inherently ambiguous [10]; natural language is used for the ease in comprehension and the ease in sharing [1].

Numerous efforts in software engineering have attempted to convert requirements into the next phases of system implementation artifacts automatically. In particular, as test planning and execution contribute to a large percentage (40-70%) of overall project cost [7]; automatic creation of Test Cases is expected to give significant benefits.

In industrial software development, a Functional Requirement Document (FRD) is typically the first level of documentation of a project. The FRD, also known as a

G. Bouma et al. (Eds.): NLDB 2012, LNCS 7337, pp. 58–69, 2012.
© Springer-Verlag Berlin Heidelberg 2012

Business Requirement Document (BRD), is generally created by Business Analysts to capture requirements as specified by the client. The corresponding testing of the Functional Requirements is known as Functional or User Acceptance Testing [4]. In Functional Testing, the System under test is treated as a black box [9] and is subjected to a series of user inputs as dictated by the Test Conditions and conformance to or departures from Expected Results are observed. In contrast, a Software Requirement Specification (SRS) document contains detailed system-level description of the requirements and use-cases. The information contained in SRS is used to create detailed system level Test Cases for Unit Testing and Interface Testing.

In large software projects, especially those involving distributed or offshore development and testing teams, precise understanding of the high level requirement sentences in the FRD is crucial. In particular, the understanding of the requirements should be consistent across the different teams. Automated analysis of requirements helps achieve this consistency.

In this paper, we present the auto-generation of Functional Test Cases from FRDs written in English. In our case, each functional requirement is expressed in a single sentence. The automated generation of Test Cases from these single sentence requirements is implemented in a tool called Litmus. The tool considers one requirement sentence at a time and generates one or more Test Cases for each requirement. To analyze free-form English, we leverage a syntactic natural language parser - Link Grammar [2]. Litmus is implemented as a plug-in into Microsoft Word and Excel and the Test Cases generated are populated in an Excel document.

This paper is structured as follows. Section 2 provides the details of the tool. We provide experimental results including the strengths and weaknesses of our approach in Section 3. The related work is presented in Section 4 and we conclude in Section 5.

2   Litmus

Litmus identifies a requirement as a sentence that has a label (given for traceability) in a requirement document and generates one or more Test Cases. A Test Case includes – the Test Condition, the Test Sequence and the Expected Result. A "Test Condition" is defined as the entry criterion for the Test Case or the particular condition which is being tested. The "Test Sequence" is the ordered sequence of steps a tester would have to execute to perform the test. The corresponding output from a correctly implement system is denoted as the "Expected Result". The Test Cases are generated through a six step process as shown in Fig. 1. Each module is explained in the coming sections. Fig. 5 shows an input requirement sentence and the corresponding Test Cases generated by Litmus.

The requirement sentences are parsed using Link Grammar parser (abbreviated as LG henceforth) developed at Carnegie-Mellon University. LG provides information on the syntax of a grammatically correct English sentence by connecting pairs of words through labeled links (Fig. 2 shows the parsed output of a sentence). The links between pairs of words are labeled and each label provides specific information about the connecting pair. For example, the Subject of a sentence is indicated by a link having a label "S". LG also provides the Noun Phrases, Verb Phrases, etc. through a constituent tree output. The links are created through a Grammar maintained in a dictionary file. The interpretation of the parsed output through the label of the links, the structure of the sentence and the constituent tree helps us decipher the following information – the sentence structure (antecedent and consequent), quantity (singular and plural), Parts-of-Speech (Noun, Verb, Verb in past tense, etc.) and grammar (Subject, Object, etc.). We interpret and exploit this information to extract entities and generate the Test Cases. LG has been shown to parse 75% [6] of the Penn Tree Bank successfully. Detailed information on the parser can be obtained at [2].

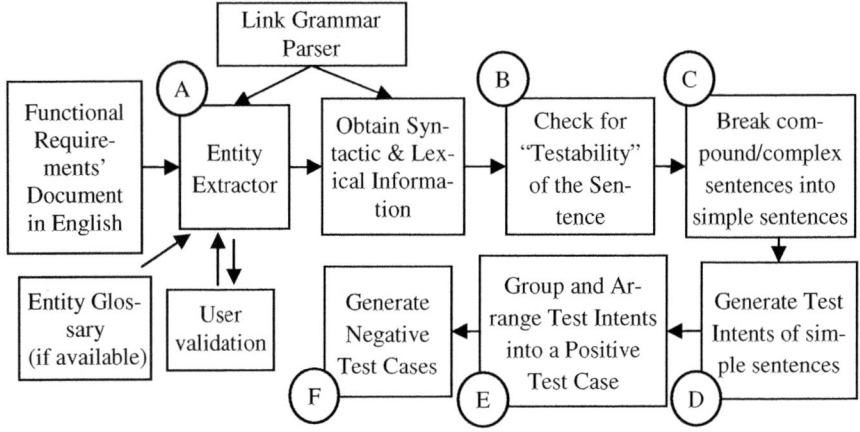

Fig. 1. Schematic of the Litmus Tool

2.1   Entity Extractor

The aim of the Entity Extractor is to pick up all entities from the requirement document. Every requirement sentence is parsed by LG and all group of words marked as a Noun Phrase (abbreviated as NP) in the constituent tree output is marked as an Entity. We have also modified the LG dictionary to ensure that words that are in title case (eg: "Order Processing System") or acronyms (eg: "ACBG") are mandatorily treated as Nouns. In addition, if the document contains a "Glossary" section, the corresponding terms are marked as entities as well.

We have often encountered Verbs being used as a Noun in software requirement documents. Two examples of such a case where the Verb "upload" is used as a Noun: (1) – "*upload data option must be enabled only if all form fields are filled*" and (2) – "*if all form fields are filled, the upload data option must be enabled*". In (1) LG is able to link the sentence, albeit linking "upload" as a Verb. In (2), LG is unable to link the sentence. To solve these cases, we use heuristics where Verbs that begin a sentence (as in (1)) or Verbs that are not linked but enclosed within Nouns or Determiners (as in (2)) are chosen as part of the entity.

The automatically identified entities are then presented to the user for validation. The user has the option to accept, reject or modify the entities. The extraction of Entities with the heuristics and the user verification helps increase the accuracy of our tool.

2.2   Identifying Testable Sentences

After marking the entities, every sentence is parsed individually using LG again incorporating the information of the entities. If the parser is unable to link the sentence, it is reported as a "failed case" and the analysis moves to the next sentence. A successfully linked sentence is tagged as either 'testable' or 'non-testable'. We define a testable sentence as one which adheres to the definition of a requirement [10]: A requirement is a contract that specifies what a user/agent does to a system (impetus) and how the system should respond. We, thus define a testable sentence as one which has a Subject, an Action and optionally an Object. The Action should contain a modal Verb (like "should", "would", etc.). We found that there are a small number of contiguous links (in terms of LG) which imply testability (a sample set is shown in Table 1). The links I and Ix correspond to modal Verbs. Fig. 2 depicts a requirement sentence and the linkages generated by LG. The sentence is testable because of Rule "S-Ix-P" (T.2 in Table 1). An example of a non-testable sentence is – "*Project Staffing Report is defined as a report containing the project name, description, employee count, staffing status*". Notice such sentences are assumptions or definitions and by themselves are not testable.

Table 1. Sample Rules to Identify Testability

Rule ID	Links in LG nomenclature
T.1	S-I-O
T.2	S-Ix-P
T.3	S-Pv-TO-I-O

Quick identification of non-testable requirements helps a Business Analyst to review and modify the requirements if needed. The aspect of testability is also critical in identifying parts of a sentence that should be tested. For example, consider a sentence – "*The maximum number of connections supported is 100, therefore the system should throw an error when more than 100 connections are requested*". Here, the underlined fragment should be inconsequential to a tester as it talks about the underlying logic. The Test Case should be based on the action of the system – which is to verify for the "error". This aspect is automatically caught by the testability module and the Test Case generated is around the action as specified in the requirement.

2.3   Breaking a Compound/Complex Sentence into Simple Sentences

In this module (shown as "C" in Fig. 1), if the testable requirement is a compound or complex sentence, it is simplified into a set of simple sentences. The idea is to develop mechanisms to analyze a simple sentence which can then be applied recursively over the set of simple sentences generated by the simplification. We define a compound sentence as one which contains two or more independent clauses or contains compound Nouns and Verbs (through coordinating conjunctions like "and", "or", ";", etc.). A complex sentence contains one or more dependent clauses (joined through subordinating conjunctions such as "if", "unless", etc., and through words such as "which", "that", etc.). We break such sentences into simple ones, each of which have no compound Nouns/Verbs and no independent/dependent clause. This simplification is made by identifying the links from the LG parse that represent compound/complex sentences and handling them appropriately. Similar to the testability rules, we have a large set of rules and associated actions to simplify compound and complex sentences. A sample set is shown in Table 2. Rule C.1 corresponds to sentences with Noun conjunctions and C.2 corresponds to complex sentences due to subordinating conjunctions.

Table 2. Sample Rules to Simplify a compund/complex sentence

Rule ID	Link in LG Nomenclature	Action
C.1	SJ	Sentence 1 = all words except those reachable through SJr Sentence 2 = all words except those reachable through SJl Ensure Number agreement between Noun and Verb Repeat Process for all SJ links.
C.2	MVs-Cs	Sentence 1 (consequent) = all words reachable through Cs+ Sentence 2 (antecedent) = all words reachable through Cs- + indicates the location of the start of the link (visually the left) - indicates the location of the end of the link (visually the right)

Consider Example (3), which contains a subordinate (*if*) and a coordinating (*or*) conjunction. Fig. 2 shows the working of the Rules. Rule C.1 results in two sentences (SENTENCE 1 – *"email must be sent if portfolio grows more than 10% during the day"* and SENTENCE 2 – *"SMS must be sent if portfolio grows more than 10% during the day"*) each of which are further broken through Rule C.2 into two sentences (SENTENCE 3 – *"portfolio grows more than 10% during the day"*, SENTENCE 4 – *"email must be sent"*, SENTENCE 5 – *"portfolio grows more than 10% during the day"* and SENTENCE 6 – *"SMS must be sent"*).

"email or SMS must be sent if portfolio grows more than 10% during the day"   (3)

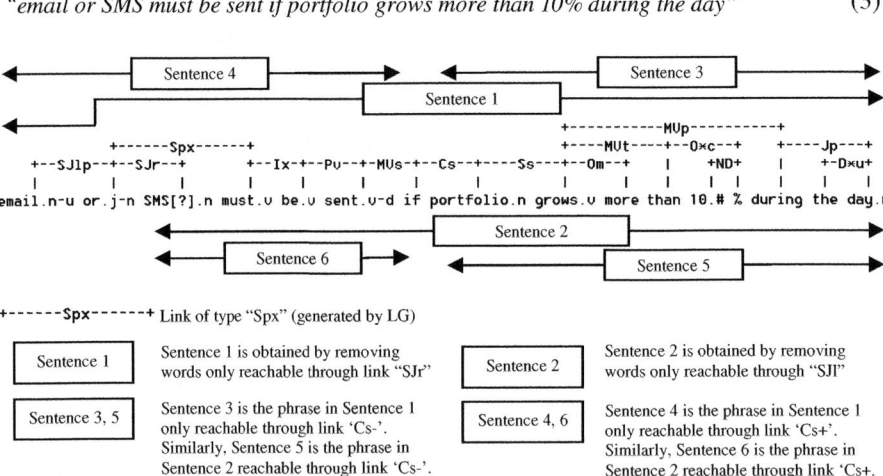

Fig. 2. Breaking a Compound/Complex Sentence into simple ones

As can be seen in the example, we maintain a particular simplification order when there are multiple compound/complex attributes – a) compound Nouns and Verbs are simplified first, b) independent clauses are simplified next, c) the dependent clauses are simplified last. The simplification of Example (3) can be represented using the following notation (and visualized as a tree as shown in Fig. 4).

$$S = (S_1 \; \emptyset_1 \; S_2) = (S_3 \; \emptyset_2 \; S_4) \; \emptyset_1 \; (S_5 \; \emptyset_3 \; S_6); \; \emptyset_1 = \text{"or"}, \emptyset_2 = \emptyset_3 = \text{"if} - \text{then"}$$

Splitting a compound sentence may break the number agreement between the Noun and the Verb. For example, *"administrator and user are required to .."* should be simplified into *"administrator is required ..."* and *"user is required..."*. We convert plural Verbs into singular for such cases through Stemming.

2.4 Generating the Test Intents of a Simple Sentence

We define a Test Intent as the smallest segment of a requirement sentence that conveys enough information about the purpose of the test. A Test Intent has a Subject, an Action and (optionally) an Object. A Test Intent is the set of words along a path of a link (from the LG parse) and bounded by NPs. Consider a sentence (4) – *"the PRTS system should print the reports selected by the user through the touch screen"* and its corresponding LG parse (Fig. 3). Referring to Fig. 3, the first test intent phrase begins from the Subject – "the PRTS system" and concludes at the occurrence of the NP – "the reports". The second test phrase begins from this point and proceeds till the next NP – "the user". Note that there are two paths that can be traversed from the Verb "selected". The second path from the NP – "the reports" till the NP – "the touch-screen" forms the third test intent phrase. Test Intents are generated from these phrases by inserting static text through a template for human comprehension. A representative set of the rule in terms of the LG links and the Test Intent templates are shown Table 3. The resulting Test Intents are shown in (5).

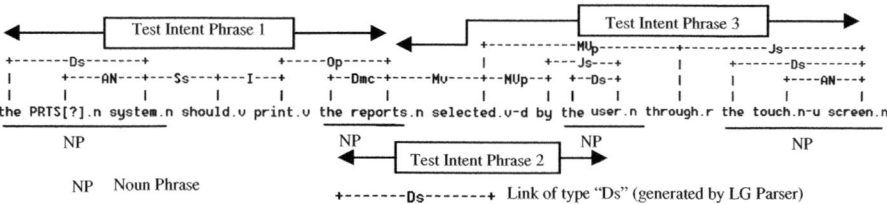

Fig. 3. Generation of Test Intents

Table 3. Sample Rules to Generate Test Intents

Rule ID	Link in LG	Action / Template
I.1	S-I-O	*Verify* NP(S+) *is/are* {<N->} *able to* <I-> NP(O-)
I.1.2	O/J-Mv	*Verify* NP(O/J) *was/were* (group of words from M till J)
I.1.3	I-MV S-MV	*Verify* <MV+> *Of* NP(O-) *happens* (group of words from MV till J)
Notations:		NP(S+) : The Noun Phrase containing the S+ link
		is/are, was/were: used according to singularity / plurality of Noun
		{<N->}: if N- link is present, insert the word 'not'
		<MV+>: Word having the MV+ link

The text from the template is in italics. Test Intent – 1 is generated due to Rule I.1 (refer the S-I-O links in Fig. 3). The template adds "Verify" to begin the Test Intent and "is able to" between the Subject and the Verb. We term this intent as "Primary" since it includes the Subject of the sentence. Similarly, Test Intent – 2 and Test Intent – 3 are generated through Rule I.1.2. Note that the Test Intents incorporate the tense and the number through "were". This is achieved from the labels of the links ("v" of "Mv" and "p" of "Op" respectively). Such intents, that do not include the Subject of the sentence, are termed "Secondary". Intuitively, the Primary Intent represents the main action conveyed by the requirement and all other actions are considered Secondary.

Test Intent – 1:	*Verify* the PRTS system *is able to* print the reports	
Test Intent – 2:	*Verify* the reports *were* selected by the user	(5)
Test Intent – 3:	*Verify* the reports *were* selected through the touch-screen	

2.5   Creating Positive Test Case(s)

A Test Case (denoted as TC) comprises of a Test Condition (C), the Test Sequence (TS) and the Expected Results (ER). The Test Condition defines the logical condition for which the test is to be performed. The Test Sequence is a set of execution steps. The corresponding output from a correctly implemented system is captured in the Expected Results. The Test Condition, the Test Sequence and the Expected Results are populated from a single requirement sentence. A Positive Test Case verifies for the action stated in the requirement sentence. Similarly, a Negative Test Case verifies that the said action does not happen when the conditions are not met. A Positive Test Case of a sentence (S) is populated according to Equation (6).

$$TC\,(S) = \begin{cases} C = prior\big(I(S)\big) + I(antecedent(S)) \\ TS = I(S) - C \\ ER = primary\,(\,I(S)) \end{cases} \tag{6}$$

Referring to (6), the Test Condition is populated with all the Test Intents (denoted as I) of the antecedent of a sentence (e.g. the "if" part of a sentence with "if-then"; shown as an operator: *antecedent(S)*) and all those secondary Test Intents that are generated through the rule "O/J-Mv" (Rule I.1.2 of Table 3). This rule indicates past events through Verbs in past tense (e.g. Test Intent – 2 and Test Intent – 3 of (5); shown as operator: *prior()*). The word "Verify" is stripped from the Test Intents which are placed in the Test Condition. The Test Sequence comprises of the Test Intents of the consequent (e.g. the "then" part of an "if-then" sentence). Any Test Intent already included in the Test Condition is removed. The Expected Results is populated with the Primary Test Intent of the consequent (shown as operator: *primary()*). For a simple sentence requirement, Equation (6) would be the final positive Test Case.

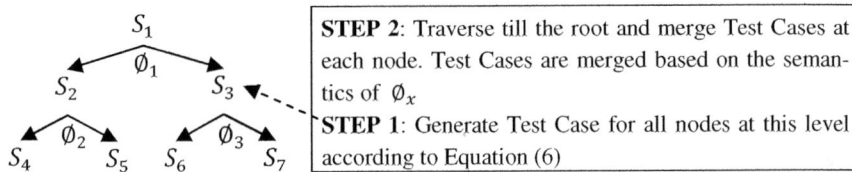

Fig. 4. Visualization of simplification of Example (3) and the Algorithm to merge Test Cases

For a compound/complex sentence, the Test Case is generated for each simplified sentence and merged according to the semantics of the conjunctions. Consider example (3), the simplification of which can be represented as a tree. The algorithm to generate the Test Case through the tree traversal is shown in Fig. 4. The merging of the Test Cases is made based on the semantics of $\emptyset$. The C, TS and ER are combined if $\emptyset =$ {"and", ","} and are kept distinct if $\emptyset =$ {"or"}.

Certain sentences linguistically imply an opposite of "if-then" even though they carry the same syntactic structure. For example, the sentence – "*Unless the user is an Admin, the edit page should be hidden*" implies - if the user is a Admin, verify for "*not* hidden". Such cases are handled by artificially inserting a "not" (thus negating the consequent) and following the same procedure for generating the Positive Test Case.

2.6   Creating Negative Test Case(s)

For the Negative Test Case, we need to identify all conditions where the affirmative action of the system should not happen. The Negative Test Case ($\overline{TC}$) comprises of the Test Condition (C), the Test Sequence (TS) and the Expected Result (ER).

$$\overline{TC}\,(S) = \begin{cases} C = neg(prior(I(S)) + I(antecedent(S))) \\ TS = Invert\ Action(primary\left(\,I(S)\right)) \\ ER = Invert\ Action(primary\left(\,I(S)\right)) \end{cases} \tag{7}$$

The Test Condition includes the Test Intents (I) of the antecedent of the sentence (S) and Test Intents of past events, similar to how the Test Condition of the Positive Test Case was generated. However, we need to identify the logical inverse of these conditions. This is achieved through a logical negation operation - $neg()$. The set of Test Intents from the two sources (*prior* and *antecedent*) are treated as mutually independent and the negative conditions are the combinations of the negations of each set.

$$\begin{aligned} neg(prior(I(S)) + I(antecedent(S))) \quad &= \quad neg\left(Prior(I(S))\right) + I(antecedent(S)) \\ &+ \quad prior(I(S)) + neg(I(antecedent(S))) \\ &+ \quad neg(Prior(I(S))) + neg(I(antecedent(S))) \end{aligned}$$

The individual Test Intents are treated similar to a parent-child relationship. For example, if *prior(I(S))* contains three Test Intents, the negation is as shown below. *neg(I(antecedent(S))*) is calculated in similar fashion.

$$neg\left(prior(I(S))\right) = neg(I_1 + I_2 + I_3) = \{neg(I_1)\} + \{I_1 + neg(I_2)\} + \{I_1 + I_2 + neg(I_3)\}$$

If any of the Test Intents contain data, we use Boundary Value Analysis (BVA) [5] to generate range of data conditions. We modified the LG dictionary to annotate data as special elements (Refer ".#" in Fig. 2). Data includes numbers (e.g. "1"), textual numerals (e.g. "one") and Boolean values (e.g. "ON", "true"). Rules were developed to identify symbolic or textual data comparators (e.g. ">", "less than") and the data units by looking at the pre and post-nominal connectors to the data. For example, the data in Example (3) is – absolute number: 10, comparator: >, units - %. The negative data range using BVA, then, would be: "=10%" and "<10%". In case the data had been "<=10%", the negative data range would have been ">10%".

$$neg(I) = \begin{cases} if\ I\ contains\ data, replace\ data\ with\ data\ points\ from\ BVA \\ else, Invert\ Action(I) \end{cases}$$

Finally, the "Invert Action" is achieved as follows. If two or more contiguous Verbs exist after the Subject, a "not" is inserted after the first Verb (and removed if present). For example, "*user should upload ...*" becomes "*user should not upload ...*". In case of a single Verb, the Verb is made singular using stemming and negated with "does not". For example, "*user uploads ...*" becomes "*user does not upload...*".

The Test Sequence of the Negative Test Case should verify that the primary action does not happen and this is achieved by inverting the Verb of the Primary Test Intent. No other Test Intent is kept. The Expected Result is calculated in similar fashion.

The Negative Test Cases generated above are for a simple sentence. For a compound/complex sentence, the merging of Test Cases of each simple sentence is done in similar fashion to that of the Positive Test Case (Fig. 4). The only difference being, the semantics for "and" and "or" are reversed. The C, TS and ER are combined if $\emptyset = \{$"or"$\}$ and are kept distinct if $\emptyset = \{$"and", ",", ";"$\}$. The Test Cases for Example (3) are provided in Fig. 5. There are two Positive Test Cases because of "*or*". The Test Conditions consist of two Test Intents (the Primary followed by the Secondary). The negation of these Intents gives the Negative Test Cases. The first two Negative Test Cases arise due to the negation of the Primary Test Intent and the use of BVA. The third Negative Test Case is generated due to the negation of the secondary Test Intent.

Requirement: *email or SMS must be sent if portfolio grows more than 10% during the day*

Positive Test Cases	

Test Condition:
  portfolio grows > 10 %
and
grow of > 10 % happens during the day
Test Sequence:
  1) Verify email is sent
Expected Result:
  1) email is sent
	Test Condition:
	  portfolio grows > 10 %
	and
	grow of > 10 % happens during the day
	Test Sequence:
	  1) Verify SMS is sent
	Expected Result:
	  1) SMS is sent

Negative Test Cases		

Test Condition:
  portfolio grows = 10 %
Test Sequence:
  1) Verify email is not sent
  2) Verify SMS is not sent
Expected Result:
  1) email is not sent
  2) SMS is not sent
	Test Condition:
	  portfolio grows < 10 %
	Test Sequence:
	  1) Verify email is not sent
	  2) Verify SMS is not sent
	Expected Result:
	  1) email is not sent
	  2) SMS is not sent
		Test Condition:
		  portfolio grows > 10 %
		and
		grow of > 10 % does not happen during the day
		Test Sequence:
		  1) Verify email is not sent
		  2) Verify SMS is not sent
		Expected Result:
		  1) email is not sent
		  2) SMS is not sent

Fig. 5. The Positive and Negative Test Cases generated by Litmus for a sentence

3   Experimental Validation and Discussion

Litmus has been implemented as a plug-in into Microsoft Word and Excel using .NET. The implementation calls the LG parser run-time (in C) as a managed code library from the .NET environment. Litmus was tested over actual requirement documents from various projects in the industry. Litmus was run on the requirement documents and the generated Test Cases were analyzed manually for accuracy. The results are in Table 4. The accuracy of LG is measured by its ability to link grammatically correct sentences. The accuracy of Litmus is measured by the number of correct (and intelligible) Test Cases as a percentage of the total number of Test Cases generated.

Table 4. Experimental Results of Litmus on Industrial Requirement Documents

Domain	Num of Requirements	Accuracy of LG	No. of Positive Test Cases	No. of Negative Test Cases	No. of Test Steps	Accuracy of Litmus
Pharma	864	89%	778	205	1249	84.2%
Pharma	42	93%	39	9	93	79.2%
Pharma	37	84%	30	24	87	55.6%
Pharma	322	80%	242	137	741	70.7%
IT	178	75%	144	44	324	63.3%
IT	183	81%	159	43	431	64.4%
Telecom	215	86%	190	39	432	80.8%
Total	1841	85%	1582	501	3357	76.7%

The overall accuracy of Litmus was seen at 77% and varied from a low of 56% to a high of 84%. We observed that most requirement documents have a small set of sentence structures which repeat across the document. Hence, any inaccuracy in handling a particular structure results in multiple similar errors. This was the primary reason for the low accuracy of 56%. The results seen are from the deployment of Litmus in Industrial projects. The feedback has been positive and Litmus is seen to provide value in the Test Design phase of the software development life-cycle. The benefits from the use of Litmus includes better test granularity (through Test Intents) which helps ensure all aspects are tested and better coverage (through negative conditions and BVA for test data). During the evaluation, it was found that Litmus could identify Test Cases that were missed by the human analysts. Litmus has applicability to Business Analysts too, who can estimate the complexity of projects through the number of Positive & Negative Test Cases generated.

We have also investigated the cases where Litmus fails to generate accurate Test Cases. We briefly discuss some of the issues. The methodology to generate Test Intents works well in most cases, however, in a few instances the Test Intents seem incomplete. Consider – *"system should have the ability to ensure all fields are entered"*. Here, the NP boundaries cause the first Test Intent to be – "Verify the system has the ability". While this Intent is not wrong, it is incomplete and not understandable. Similarly for – *"system should send one-second block of 8 frames"*, the first Test Intent is – "Verify the system is able to send one-second block". Here, the Intent is unintelligible unless "block of 8 frames" is used in its entirety. Such issues can be solved by skipping NP boundaries for particular sentence structures.

Litmus is driven largely on the syntax of the sentence. Some lexical semantics is inbuilt, for handling coordinating and subordinating conjunctions. Such semantics are context and domain agnostic. However, consider – *"Product code and lot number combinations should be unique"*. Here, the Test Intents from Litmus are – "Verify product code is unique" and "Verify lot number combinations are unique". The semantics of "combination" is lost in the Test Intents. Other examples include: "between A and B", "one or more", etc. The semantics for words like "only", "all", etc. can be handled in this way as well. Other forms of semantics are dictated by domain knowledge. The example – *"system shall maintain an audit trail of the database updates"* requires the understanding that "audit trail" is concerned with "updates". The ideal Test Condition would have been to perform certain updates and check for the same through the audit trail. This Test Condition, to be generated, needs semantics.

In our tests, the LG parser showed an accuracy of 85%. LG generates multiple link structures for a given sentence and uses a "ranking" mechanism to order these structures through link weights. In most cases, the highest ranked structure is also the correct interpretation. However, the ranking mechanism was found to be inadequate in few cases. Consider – *"system shall update profile to JPEG and file to DOC"*. While the correct structure links *"and"* with *"profile"* and *"file"*, the first link generated by LG connects *"and"* with *"JPEG"* and *"file"* since LG is designed to favor links of shorter length. This can be solved by incorporating statistical parsing techniques.

4   Related Work

In requirements engineering, research has been in three areas – requirement critiquing, auto-generation of design artifacts and auto-generation of test artifacts from requirements. Most research takes recourse to supporting only a subset of natural language or expects formal specifications. We will focus on work that handles unconstraint natural language.

The generation of Test Cases from unconstrained English requirements is studied in [3]. A requirement with an Object is considered testable. It is then classified into: 1) a "condition" if it contains *"if"*; 2) a "statement" if the Subject is assigned a value; and 3) an "action" if a Verb acts on the Subject. A conditional sentence creates two Test Cases while others create one. Our work, in contrast, provides richer test conditions (from subordinate conjunctions and past events) and provides better granularity (from Test Intents). Further, the use of BVA provides a better test coverage.

In [11], Test Cases are created from natural language use-cases through an auto-generated class model. The Actions and Entities in the model are based on a manually built list. Further, a shallow parser (thus restricting accuracy) is used to identify the grammar of the requirement. The Test Conditions are also limited to subordinate conjunctions. Besides, precise methodology of the generation of Test Cases from the class model is not provided.

A model from unconstrained natural language is discussed in [12]. Initially, a model is created manually from a set of requirements. The entities and relationships in model are used as a training set to learn rules. These rules are then used to auto-generate a model on a new requirement set. The drawbacks include the need for manual creation of the model and domain dependence. However, this approach can potentially capture semantics and tacit knowledge through the manually built model.

The work in [8] uses a dependency parser to organize the Subjects and Objects of requirement sentences into entities. The verbs are made into relations. Thus the basic sentence structure handled is the Subject-Action-Object (SAO) tuple. A class model is then generated by spotting pre-defined keywords like "consists of". The use of a dependency parser provides rich grammatical information, but the interpretation of only the SAO structure limits the coverage.

5   Conclusion

In this paper, we present Litmus – a methodology and tool to generate functional Test Cases from natural language (English) FRDs. Litmus uses a syntactic parser called Link Grammar parser and has a set of pattern matching rules to check the testability, identify Test Intents and generate Positive and Negative Test Cases in a systematic manner. We have presented the accuracy of the tool over Industrial project requirements. Litmus has received positive feedback in its practical deployment from test engineers and business analysts. Our future research is focused on improving the accuracy through handling certain sentence structures better and building a stronger set of language and domain specific semantics. Importantly, we are developing the ability to handle requirements that span multiple sentences and use cases.

References

1. Neill, C.J., Laplante, P.A.: Requirements engineering: The state of the practice. IEEE Software 20(6), 40–45 (2003)
2. Sleator, D.D.K., Temperley, D.: Parsing English with a Link Grammar. In: Third International Workshop on Parsing Technologies (1993)
3. Sneed, H.M.: Testing against Natural Language Requirements. In: Seventh International Conference on Quality Software, pp. 380–387 (2007)
4. IEEE-829-2008, IEEE Standard for Software and System Test Documentation
5. Stoer, J., Bulirsch, R.: Introduction to Numerical Analysis. Springer (2002)
6. Link Grammar,
   http://www.abisource.com/projects/link-grammar/dict/ph-explanation.html
7. Luisa, M., Mariangela, F., Pierluigi, I.: Market Research for Requirements Analysis using Linguistic Tools. Requirements Engineering 9(1), 40–56 (2004)
8. Ilieva, M.G., Ormandjieva, O.: Automatic Transition of Natural Language Software Requirements Specification into Formal Presentation. In: Montoyo, A., Muñoz, R., Métais, E. (eds.) NLDB 2005. LNCS, vol. 3513, pp. 392–397. Springer, Heidelberg (2005)
9. Howden, W.: Functional Program Testing & Analysis. McGraw Hill, New York (1987)
10. IEEE-830-1998: IEEE Recommended Practice for Software Requirements Specifications
11. Sutton, S.M., Sinha, A., Paradkar, A.: Text2Test: Automated Inspection of Natural Language Use Cases. In: Third International Conference on Software Testing, Verification and Validation, pp. 155–164 (2010)
12. Kof, L.: Faster from Requirements Documents to System Models: Interactive Semi-Automatic Translation. In: First International Requirements Engineering Efficiency Workshop (2011)

Extracting Multi-document Summaries with a Double Clustering Approach

Sara Botelho Silveira and António Branco

University of Lisbon, Portugal
Edifício C6, Departamento de Informática
Faculdade de Ciências, Universidade de Lisboa
Campo Grande, 1749-016 Lisboa, Portugal
{sara.silveira,antonio.branco}@di.fc.ul.pt

Abstract. This paper presents a method for extractive multi-document summarization that explores a two-phase clustering approach. First, sentences are clustered by similarity, and one sentence per cluster is selected, to reduce redundancy. Then, in order to group them according to topics, those sentences are clustered considering the collection of keywords. Additionally, the summarization process further includes a sentence simplification step, which aims not only to create simpler and more incisive sentences, but also to make room for the inclusion of relevant content in the summary as much as possible.

Keywords: Multi-document summarization, sentence clustering, sentence simplification.

1 Introduction

Automatic text summarization is the process of creating a summary from one or more input text(s) through a computer program. It seeks to combine several goals: (1) the preservation of the idea of the input texts; (2) the selection of the most relevant content of the texts; (3) the reduction of eventual redundancy; and (4) the organization of the final summary. While meeting these demands, it must be ensured that the final summary complies with the desired compression rate. This is, thus, a complex task to be accomplished by a human let alone a computer.

This paper presents a multi-document summarization system, SIMBA, that receives a collection of texts and retrieves an extract summary, that is, a text composed by sentences obtained from the source texts. The main goals of multi-document summarization are tackled through a double clustering approach, which includes a similarity clustering phase and a keyword clustering phase. Redundancy is addressed by clustering all the sentences based on a measure of similarity. Afterwards, the sentences are assembled by topics, using the keywords retrieved from the collection of texts. This approach impacts on the content of the final summary. On the one hand, the similarity clustering ensures that this content is not repetitive. On the other hand, the keyword clustering assures the

G. Bouma et al. (Eds.): NLDB 2012, LNCS 7337, pp. 70–81, 2012.

selection of the most relevant content, the preservation of the idea of the input texts, and the organization of the final summary. Furthermore, this system includes a step of sentence simplification, at the end of its processing pipeline, which aims at producing simpler and more incisive sentences, thus allowing that more relevant content enters the summary. Finally, an automatic evaluation of SIMBA is presented.

From now on, this paper is organized as follows: Section 2 provides an overview of automatic summarization systems that inspired our work; Section 3 describes our multi-document summarization system; Section 4 reports on the system evaluation; and, finally, the conclusions in Section 5.

2   Related Work

The summarization process comprises typically three stages: analysis, transformation and synthesis [11].

The **analysis** phase processes the input text(s) and builds an internal representation of those text(s). Typical steps include tokenization, part-of-speech annotation, lemmatization, or stop-words removal. Also, in this phase, the information units (sentences) to be handled in the next phases are identified.

The **transformation** phase aims to handle the sentences and to identify the ones that are the most relevant within the input text(s). Each sentence is assigned a score. In many summarization systems, measures of significance of the sentence are calculated over the text(s) to determine each sentence score.

These measures of significance can be in the form of reward metrics or penalty metrics, that take into account features related to the sentence or the words composing it. The reward ones assign positive scores to the sentences, while penalty metrics remove values from the sentence scores.

The keyword method was proposed in [9], and assumes that more frequent words in a document indicate the topic discussed, so that more frequent words are assigned with reward values. [5] experimented the cue method (sentences with cue words are relevant); the location method (sentences in specific positions are important); finally, the title method (words in the title are keywords of the text).

In fact, the most used score measure has been `tf-idf` (Term Frequency × Inverse Document Frequency). Sentences containing frequent or highly infrequent terms are more likely to be significant in the global context of the collection of texts.

Vocabulary overlap measures are also used to define sentence scores. The Dice coefficient [4], the Jaccard index [6], and the Cosine similarity coefficient compute a similarity metric between pairs of sentences, determining a relation between them. For instance, to, along with other features, define the sentence score, [13] computes the overlap between each sentence and the first sentence of the text, which, in news articles, is considered one of the most important sentences. Yet, [15] computes the overlap between each sentence and the document title, in order to reward the sentences that have a high degree of similarity with the title.

The reward metrics are both suitable for single- or multi-document summarization. However, multi-document summarization represents new challenges. Beyond selecting the most salient information which represents the collection of documents, removing the redundancy present in the texts is another challenging issue to overcome.

Many works addressed these challenges by trying new metrics of word or sentence significance. The Maximal Marginal Relevance (MMR) [3] is a linear combination metric that relates query-relevance with information-novelty. It strives to reduce redundancy while considers query relevance to select the appropriate passages to be part of the summary.

After determining the sentence scores, **penalty metrics** can be computed to refine each score. The most common penalty metric is sentence length. Sentences with less than ten words are considered too small and conveying limited information, so that a predefined score value is removed from the sentence score. [8] used another metric, the stigma words penalty. Sentences starting with conjunctions, question marks, pronouns such as "he"/"she"/"they", and the verb "say" and its derivatives are penalized. Finally, the text publication date can either be a reward or a penalty metric, whether the text is a recent one or not.

Once all the sentences in the text(s) have been scored, the **generation** phase aims to order those sentences and to select the ones that will compose the final summary, seeking to fulfill the compression rate. Typically, the sentences are ordered considering their score. The ones with the highest score are selected until the compression rate is attained.

3   The SIMBA System

SIMBA is an extractive multi-document summarizer for the Portuguese language. It receives a collection of Portuguese texts, from any domain, and produces informative summaries, for a generic audience. The length of the summaries is determined by a compression rate value that is submitted by the user. It performs summarization by using a shallow yet efficient approach that relies on statistical features computed over the text elements.

Summarization is performed by executing five main stages: identification, matching, filtering, reduction, and presentation. These stages are described in the following sections. These five stages can be enclosed in the three generic phases of summarization mentioned above, as identification is a step of the analysis phase; matching and filtering are stages of the transformation phase; and, finally, reduction and presentation are included in the synthesis phase.

3.1   Identification

The identification stage is executed in two phases. The first phase handles the documents submitted by the user, converts them into the same format and removes the existing noise.

Once the documents are in the same format, texts are accessible to be processed automatically. Afterwards, a set of shallow processing tools for Portuguese, LX-Suite [2], is used to annotate the texts. Sentence and paragraph boundaries are identified and words are tagged, with its corresponding POS and lemmata. Also, for each sentence, a parse tree, representing the sentence syntactic structure, is built by LX-Parser [14].

Henceforth, the collection of texts is handled as a set of sentences.

3.2 Matching

The matching stage aims to identify relevant information in the collection of texts. First, sentence scores are computed. Then, sentences are clustered by similarity to remove redundancy within the collection. Finally, sentences are clustered by keywords to identify the ones that have the most relevant information.

It is important to note here that in our summarization process we consider three types of scores: the main score, the extra score and the complete score. The main score reflects the sentence relevance in the overall collection of sentences. The extra score is used in the summarization process to reward or penalize the sentences, by adding or removing predefined score values.[1] The complete score is the sum of these two scores.

Computing Sentence Main Score. Once the sentences and the words have been identified, `tf-idf` score is computed for each word, by considering its lemma. The sentence main score is then the sum of the `tf-idf` score of each word, smoothed by the total number of words in the sentence.

Clustering Sentences by Similarity. In order to identify redundant sentences, conveying the same information, the next step aims to cluster sentences, considering their degree of similarity.

The similarity between two sentences (Equation 3) comprises two dimensions, computed considering the word lemmas: the sentences subsequences (Equation 1) and the word overlap (Equation 2).

The subsequences value is inspired in ROUGE-L and consists in the sum of the number of words in all the subsequences common to each sentence, smoothed by the total number of words of each sentence being considered, and divided by the total number of subsequences found between the two sentences. The overlap value is computed using the Jaccard index [6].

The similarity value is the average of both these values: the overlap and the subsequences value. It is then confronted with a predefined threshold – similarity threshold[2] –, initially set to 0.75, determining that sentences must have at least 75% of common words or subsequences to be considered as conveying the same information.

---

[1] The predefined value is set to 0.1, both for the reward and the penalty values. This value has been determined empirically, through a set of experiments.

[2] This threshold was determined empirically, using a set of experiments, since there is no reference for such a value for the Portuguese language.

$$subsequences(s_1, s_2) = \frac{\sum_i \left(\frac{\text{subsequence}_i}{\text{totalWords}_{s_1}} + \frac{\text{subsequence}_i}{\text{totalWords}_{s_2}} \right)}{totalSubsequences} \qquad (1)$$

$$overlap(s_1, s_2) = \frac{\sum \text{commonWords}(s_1, s_2)}{\text{totalWords}_{s_1} + \text{totalWords}_{s_2} - \sum \text{commonWords}(s_1, s_2)} \qquad (2)$$

$$similarity(s_1, s_2) = \frac{\text{subsequences}(s_1, s_2) + \text{overlap}(s_1, s_2)}{2} \qquad (3)$$

$overlap(s_1, s_2)$ − number of overlapping words between the two sentences.

$subsequence(s_1, s_2)$ − number of overlapping words in the subsequences between the two sentences.

$commonWords(s_1, s_2)$ − common words between the two sentences.

$totalWords_{s_i}$ − total words in the sentence i.

$subsequence_i$ − number of words of the subsequence i.

$totalSubsequences$ − number of subsequences between the two sentences.

Two examples are discussed below. Taking into account this threshold, the sentences in the following example are considered to be similar.

Sentence#1:

A casa que os Maias vieram habitar em Lisboa, no outono de 1875, era conhecida pela casa do Ramalhete.

The house in Lisbon to which the Maias moved in the autumn of 1875, was known as the Casa do Ramalhete.

Sentence#2:

A casa que os Maias vieram habitar, no outono de 1875, era conhecida pela casa do Ramalhete.

The house to which the Maias moved in the autumn of 1875, was known as the Casa do Ramalhete.

Overlap	Subsequences	Similarity Value
0.89	0.95	0.92

These sentences share most of the words, but there is a leap (*"em Lisboa"*) between Sentence#1 and Sentence#2. Both the overlap and the subsequences values are high, because these two sentences share most of their words. So, as the similarity value is high (0.92), the sentences are considered similar.

The two sentences in the following example are not similar though having many words in common.

Sentence#1:

A casa que os Maias vieram habitar em Lisboa, no outono de 1875, era conhecida pela casa do Ramalhete.

The house in Lisbon to which the Maias moved in the autumn of 1875, was known as the Casa do Ramalhete.

Sentence#2:

A casa que os Maias vieram habitar em Lisboa, no outono de 1875, era conhecida na vizinhança da Rua de S. Francisco de Paula, pela casa do Ramalhete ou simplesmente o Ramalhete.

The house in Lisbon to which the Maias moved in the autumn of 1875, was known in Rua S. Francisco de Paula, as the Casa do Ramalhete or, more simply, as Ramalhete.

Overlap	Subsequences	Similarity Value
0.59	0.79	0.69

Despite Sentence#1 is contained in Sentence#2, both sentences are considered not to be similar, since their similarity value is below the threshold.

Afterwards, the sentences are actually clustered considering their similarity value. A cluster is composed by a collection of sentences, a similarity value, and a centroid – the highest scored sentence in the collection of values.

The algorithm starts with a empty set of clusters. All sentences in the collection of texts are considered. The first sentence of the collection creates the first cluster. Then, each sentence in the collection of sentences is compared with the sentences already clustered. For each cluster, the similarity value is computed between the current sentence being compared and all the sentences in the collection of sentences of each cluster. The similarity value considered is the one that is the highest between the current sentence and all the sentences in the collection of sentences. Then, if the similarity value is higher than the already mentioned similarity threshold, the sentence will be added to the current cluster.

When a sentence is added to a cluster, its centroid must be updated. If the score of the sentence being added is higher than the centroid one, the newly added sentence becomes the centroid of the cluster. Also, each centroid is given an extra score value (0.1), which is subtracted from the sentences which are replaced as centroids.

Finally, if all the clusters have been considered, and the sentence was not added to any cluster, a new cluster with this sentence is created.

Once the procedure is finished, sentences with redundant information are grouped in the same cluster and the one with the highest score (the centroid) represents all sentences in the cluster.

Clustering Sentences by Keywords. After identifying similar sentences, the centroids of each similarity cluster are selected to be clustered by keywords. The sentences contained in the collection of values of each cluster are ignored, and will not be clustered in this phase, since they have been considered redundant.

Our system produces a generic summary, so it is not focused on a specific matter. Thus, the keywords that represent the global topic within the collection of texts are identified.

Keywords are determined in three steps. The first step selects the potential keywords in the complete collection of words present in all texts. Words such as clitics, and contractions are ignored, filtering out words with little discriminative power. The list with candidate keywords is constructed containing common and proper names, since these words are the ones that identify ideas or themes. To be added to this list, words are compared considering their lemmas, to ensure that the words in the collection are unique. The second step orders this list, containing the words which are candidates to be keywords, by the word score. Finally, a predefined number of keywords is retrieved in order to build the final set of keywords. We define k, the number of keywords, as $k = \sqrt{\frac{N}{2}}$, where N is the total number of words in the collection of documents.

Afterwards, sentences are clustered based on these keywords. The procedure starts by retrieving from the similarity clusters all its centroids. In this phase, a cluster is identified by a keyword, and contains a centroid (a sentence), and a collection of values (the sentences related to the keyword). The algorithm that clusters sentences by keywords is an adapted version of the K-means algorithm [10], and follows the steps described below:

1. Choose the number of clusters, k, defined by the number of keywords previously selected;
2. Create the initial clusters, represented by each keyword obtained;
3. Consider each sentence:
   (a) Compute the occurrences of each keyword in the sentence;
   (b) Assign the current sentence to the cluster whose keyword occurs more often;
4. Recompute the cluster centroid. If the current sentence has more occurrences of the keyword than the previous centroid sentence had, the newly added sentence becomes the cluster centroid;
5. If the sentence does not contain any keywords, it is added to a specific set of sentences which do not have any keyword ("no-keyword" set);
6. Recompute the set of keywords if:
   (a) All the sentences have been considered;
   (b) The "no-keyword" set contains new sentences.
7. Repeat previous steps (2 – 6) while the "no-keyword" set of sentences remains different in consecutive iterations.

As in the similarity algorithm, each centroid is assigned with an extra score value. When the centroid is changed, the extra scores of the current centroid and of the previous centroid are updated. In addition, an extra score is also assigned to the sentences in the clusters that represent the original set of keywords.

Finally, sentences in the "no-keyword" set are ignored, while the ones that have been clustered are considered in the next phases.

3.3   Filtering

In this phase, the sentences that have been clustered by keywords are considered. The ones that have less than ten words are penalized and an extra score value is subtracted from their extra score. However, the sentences that have more than ten words are assigned with an extra score value.

The complete score, defined in Equation 4, is used to rank all the sentences:

$$completeScore = score_s + extraScore_s \qquad (4)$$

So, sentences are ranked by their complete score, defining the order of the sentences to be chosen to be part of the final summary.

3.4   Reduction

The reduction process aims at reducing the original content to produce a summary containing simpler and more informative sentences. This phase comprises two steps: simplification and compression.

Simplification. Sentence simplification is performed by removing expressions or phrases whose removal is less detrimental to the comprehension of the text. Three types of structures are identified: sentence clauses, parenthetical phrases, explanatory or qualifying phrases; and apposition phrases, which are a specific type of parentheticals, composed by a noun phrase that describes, details or modifies its antecedent (also a noun phrase).

Parentheticals are typically enclosed either by parenthesis, or by commas or dashes. Appositions, in turn, can only be enclosed by commas or dashes and consist of a noun phrase. These phrases are candidates to removal.

In order to identify the passages, the sentence parse tree (built in the identification phase) is used. Once have been identified, all those passages are candidates to be removed.

Apposition and parenthetical phrases are removed from the sentence without further constraints, since they are considered less important than the rest of the sentences. Concerning sentence clauses, for each sentence, the simplification score (Equation 5) is computed. The simplification score is the sum of each word score entering the sentence, divided by the length of the sentence.

$$simplificationScore_s = \frac{\sum_{w \in s} score_w}{length_s} \tag{5}$$

For each sentence, the algorithm selects each clause separately. Afterwards, the simplification score is computed for both the main sentence and the new sentence. If the new sentence has a higher simplification score than the one of the main sentence, the main sentence is marked to be replaced by the sentence without this clause.

Compression. The compression rate previously defined (either given by the user or defined by default as 70% – the most commonly used default value – , that is the summary will contain 30% of the words in the collection of texts) is then applied to the collection of sentences, which are added to the final summary based on its total number of words. If the total words of the already added sentences reaches or surpasses the maximum number of words determined by the compression rate, no more sentences will be added, and the summary is created.

3.5   Presentation

Once the summary sentences have been identified, the summary is delivered to the user in the form of a text file.

4   Evaluation

In order to perform evaluation we used the *CSTNews* corpora [1] (described below), containing sets of texts in Portuguese and its corresponding ideal summaries. Concerning the evaluation itself, we compared SIMBA with GISTSUMM [12].

GISTSUMM is a summarizer built to deal with texts in Portuguese. It is based on the notion of gist, which is the most important passage of the text, conveyed by just one sentence, the one that best expresses the text's main topic. The system algorithm relies on this sentence to produce extracts. GISTSUMM is the only summarizer for Portuguese available on-line. Despite it has been built to produce summaries from a single-document, it also performs multi-document summarization by means of an option in its interface that allows to produce a summary from a collection of texts. GISTSUMM is used as a baseline for our work.

4.1   Corpus

CSTNews is an annotated corpus, whose texts were collected from five Brazilian newspapers. It contains 50 sets of news texts from several domains. Each set contains in average 3 documents which address the same subject, accompanied by its ideal summary. Table 1 summarizes the corpus data.

Table 1. Corpus statistics

Source texts statistics:

Total documents	Average documents	Total sentences	Average sentences	Total words	Average words
140	2.8	2,234	15.9	47,350	338.2

Ideal summaries statistics:

Total words	Average words	Average compression rate
6,859	137.18	85%

4.2   Results

First, for each set of the *CSTNews* corpora two summaries were created: one by SIMBA, and another by GISTSUMM.

In order to understand the differences that may lie between both summaries – the GISTSUMM and the SIMBA summary –, Table 2 details the phrases considered in the simplification process.

Hence, 35% of the sentences have an apposition phrase, and 17% have a parenthetical phrase. Thus, in the simplification process, we can work with around half of the sentences of the corpora. Appositions have in average five words, while parentheticals have three words in average. In fact, there is a large number of phrases to be examined. Considering both the apposition and the parenthetical phrases, they provide a total of 4,437 words.

The summaries have a compression rate of 85%, which means that the summary contains 85% of the words contained in the set of texts. This is the chosen value because it is the average compression rate of the ideal summaries.

Table 2. Phrases that are candidates to removal in the simplification process

Apposition phrases:

Tot. Sentences	Avg./document	Avg./sentence	Tot. Words	Avg. Words
774	15	0.35	3457	5

Parenthetical phrases:

Tot. Sentences	Avg./document	Avg./sentence	Tot. Words	Avg. Words
369	3	0.17	980	3

After both summaries have been built, they were compared with the corpus ideal summaries, using ROUGE [7]. In fact, a more precise metric of ROUGE was used, ROUGE-L (longest common subsequence), since it identifies the common subsequences between two sequences. The simplification process introduces gaps in the extracted sentences. This metric does not require consecutive matches but in-sequence matches, which reflect sentence level word order. This is a fairer metric, considering the type of arrangements made in the text. Precision, recall and f-measure values for each summarizer are detailed in Table 3.

Table 3. Multi-document evaluation metrics

	GISTSUMM	SIMBA
Precision	0.38469	0.47375
Recall	0.43616	0.45542
F-measure	0.40398	0.45980

By observing graph in Figure 1, we can see that SIMBA has a better performance than GISTSUMM.

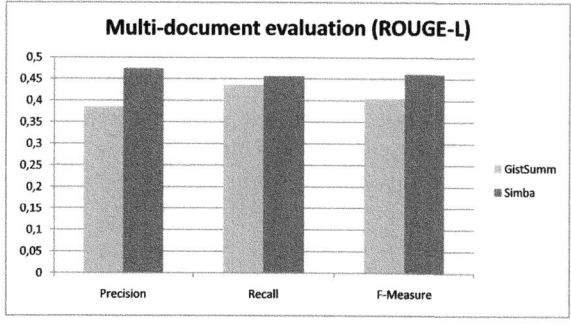

Fig. 1. ROUGE-L metric for GISTSUMM and SIMBA summaries

The complete summarization process has an overall better performance than GISTSUMM. SIMBA f-measure value overcomes the one of GISTSUMM in five percentage points, meaning that SIMBA summaries have more significant information than the ones of GISTSUMM.

The precision value obtained by SIMBA is very interesting. A high precision value means that considering all the information in the input texts, the retrieved information is relevant. Thus, obtaining the most relevant information in the sentences by discarding their less relevant data ensures that the summary contains indeed the most important information conveyed by each of its sentences.

The recall values of the two systems are closer than the ones concerning precision. Due to the simplification process, less in-sequence matches are likely to be found in SIMBA summaries, when compared to the ideal summaries. Thus, its recall values should be similar or even decrease. Still, considering both recall values, we can conclude that SIMBA summaries cover more significant topics than the summaries produced by GISTSUMM.

Thus, considering the evaluation results, we can conclude that this approach produces better summaries when compared to the one used by GISTSUMM.

5   Concluding Remarks

The quality of an automatic summary can be improved by (1) performing specific multi-document tasks – as removing the redundant information, or considering all the texts in each set as a single information source; and (2) executing an algorithm that seeks to optimize the content selection, combined with a simplification process that removes less relevant content, allowing the addition of new relevant information.

Despite the core algorithm is language independent, this system uses language specific tools that aim to improve not only the content selection, but also the general quality of a summary produced from a collection of texts written in Portuguese.

The multi-document summarizer presented relies on statistical features to perform summarization of a collection of texts in Portuguese. On the one hand, the double-clustering approach identifies the most relevant sentences in the texts. On the other, the simplification process removes from those sentences the information that adds no crucial content to the final summary. As the encouraging results, shown in the final evaluation, state, the combination of these two approaches produces highly informative summaries. In fact, these summaries not only preserve the idea conveyed by the collection of texts to be summarized, but also they cover the most significant topics mentioned in that collection.

References

1. Aleixo, P., Pardo, T.A.S.: Cstnews: Um córpus de textos jornalísticos anotados segundo a teoria discursiva multidocumento cst (cross-document structure theory). Tech. rep., Universidade de São Paulo (2008)

2. Branco, A., Silva, J.: A suite of shallow processing tools for portuguese: Lx-suite. In: Proceedings of the 11th Conference of the European Chapter of the Association for Computational Linguistics, EACL 2006 (2006)

3. Carbonell, J.G., Goldstein, J.: The use of MMR, diversity-based reranking for reordering documents and producing summaries. In: Research and Development in Information Retrieval, pp. 335–336 (1998)

4. Dice, L.R.: Measures of the amount of ecologic association between species. Ecology 26(3), 297–302 (1945)

5. Edmundson, H.P.: New methods in automatic extracting. J. ACM 16(2), 264–285 (1969)

6. Jaccard, P.: Nouvelles recherches sur la distribution florale. Bulletin de la Sociète Vaudense des Sciences Naturelles 44, 223–270 (1908)

7. Lin, C.Y.: Rouge: A package for automatic evaluation of summaries. In: Marie-Francine Moens, S.S. (ed.) Text Summarization Branches Out: Proceedings of the ACL 2004 Workshop, pp. 74–81. ACL, Barcelona (2004)

8. Lin, C.Y., Hovy, E.: From single to multi-document summarization: A prototype system and its evaluation. In: Proceedings of the ACL, pp. 457–464. MIT Press (2002)

9. Luhn, H.P.: The automatic creation of literature abstracts. IBM Journal of Research and Development 2 (1958)

10. MacQueen, J.B.: Some methods for classification and analysis of multivariate observations. In: Proceedings of the 5th Berkeley Symposium on Mathematical Statistics and Probability, vol. 1, pp. 281–297. University of California Press (1967)

11. Mani, I.: Automatic Summarization. Benjamins Pub. Co., Amsterdam (2001)

12. Pardo, T.A.S., Rino, L.H.M., das Graças Volpe Nunes, M.: GistSumm: A Summarization Tool Based on a New Extractive Method. In: Mamede, N.J., Baptista, J., Trancoso, I., Nunes, M.d.G.V. (eds.) PROPOR 2003. LNCS, vol. 2721, pp. 210–218. Springer, Heidelberg (2003)

13. Radev, D.R., Jing, H., Budzikowska, M.: Centroid-based summarization of multiple documents: sentence extraction, utility-based evaluation, and user studies. In: Proceedings of the 2000 NAACL-ANLP Workshop on Automatic Summarization, NAACL-ANLP-AutoSum 2000, pp. 21–30. ACL (2000)

14. Silva, J., Branco, A., Castro, S., Reis, R.: Out-of-the-Box Robust Parsing of Portuguese. In: Pardo, T.A.S., Branco, A., Klautau, A., Vieira, R., de Lima, V.L.S. (eds.) PROPOR 2010. LNCS, vol. 6001, pp. 75–85. Springer, Heidelberg (2010)

15. White, M., Korelsky, T., Cardie, C., Ng, V., Pierce, D., Wagstaff, K.: Multidocument summarization via information extraction. In: HLT 2001: Proceedings of the First International Conference on Human Language Technology Research, pp. 1–7 (2001)

Developing Multilingual Text Mining Workflows in UIMA and U-Compare

Georgios Kontonasios, Ioannis Korkontzelos, and Sophia Ananiadou

National Centre for Text Mining, School of Computer Science
The University of Manchester
georgios.kontonatsios@cs.man.ac.uk,
{Ioannis.Korkontzelos,Sophia.Ananiadou}@manchester.ac.uk

Abstract. We present a generic, language-independent method for the construction of multilingual text mining workflows. The proposed mechanism is implemented as an extension of U-Compare, a platform built on top of the Unstructured Information Management Architecture (UIMA) that allows the construction, comparison and evaluation of interoperable text mining workflows. UIMA was previously supporting strictly monolingual workflows. Building multilingual workflows exhibits challenging problems, such as representing multilingual document collections and executing language-dependent components in parallel. As an application of our method, we develop a multilingual workflow that extracts terms from a parallel collection using a new heuristic. For our experiments, we construct a parallel corpus consisting of approximately 188.000 PubMed article titles for French and English. Our application is evaluated against a popular monolingual term extraction method, C Value.

Keywords: text mining, multilingual text mining workflows, UIMA, U-Compare, multilingual term extraction.

1 Introduction

Recently, an increasing number of platforms are capable of developing text mining workflows. Workflows cleverly abstract the need of programming and technical skills and thus minimize the effort required to conduct *in silico* experiments. For this reason, platforms that support workflows are highly popular and widely applicable. Examples of such platforms in bioinformatics and chemoinformatics are *Taverna* [1] and *Galaxy* [2]. *Discovery Net* [3] is intended for molecular biology, *Kepler* [4] for environmental analysis, while *Pipeline-Pilot*[1] is a commercial system for business intelligence. *UIMA* [5] and *UCompare* [6] are designed for text mining and Natural Language Processing[2].

---

[1] accelrys.com/products/pipeline-pilot

[2] *GATE* [7] is also a popular platform, but is it more pipeline than workflow oriented, since it does not integrate a workflow management mechanism.

G. Bouma et al. (Eds.): NLDB 2012, LNCS 7337, pp. 82–93, 2012.
© Springer-Verlag Berlin Heidelberg 2012

Most of text analysis tools offered by the platforms are language specific (usually applicable to English, only) and therefore the workflows that contain these tools are monolingual. Exceptions are LIMA [8] and ATLAS [9], both built on top of UIMA. Although they offer components that support more than one language, each workflow instance can only handle a single language.

In this paper, we propose a method for developing multilingual workflows within the UIMA/U-Compare infrastructure. Resulting workflows are able to handle documents that contain multilingual content, while workflows of other platforms are able to handle documents of monolingual content, only. As an application, we present a UIMA/U-Compare workflow for multilingual term extraction from parallel biomedical corpora, based on the intuition that biomedical terms are invariant to translation. Our application workflow was tested on a bilingual parallel corpus consisting of approximately 188K pairs of PubMed article titles in French and English. The resulting set of bilingual terms was evaluated against large collections of biomedical ontologies: the Unified Medical Language System (UMLS) for English and part of the Unified Medical lexicon for French (UMLF). A popular monolingual term extraction method, *C Value* [10], was used as comparative baseline. Our experimental results show that existing monolingual term extraction techniques outperform our method.

The remaining of this paper is structured as follows: in section 2 we discuss literature related both to our method of developing multilingual text mining workflows (subsection 2.1) and to the application workflow for multilingual term extraction (subsection 2.2). In section 3 we present our methodology for constructing multilingual text mining workflows within the UIMA/U-Compare infrastructure. Based on our methodology we develop a multilingual workflow application that we demonstrate in section 4. Our application deploys a word-to-word alignment method for multilingual term extraction. Finally, we discuss the effectiveness of our methodology for developing complex multilingual workflows (subsection 4.5) and we report the results from the experiments that we have conducted on our application workflow(subsection 4.6).

2   Related Work

This section reviews text mining platforms that support the integration of multilingual components into their workflows or applications. Then, it briefly discusses methods for multilingual term extraction, relevant to our application workflow.

2.1   Platforms Supporting Multilinguality

ATLAS [9] is a language processing platform that focuses on the reusability and interoperability of NLP tools for a range of European languages. It is workflow-oriented and offers various levels of text analysis. The core processing level of ATLAS is built on top of the UIMA annotation model that allows components to be organized into workflows. Since UIMA offers only an abstract *Type System*, a hierarchy of data types, ATLAS extends it to enable interoperability. In terms of multilingual components, ATLAS offers a Machine Translation API and a language detection component, nonetheless, the vast majority of its core text mining tools are wrapped OpenNLP[3] modules tuned for English, only.

GATE is a mature framework that has been used for developing many text mining applications. It incorporates java libraries that handle character encodings for approximately 100 languages and makes a clean separation between algorithms and data [11]. Consequently, multilingual tools (e.g. POS taggers) can be reconfigured for different languages by modifying their settings. The multilingual characteristics of GATE aim to assist in developing new applications but do not support interoperability of existing tools.

LIMA [8] is another framework for developing text mining applications that supports multilingual components. LIMA adopts a language-oriented model, in which workflows are reconfigured in order to support different languages. For instance, a clitic stemmer pipeline, initially developed for Arabic is extended for Spanish by modifying the existing workflow. In terms of interoperability, the Lexical Markup Framework(LMF) [12] defines a standard within ISO for multilingual lexica. The components of LIMA support many languages, but each component in a workflow can handle text in one language, only.

2.2 Multilingual Term Extraction

Approaches to multilingual term extraction can be classified in distributional and heuristic-based ones. The former use the *Distributional Hypothesis* [13], which states that words occurring in similar context, tend to have similar meaning. This assumption was firstly used for compiling bilingual lexica automatically in [14] and [15]. Source and target words are encoded as frequency vectors whose dimensions represent their most likely neighbour words. A general-purpose bilingual dictionary is used to translate the dimensions of the source word vector in the language of the target word vector. Hence, the source and the target word vectors are comparable and their similarity according to some metric (e.g. cosine) is measurable. Similarity scores largely depend on the coverage of the general-purpose bilingual dictionary. Unavailability of such a dictionary for a given pair of languages or even poor coverage can affect the result detrimentally and make this method inapplicable. Several solutions have been proposed to deal with these problems. The *backing-off method* [16] suggests to map unknown word sequences to their components that occur in the bilingual dictionary. The *morphology-based compositional method* [17] attempts to map unknown words to words that occur in the dictionary using morphological information. However, in the biomedical domain, new terms are introduced with an exponential rate [18]. As a result a general-purpose bilingual dictionary would exhibit a rapidly decreasing coverage. The *backing-off method* would fail to translate multiword terms whose components do not occur in the dictionary (e.g. *beta-hydroxysteroid dehydrogenase*) and the *morphology-based compositional method* would still be inadequate for terms derived from rare or unknown morphemes (e.g. the metabolite *12-hete*).

---

[3] incubator.apache.org/opennlp/

Heuristic-based approaches to multilingual term extraction apply rules to parallel text collections. For example, [19] identifies patterns in the prefixes and suffixes of biomedical terms, which consist of Greek and Latin characters. The method suffers low recall, since it cannot retrieve terms on which the patterns do not apply (e.g. *disease, blood, x-ray*). [20] and [21] employ morpho-syntactic patterns (e.g. N_1 *Adj* N_2) to identify terms.

3   Methodology

In this section, we present our method that makes UIMA and U-Compare able to process multilingual documents and discuss how it supports workflow interoperability. We chose the UIMA/U-Compare infrastructure for a number of reasons. Firstly, UIMA offers a mechanism that can be employed to analyze multilingual content. Each document is associated with one or more independent views each of which refers to a part of the document. Thus, a bilingual document is represented by two UIMA views, one for each language. Secondly, U-Compare comes with a sharable Type System than enables interoperability. Components compatible with U-Compare's Type System are sharable within any U-Compare workflow. Thirdly, U-Compare currently offers the largest library of UIMA tools for text mining available for developing new multilingual workflows [6].

3.1   Processing Multilingual Documents in UIMA/U-Compare

Multilingual text mining usually processes parallel documents, i.e. documents with text in more than one language. Thus, multilingual workflows require a mechanism to separate text in different languages within a single document. Then, each part can be processed by language-dependent components, without affecting other parts. For instance, an English POS tagger will process the French part of the same document. Moreover, it is also desired that language-specific components share the same output data types so that results can be further processed in a unified manner. At the same time, different languages should remain identifiable, so that successor components in the workflow able to distinguish them. Moreover, distinguishing text and annotations of different languages is desirable for representational purposes. Inability to fulfil these requirements would harm interoperability or produce to duplicate and incorrect annotations.

To address the above challenges, we employ *the Subject of Analysis (SOFA)*, a mechanism embedded in UIMA and able to accommodate different views of the same document. SOFAs are applicable to cases that various workflow components need to access different parts of a single document, e.g. tagged or detagged parts of HTML pages, monolingual snippets of a multilingual documents, *gold standard* and *test* annotations etc. Each UIMA workflow component is linked with a specific view (SOFA) and processes the corresponding textual fragment.

Figure 1(a) illustrates a generic UIMA approach to processing parallel documents. Firstly, each monolingual fragment of a parallel document is assigned to a SOFA, so that the preceding language-dependent components process the

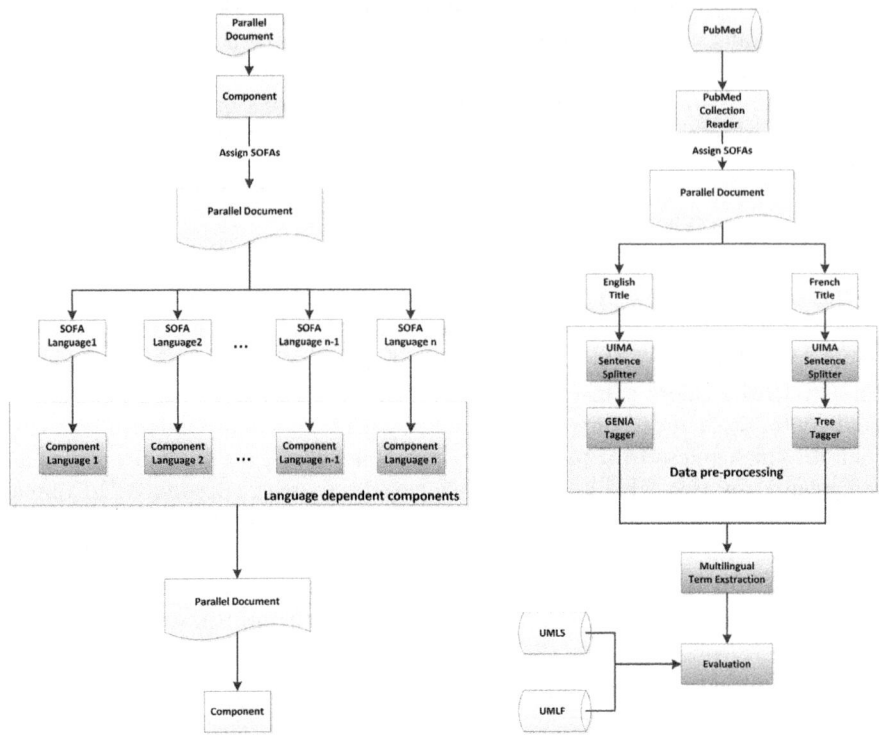

(a) Processing a multilingual document using UIMA's SOFAs

(b) Multilingual term extraction

Fig. 1. Block diagram architectures

correct fragment, only. SOFAs not only distinguish different regions of a single document but also can carry their own annotations. Annotations generated by the language-dependent components are linked with the corresponding SOFA and thus annotations for different languages remain distinct.

3.2 Enhancing Interoperability

Adding SOFAs to U-Compare grants multilingual functionality without modifying the U-Compare Type System. This ensures that U-Compare remains compatible with the existing monolingual U-Compare components, workflows and resources. Interoperability of text mining tools is essential. Firstly, compliance of tools speeds up development of new applications by allowing reuse. Secondly, comparison and evaluation of different workflows can be simplified given that workflows are interoperable. This is in text mining [6], since it helps developers locate problems and optimise their workflows. Using interoperable parallel or comparable corpora strengthens evaluations, since experiments can be repeated and validated with minimum effort by third-parties on a common basis.

4   Application Workflow: Multilingual Biomedical Term Extraction

In this section, we present an application of the proposed framework for multilingual term extraction from a parallel document collection. Our goal is to assess the effectiveness of our methodology for developing real life multilingual applications. We implement a new multilingual term extraction component and we apply it on a new biomedical, parallel corpus. Identifying terms, i.e. sequences related to a domain concept, is important towards extracting knowledge from published literature. Our term extraction workflow can be particularly useful for translators unfamiliar with medical terminology. The accuracy of human translators is reported to double when using multilingual term extraction tools [22]. In addition, the application can be used as a component in a Cross-Lingual Information Retrieval (CLIR) workflow. Experimentation has shown that query expansion or translation using multilingual resources, e.g. lexica, can considerably enhance the performance of CLIR systems [23]. Our workflow can also be used for updating medical bilingual dictionaries automatically as manual updating is laborious, due to the vast number of neologisms in this domain [18].

We exploit the assumption that the realizations of a term in different languages exhibit limited string variability. Given a parallel document, our approach considers as candidate terms sequences that are lexically similar in different languages. Our multilingual term extraction workflow can be configured for many languages including Dutch, French, Italian, German, Spanish, Greek and Bulgarian. However, we have only experimented with an English/French collection.

4.1   Overview of the Application Workflow

Figure 1(b) shows the architecture of our multilingual term extraction workflow. All its components are UIMA compliant and inherit data types of the U-Compare type system. The PubMed collection reader, the multilingual term extraction and the evaluation component were implemented from scratch. For POS-tagging text in English, we used the GENIA tagger [24], already existing in U-Compare's library. For text in French we used TreeTagger [25], after wrapping it into the infrastructure. The gold standard of English terms was constructed using the UMLS metathesaurus[4]. In French, we used UMLF [26] that includes French translations of most of the resources of UMLS (e.g. WHOs ICD-10 [5], SNOMED CT[6]) and an additional online medical dictionary[7].

4.2   Parallel Corpus

We created a parallel corpus using the titles of PubMed articles in English and French. Although, PubMed does not index translation of abstracts, it holds the

---

[4] nlm.nih.gov/research/umls
[5] who.int/classifications/icd/en
[6] nlm.nih.gov/research/umls/Snomed/snomed_main.html
[7] debussy.hon.ch/cgi-bin/HONselect

original titles of articles published in a language other than English under its *transliterated/vernacular* title field and also the corresponding translation in English[8]. Our parallel documents consist of these title pairs. The PubMed collection reader is responsible for creating the workflow SOFAs, the UIMA mechanism that we exploit so as to process parallel documents. Since our workflow is bilingual, we need to assign two SOFAs to the initial document. The English and French title are assigned to different SOFAs so that the language dependent components access the correct parts of the document.

4.3   Data Pre-processing

At this point, the workflow has retrieved relevant article titles from PubMed, has constructed the local collection, and has created the required SOFAs. Data preprocessing consists of a suite of components: two sentence splitters and two POS-taggers. It aims to generate all annotations required by the main term extraction component. POS-tagging is used to filter out noisy tokens, i.e. tokens that should not be allowed to be part of terms and may decrease performance if they do: punctuation marks, determiners, pronouns and prepositions are removed in both languages. We also remove single character tokens, assuming that they cannot be (part of) terms.

4.4   Multilingual Term Extraction Using String Similarity

We exploit the assumption that textual fragments that are invariant to translation are likely to be terms. Applying a strict invariant-to-translation criterion prevents the retrieval of terms that only differ in a few characters. For example the English biomedical term *hepatitis* with the French *hepatite*, cannot be retrieved although they are semantically identical. Introducing a similarity threshold to relax this strictly binary criterion allows retrieving such terms. Pairs whose similarity exceeds the threshold are accepted as pairs of biomedical terms. Selecting a good threshold value is vitally important because it directly affects classification performance. The threshold is a trade-off between identical and loosely similar strings and can be thought as the possibility that our hypothesis holds. The lower the threshold, the less possible it is that our hypothesis holds.

String similarity can be computed using sequence alignment, a technique used widely in bioinformatics [27,28] for identifying similarities between DNA, RNA or protein sequences and in text mining [29,30] for dictionary matching or Machine Translation applications. Sequence alignment aims to find the highest scoring alignment of two input strings. From a variety of string alignment methods, we choose to deploy the Needleman-Wunsch algorithm [28] because it is simple, and easy to implement and adjust. The maximum alignment score is 5, achieved by identical strings. A minimum score is not defined by the algorithm.

The strict binary version of our hypothesis is verified by very few candidates, and, unfortunately, not all of them are medical terms, e.g. *role*. In addition, for

---

[8] `nlm.nih.gov/pubs/techbull/jf06/jf06_trans_title.html`

high threshold values, our method is not able to match n-grams which consist of the same tokens in a different order. For example, *pelvic actinomycosis*, a rare disease in English, was not matched with its French equivalent, *actynomycose pervienne*. In the diametric case of low threshold values, our method matches more actual biomedical terms but also introduces more noise.

4.5   Evaluation of the Methodology

Although our string-alignment application is simple, the *C Value* method [10] for term extraction that we wrap into U-Compare and deploy as a multilingual workflow is complex. UIMA NLP applications by default operate on a per-document level, while C Value operates on both single documents and document collections. C Value a statistical approach consisting of two stages: firstly, the frequency of every candidate term is measured and then the number of times a candidate term occurs as nested in longer terms is calculated. To deploy these stages in U-Compare we had to incorporate other UIMA mechanisms (CAS multipliers) that allow pausing the workflow execution after all documents are seen once and statistics are computed, run the C Value algorithm on these statistics, and then resume the workflow execution so as to revisit all documents to annotate candidate terms. It is worth noting that the C Value workflow was developed according to the proposed framework, i.e. SOFAs, so as to support multilinguality. Thus, the framework is useful for building complex state-of-the art multilingual term extraction workflows.

The extension that we propose in this paper could not possibly be verified following a standard evaluation process, i.e. using accuracy and performance measures. However, using the comparison mechanism embedded within U-Compare, we can evaluate the effectiveness of the multilingual workflow application against state-of-the-art multilingual term extraction algorithms. U-Compare extended with our framework allows easily comparing term extraction tools in a multilingual setting, where the algorithmic part is separated from the language components required.

4.6   Evaluation of the Multilingual Workflow Application

We experimented with 250 alignment threshold values lying in the range $[-0.1, 5]$. For each execution we computed the standard information retrieval precision, recall and F-Score metrics for English and French documents separately and for the parallel collection as a whole. Figure 2 illustrates the precision and recall for different values of the threshold.

Starting from low threshold values, our method accepts most candidate tokens as biomedical terms. As a result, the recall is very high, almost perfect, but the precision is diametrically low. While the threshold increases, we accept less and less candidate tokens and therefore the recall constantly decreases. Our results show that the optimal threshold values are different for English and French but still close. We obtained a maximum precision of 67% for a threshold value of 3.2 for English, 62% at 4.22 for French and an overall maximum precision of

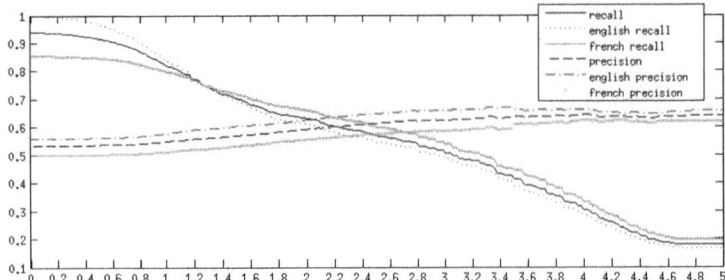

Fig. 2. Precision and Recall for different threshold values

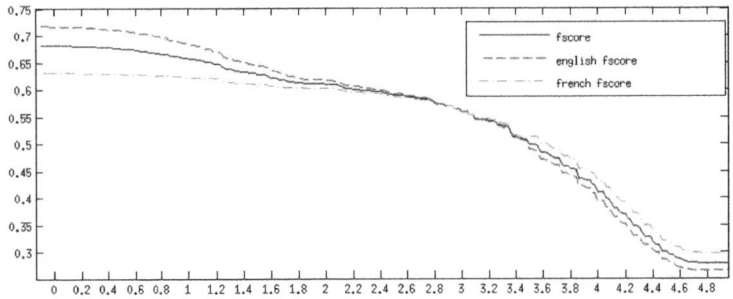

Fig. 3. F-Score measures for different threshold values

64% at a 3.84 threshold for the whole parallel collection. The performance of our workflow is lower on French documents, because the UMLF gold standard is less complete than the UMLS [26] that we use for English.

Figure 3 shows the F-Score metrics with respect to threshold values. We observe that the curves are continuously decreasing. This is an indication of poor performance, which is expected due to the simplicity and naiveness of our approach. However, we emphasize that we do not intend to develop a state-of-the-art multilingual term extraction system but we rather focus in demonstrating how the proposed methodology allows developing multilingual and interoperable text mining workflows for term extraction in a straightforward manner. Building workflows within the U-Compare platform supports easy comparison and evaluation depending on the selection of individual components in the pipeline.

To verify our results further, we compare our multilingual workflow with *C Value* [10]. C Value extracts candidate terms based on POS patterns and assigns a score that represents the likelihood that a sequence of words is a valid term. Since, it is a monolingual term extraction method, we need to make different configurations for the two languages of our parallel collection. C Value is minimally language dependent, because it uses POS patterns, which vary for different languages. Hence, after implementing C Value and wrapping it into

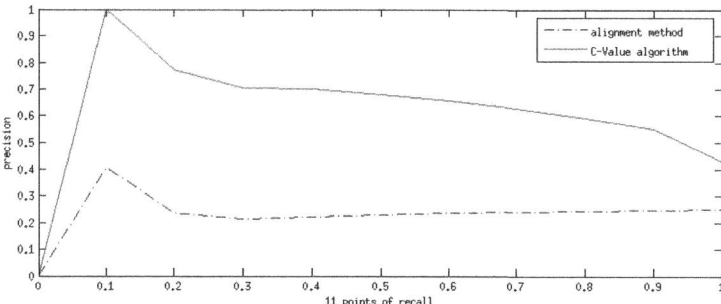

Fig. 4. Comparison of C Value with our multilingual term extraction component for bigram terms

the UIMA/U-Compare infrastructure, we defined POS patterns for English and French term candidates. The former accept any adjective and noun sequences that end with a noun while the latter any adjective and noun sequence.

While comparing the two methods we faced two inconsistencies: firstly, C Value only identifies multiword terms and, thus, the comparison was performed for bigrams, only. Secondly, C Value is a statistical-based approach and generates a ranked list of candidate terms. Therefore, thresholds for the two methods are different. In order to make the two methods comparable we calculated the best achieved precision at 11 points of recall.

Figure 4 illustrates our experimental results and verifies that our system performs poorly compared the popular term extraction technique C Value. The C Value algorithm outperformed our approach for all 11 points of recall.

5 Conclusions and Future Work

In this paper we presented a method for constructing multilingual workflows using the UIMA/U-Compare infrastructure. We examined the problems related to multilingual workflows and ways to address them effectively using different views of parallel documents as implemented by UIMA's SOFAs. Furthermore, we implemented a series of multilingual components for U-Compare, wrapped existing multilingual resources into the infrastructure and explained how U-Compare supports interoperability of these resources. As an application of our method, we coordinated the above components into a multilingual workflow that extracts terms from parallel documents in the biomedical domain.

The experiment showed that our multilingual workflow does not perform as robustly an existing term extraction algorithm, C Value. Therefore, as future work we propose a hybrid model that combines the existing string alignment algorithm with a statistical-based approach in order to rank candidate terms and quantify the importance of a candidate term for a specific domain.

Acknowledgements. This work was supported by the UK Joint Information Systems Committee (JISC) as part of the ISHER project, awardee of the Digging into Data Challenge.

References

1. Oinn, T., Greenwood, M., Addis, M., Alpdemir, M., Ferris, J., Glover, K., Goble, C., Goderis, A., Hull, D., Marvin, D., et al.: Taverna: lessons in creating a workflow environment for the life sciences. Concurrency and Computation: Practice and Experience 18(10) (2006)
2. Goecks, J., Nekrutenko, A., Taylor, J., Team, T.: Galaxy: a comprehensive approach for supporting accessible, reproducible, and transparent computational research in the life sciences. Genome Biology 11(8), R86 (2010)
3. Rowe, A., Kalaitzopoulos, D., Osmond, M., Ghanem, M., Guo, Y.: The discovery net system for high throughput bioinformatics. Bioinformatics 19(suppl. 1) (2003)
4. Barseghian, D., Altintas, I., Jones, M., Crawl, D., Potter, N., Gallagher, J., Cornillon, P., Schildhauer, M., Borer, E., Seabloom, E.: Workflows and extensions to the kepler scientific workflow system to support environmental sensor data access and analysis. Ecological Informatics 5(1) (2010)
5. Ferrucci, D., Lally, A.: Building an example application with the unstructured information management architecture. IBM Systems Journal 43(3) (2004)
6. Kano, Y., Miwa, M., Cohen, B., Hunter, L., Ananiadou, S., Tsujii, J.: U-Compare: A modular nlp workflow construction and evaluation system. IBM Journal of Research and Development 55(3) (2011)
7. Cunningham, H., Maynard, D., Bontcheva, K., Tablan, V.: GATE: A Framework and Graphical Development Environment for Robust NLP Tools and Applications. In: Proceedings of the 40th Anniversary Meeting of the Association for Computational Linguistics, ACL 2002 (2002)
8. Besançon, R., de Chalendar, G., Ferret, O., Gara, F., Semmar, N.: Lima: A multilingual framework for linguistic analysis and linguistic resources development and evaluation. In: 7th Conference on Language Resources and Evaluation (LREC 2010), Malta (2010)
9. Ogrodniczuk, M., Karagiozov, D.: Atlas multilingual language processing platform. Procesamiento de Lenguaje Natural 47 (2011)
10. Frantzi, K., Ananiadou, S., Mima, H.: Automatic recognition of multi-word terms: the C-value/NC-value method. International Journal on Digital Libraries 3(2) (2000)
11. Bontcheva, K., Maynard, D., Tablan, V., Cunningham, H.: Gate: A unicode-based infrastructure supporting multilingual information extraction. In: Proceedings on Information Extraction for Slavonic and other Central and Eastern European Languages, Borovets, Bulgaria (2003)
12. Francopoulo, G., George, M., Calzolari, N., Monachini, M., Bel, N., Pet, M., Soria, C.: Lexical markup framework (lmf). In: International Conference on Language Resources and Evaluation-LREC. Number 2006 (2006)
13. Harris, Z.: Distributional structure. Word (1954)
14. Rapp, R.: Identifying word translations in non-parallel texts. In: Proceedings of the 33rd Annual Meeting on Association for Computational Linguistics. Association for Computational Linguistics (1995)

15. Fung, P., McKeown, K.: Finding terminology translations from non-parallel corpora. In: Proceedings of the 5th Annual Workshop on Very Large Corpora. (1997)
16. Robitaille, X., Sasaki, Y., Tonoike, M., Sato, S., Utsuro, T.: Compiling french-japanese terminologies from the web. In: Proceedings of the 11st Conference of the European Chapter of the Association for Computational Linguistics (2006)
17. Morin, E., Daille, B.: Compositionality and lexical alignment of multi-word terms. Language Resources and Evaluation 44(1) (2010)
18. Ananiadou, S., Mcnaught, J.: Text mining for biology and biomedicine. Artech House Publishers (2006)
19. Bernhard, D.: Multilingual term extraction from domain-specific corpora using morphological structure. In: Proceedings of the Eleventh Conference of the European Chapter of the Association for Computational Linguistics: Posters & Demonstrations. Association for Computational Linguistics (2006)
20. Daille, B.: Study and implementation of combined techniques for automatic extraction of terminology. In: Klavans, J., Resnik, P. (eds.) The Balancing Act: Combining Symbolic and Statistical Approaches to Language. MIT Press, Cambridge (1996)
21. Fidelia, I.: Terminological variation, a means of identifying research topics from texts. In: Proceedings of the 17th International Conference on Computational Linguistics, COLING 1998, vol. 1. Association for Computational Linguistics, Stroudsburg (1998)
22. Fung, P., McKeown, K.: A technical word-and term-translation aid using noisy parallel corpora across language groups. Machine Translation 12(1) (1997)
23. Sheridan, P., Braschlert, M., Schäuble, P.: Cross-language Information Retrieval in a Multilingual Legal Domain. In: Peters, C., Thanos, C. (eds.) ECDL 1997. LNCS, vol. 1324, pp. 253–268. Springer, Heidelberg (1997)
24. Tsuruoka, Y., Tateishi, Y., Kim, J.-D., Ohta, T., McNaught, J., Ananiadou, S., Tsujii, J.: Developing a Robust Part-of-Speech Tagger for Biomedical Text. In: Bozanis, P., Houstis, E.N. (eds.) PCI 2005. LNCS, vol. 3746, pp. 382–392. Springer, Heidelberg (2005)
25. Schmid, H.: Treetagger: a language independent part-of-speech tagger (1995), www.ims.uni-stuttgart.de/projekte/corplex/TreeTagger/
26. Zweigenbaum, P., Baud, R., Burgun, A., Namer, F., Jarrousse, E., Grabar, N., Ruch, P., Le Duff, F., Forget, J., Douyere, M.: Umlf: a unified medical lexicon for french. International Journal of Medical Informatics 74(2-4) (2005)
27. Altschul, S., Gish, W., Miller, W., Myers, E., Lipman, D.: Basic local alignment search tool. Journal of Molecular Biology 215(3) (1990)
28. Needleman, S., Wunsch, C.: A general method applicable to the search for similarities in the amino acid sequence of two proteins. Journal of Molecular Biology 48(3) (1970)
29. Wu, X., Matsuzaki, T., Tsujii, J.: Fine-grained tree-to-string translation rule extraction. In: Proceedings of the 48th Annual Meeting of the Association for Computational Linguistics. Association for Computational Linguistics (2010)
30. Okazaki, N., Tsujii, J.: Simple and efficient algorithm for approximate dictionary matching. In: Proceedings of the 23rd International Conference on Computational Linguistics. Association for Computational Linguistics (2010)

Geographic Expansion of Queries to Improve the Geographic Information Retrieval Task

José M. Perea-Ortega[1] and L. Alfonso Ureña-López[2]

[1] Languages and Information Systems Department
University of Sevilla, Spain
jmperea@us.es
[2] Computer Science Department
University of Jaén, Spain
laurena@ujaen.es

Abstract. Geographic Information Retrieval (GIR) is concerned with improving the quality of geographically-specific Information Retrieval (IR), focusing on access to unstructured documents. Since GIR can be considered as an extension of IR, the application of Natural Language Processing (NLP) techniques, such as query expansion, can lead to significant improvements. In this paper we propose two NLP techniques of query expansion related to the augmentation of the geospatial part that is usually identified in a geographic query. The aim of both approaches is to retrieve possible relevant documents that are not retrieved using the original query. Then, we propose to add such new documents to the list of documents retrieved using the original query. In this way, the geo-reranking process takes into account more possible relevant documents. We have evaluated the proposed approaches using GeoCLEF as evaluation framework for GIR systems. The results obtained show that the use of proposed query expansion techniques can be a good strategy to improve the overall performance of a GIR system.

Keywords: Geographic Information Retrieval, Query Expansion, GeoCLEF.

1 Introduction

Natural Language Processing (NLP) techniques, such as query expansion, are an integral part of most Information Retrieval (IR) architectures. Since Geographic Information Retrieval (GIR) can be considered as an extension of IR [11], the application of these techniques in GIR systems can lead to significant improvements. While in classic IR retrieved documents are ranked by their similarity to the text of the query, in a search engine with geographic capabilities, the semantics of geographic terms should be considered as one of the ranking criteria [2].

G. Bouma et al. (Eds.): NLDB 2012, LNCS 7337, pp. 94–103, 2012.
© Springer-Verlag Berlin Heidelberg 2012

Specifically, GIR is concerned with improving the quality of geographically-specific information retrieval, focusing on access to unstructured documents [11,14]. The IR community has primarily been responsible for research in the GIR field, rather than the Geographic Information Science (GIS) community. The type of query in a IR engine is based usually on natural language, in contrast to the more formal approach common in GIS, where specific geo-referenced objects are retrieved from a structured database. In a GIR system, a geographic query can be structured as a triplet of $<theme><spatial\ relationship><location>$, where $<theme>$ is the main subject of the query, $<location>$ represents the geographic scope of the query and $<spatial\ relationship>$ determines the relationship between the subject and the geographic scope. For example, the triplet for the geographic query *"airplane crashes close to Russian cities"* would be $<airplane\ crashes><close\ to><Russian\ cities>$. Thus, a search for *"castles in Spain"* should return not only documents that contain the word *"castle"*, also those documents which have some geographical entity related to Spain. For this reason, it is important to pay attention to finding effective methods for query expansion to improve the quality of the retrieved documents. From an IR point of view, query expansion refers to the process of automatically adding additional terms to the query, in an effort to improve the relevance of the retrieved results. From a GIR point of view, geographic expansion techniques can be used to augment any geographic term identified in queries, thereby increasing the likelihood of finding relevant documents with geographic entities that match the geographic scope identified in the query.

To carry out query expansion for geographic queries, we can take into account both lexical-syntactic features and geographical aspects. In this paper we propose two query expansion techniques based on the addition of synonyms of the geospatial scope identified in the query and, on the other hand, the addition of geographic terms that match with the geospatial scope of the query. Then, we merge the documents retrieved by using the original query with those new documents retrieved by using the proposed query expansions. Finally, we apply a reranking function based on the textual and geographical similarity between each retrieved document and the query. To carry out the evaluation, we have used the most important evaluation framework in this context: GeoCLEF[1] [8,17]. The results show that the proposed query expansion techniques retrieved relevant documents that were not retrieved by using the original query, so that after applying the reranking function, our GIR system was able to improve its overall performance.

The remainder of this paper is structured as follows: in Section 2, the most important works related to the query expansion in GIR are expounded; in Section 3 we describe the GIR system used for the experiments carried out in this work; in Section 4 and Section 5, the evaluation framework is briefly described and the experiments and an analysis of the results are presented, respectively. Finally, in Section 6, some conclusions and future work are expounded.

---

[1] http://ir.shef.ac.uk/geoclef/

2   Related Work

Jansen et al. [10] define the concept of query reformulation as the process of altering a given query in order to improve search or retrieval performance. Sometimes, query reformulation is applied automatically by search engines as with *relevance feedback* technique. It is a method that allows users to judge whether a document is relevant or not, so that automatic rewritings can be generated depending on it. At other times, query reformulation is carried out analysing the top retrieved documents without the user's intervention, taking into account term statistics. However, it has been found that users rarely utilize the relevance feedback options [22] and usually reformulate their needs manually [3].

The focus of this paper is geographic queries. According to Gravano [9], search engines are criticised because of their ignorance to the geographical constraints on users' queries and, therefore, retrieve less relevant results. This could be attributed to the way search engines handle queries in general as they adopt a keywords matching approach without spatially inferring the scope of the geographic terms. However, it shall be noted that a number of services to deal with this issue have recently been proposed in major search engines, but not in the general purpose tools.

Several authors have studied what users are looking for when submitting geographic queries [21,7,12]. One of the main conclusions of these studies is that the structure of geographic queries consists of thematic and geographical parts, with the geo-part occasionally containing spatial or directional terms. From a geographical point of view, Kohler [13] provides a research about geo-reformulation of queries. She concludes that the addition of more geo terms in the query is commonly used to differentiate between places that share the same name. This is also known as query expansion using geographic entities.

In the literature, we can find various works that have addressed the spatial query expansion. Cardoso et al. [5] present an approach for geographical query expansion based on the use of feature types, readjusting the expansion strategy according to the semantics of the query. Fu et al. [6] propose an ontology-based spatial query expansion method that supports retrieval of documents that are considered to be spatially relevant. They improve search results when a query involves a fuzzy spatial relationship, showing that proposed method works efficiently using realistic ontologies in a distributed spatial search environment. Buscaldi et al. [4] use WordNet[2] during the indexing phase by adding the synonyms and the holonyms of the encountered geographical entities to each documents index terms, proving that such method is effective. Li et al.[15] describe two types of geo-query expansion: *downward expansion* and *upward expansion*. *Downward expansion* extends the influence of a geo-term to some or all of its descendants in the hierarchical gazetteer structure, to encompass locations that are part of, or subregions of, the location specified in the query. *Upward expansion* extends the influence of a geo-term to some or all of its ancestors, and then possibly downward again into other siblings of the original node. This facilitates

---

[2] http://wordnet.princeton.edu/

the expansion of geo-terms in the query to their nearby locations. Finally, Stokes et al. [23] conclude that significant gains in GIR will only be made if all query concepts (not just geospatial ones) are expanded.

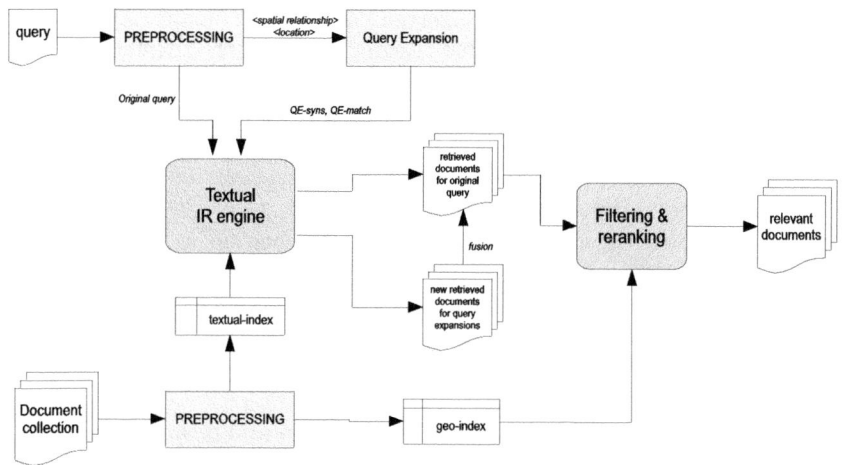

Fig. 1. Overview of the SINAI-GIR system

3   GIR System Overview

In this Section we describe an example of a GIR system. Specifically, we have used our own GIR system called SINAI-GIR [20]. GIR systems are usually composed of three main stages: preprocessing of the document collection and queries, textual-geographical indexing and searching and, finally, reranking of the retrieved results using a particular relevance formula that combines textual and geographical similarity between the query and the retrieved document. The GIR system used in this work follows a similar approach, as can be seen in Figure 1.

On the one hand, each query is preprocessed and analyzed, identifying the geographic scope and the spatial relationship that may contain. On the other hand, the document collection is also preprocessed, detecting all the geographic entities and generating a geo-index with them. In this phase, the stop words are removed and the stem of each word is taken into account. Then, each preprocessed query (including their geographic entities) is run against the search engine.

Regarding query processing, it is mainly based on detecting the geographic entities. We have used Geo-NER[19] to recognize spatial entities in the collection and queries, a Named Entity Recognizer (NER) for geographic entities based on GeoNames[3] and Wikipedia. This phase also involves specifying the triplet explained in Section 1, which will be used later during the filtering and reranking

---

[3] http://www.geonames.org

process. To detect such triplet, we have used a Part Of Speech tagger (POS tagger) like TreeTagger[4], taking into account some lexical syntactic rules such as *preposition + proper noun*, for example. Moreover, the stop words are removed and the Snowball stemmer[5] is applied to each word of the query, except for the geographical entities. During the text retrieval process, we obtain 1,000 documents for each query. We have used Terrier[6] as a search engine. According to a previous work [18], it was shown that Terrier is one of the most used IR tools in IR systems in general and GIR systems in particular, obtaining promising results. The weighting scheme used has been *inL2*, which is implemented by default in Terrier. This scheme is the Inverse Document Frequency (IDF) model for randomness, Laplace succession for first normalization, and Normalization 2 for term frequency normalization [1].

In addition to the original preprocessed query, each query expansion is also launched against the search engine. Then, the new documents retrieved by each expansion are added to the list of documents retrieved for the original query with the lowest Retrieval Status Value (RSV) found in such list. In this way, the reranking process will take into account more possible relevant documents to rerank.

Finally, in the last phase, the fusion list is filtered and reranked, leaving only 1,000 documents to return and making use of the reranking function that combines both similarities textual and geographical between the query and each document. Many GIR systems, for example those of Li et al. [16] and Andrade and Silva [2], combine the scores of textual terms and geographic terms using linear combinations of the form:

$$sim(Q, D) = \alpha \times sim_{text}(Q, D) + (1 - \alpha) \times sim_{geo}(Q, D) \qquad (1)$$

where $sim_{text}(Q, D)$ is the score assigned by the search engine to each document, i.e. the RSV score. For the experiments carried out in this work, we have tried several values of α, obtaining the best performance with $\alpha = 0.5$. On the other hand, $sim_{geo}(Q, D)$ is the geographic similarity between a document (D) and a query (Q) and it is calculated using the following formula:

$$sim_{geo}(Q, D) = \frac{\sum_{i \in geoEnts(D)} match(i, GS, SR) \cdot freq(i, D)}{|geoEnts(D)|} \qquad (2)$$

where the function $match(i, GS, SR)$ returns 1 if the geographic entity i satisfies the geographic scope GS for the spatial relationship SR and 0 otherwise. $freq(i, D)$ means frequency of the geographic entity i in document D, and $|geoEnts(D)|$ represents the total number of geographic entities identified in the document D. To explain the performance of the *match* function, we can

---

[4] TreeTagger v.3.2 for Linux. Available in http://www.ims.uni-stuttgart.de/projekte/corplex/TreeTagger/DecisionTreeTagger.html
[5] Available in http://snowball.tartarus.org
[6] Version 2.2.1, available in http://terrier.org

use the following query: *"Hurricane Katrina in the United States"*. In this case, it is a geographic query because we can recognize a geographical scope (*United States*) and a spatial relationship (*in*). The theme or subject of the query would be *Hurricane Katrina*. Therefore, when the system finds a geographic entity i (for example, *New York*) in a retrieved document (D) which belongs to United States, then $match(NewYork, UnitedStates, in) = 1$. If the geographic entity did not belong to the geographic scope (GS), then the *match* function would return 0 (for example, $match(Madrid, UnitedStates, in) = 0$). In short, the $match(i, GS, SR)$ function receives as input the geographic entity i of the document, the geographic scope (GS) of the query and the spatial relationship (SR) identified in the query. This function is based on manual rules such as *"if $SR = in$ and $i \in GS$ then return 1, else return 0"*. Obviously, this function makes use of an external geographical database like GeoNames in order to check if a city belongs to a country or a continent, for example.

4   GeoCLEF: The Evaluation Framework

In order to evaluate the proposed query expansions, we have used the GeoCLEF framework [8,17], an evaluation forum for GIR systems held between 2005 and 2008 under the CLEF[7] conferences. GeoCLEF provides a document collection that consists of 169,477 documents, composed of stories and newswires from the British newspaper *Glasgow Herald* (1995) and the American newspaper *Los Angeles Times* (1994), representing a wide variety of geographical regions and places. On the other hand, there are a total of 100 textual queries or topics provided by GeoCLEF organizers (25 per year). They are composed of three main fields: *title* (T), *description* (D) and *narrative* (N). For the experiments carried out in this work, we have only taken into account the *title* field because it represents in a similar way how a user would launch a geographic query to a search engine. Some examples of GeoCLEF topics are: *"vegetable exporters of Europe"*, *"forest fires in north of Portugal"*, *"airplane crashes close to Russian cities"* or *"natural disasters in the Western USA"*.

Regarding the evaluation measures used, results have been evaluated using the relevance judgements provided by the GeoCLEF organizers and the TREC evaluation method. The evaluation has been accomplished by using the Mean Average Precision (MAP) that computes the average precision over all queries. The average precision is defined as the mean of the precision scores obtained after each relevant document is retrieved, using zero as the precision for relevant documents that are not retrieved.

5   Experiments and Results

As previously mentioned, two types of query expansions for the GIR task are proposed in this work. Both strategies use the geographic scope identified in the

---

[7] http://www.clef-initiative.eu/

query. The aim of these query expansions is to improve the retrieval process trying to find relevant documents that are not retrieved using the original query. Specifically, we propose the following query expansions:

- QE-syns: the geographic part is expanded using only synonyms of the geographic scope identified in the query.
- QE-match: the geographic part is expanded using locations or places that match with the geographic scope and the spatial relationship identified in the query.

Table 1. Example of query expansions generated for the query *"Visits of the American president to Germany"*

Query Expansion	Text of the query
original	visit American presid Germany
QE-syns	#and(visit American presid #or(Germany #3(Federal Republic of Germany) Deutschland FRG))
QE-match	#and(visit American presid #or(Germany Berlin Hamburg Muenchen Koeln #2(Frankfurt am Main) Essen))

Table 1 shows an example of the query expansions generated for the query *"Visits of the American president to Germany"*. As can be seen, QE-syns and QE-match expand only the geographical part of them, making use of the synonyms of the geographic scope identified in the query and with places that match with the geospatial scope of the query, respectively. Table 2 shows the results of each query expansion strategy compared with those obtained using original queries without applying any reranking process.

Table 2. Results of each query expansion strategy compared with those obtained using original queries without applying any reranking process

Query set	MAP orig query baseline $inL2$	MAP QE-syns baseline $inL2$	MAP QE-match baseline $inL2$
2005	0.3514	0.2242	0.0952
2006	0.2396	0.2064	0.1811
2007	0.2311	0.1687	0.1874
2008	0.2484	0.1619	0.1906

As has been explained in Section 3, the new documents retrieved using the query expansions proposed (QE-syns and QE-match) are added to the list of documents retrieved using the original query and, therefore, a *fusion* list of documents is generated. This list is reranked taking into account the formula expounded in 1. Table 3 shows the results obtained using this *fusion* list compared with those obtained using the original query. Moreover, it is also shown the total number of relevant documents for each query set (*Total num rel*), the number of relevant documents retrieved (*Num rel ret*) by the original query and the *fusion* list and the MAP score obtained for each experiment.

Table 3. Summary of the experiments and results

Query set	Total num rel	Num rel ret orig query	Num rel ret fusion list	MAP orig query baseline	MAP orig query reranked	MAP fusion list reranked
2005	1028	**908**	904	0.3514	**0.3608**	0.3606
2006	378	284	**291**	0.2396	0.2417	**0.2419**
2007	650	543	**570**	0.2311	0.2448	**0.2464**
2008	747	588	**614**	0.2484	0.2606	**0.2614**

Analyzing these results, we can observe that for three of the four query sets, the proposed query expansions added relevant documents that were not retrieved using the original query. Specifically, for the 2007 and 2008 query sets, the expansion techniques provided 27 and 26 new relevant documents that were not retrieved using the original query, respectively. For the 2006 query set, the *fusion* list provided 7 new relevant documents. Obviously, these results have a positive impact on the calculation of the MAP score. As can be seen in Table 3, the MAP value obtained using the reranked *fusion* list improved 2.62%, 0.96%, 6.62% and 5.23% the MAP score obtained using the original query without applying the reranking process for the 2005, 2006, 2007 and 2008 query sets, respectively. The differences achieved by the reranked *fusion* list were smaller when we applied the reranking process to the list of documents retrieved for the original query solely: +0.08%, +0.65% and +0.31% for the 2006, 2007 and 2008 query sets, respectively.

6   Conclusions and Further Work

In this paper we propose two NLP techniques of query expansion related to the augmentation of the geospatial part that is usually identified in a geographic query. The aim of both approaches is to retrieve possible relevant documents that are not retrieved using the original query. Then, we propose to add such new documents to the list of documents retrieved using the original query. In this way, the geo-reranking process takes into account more possible relevant documents. We have evaluated the proposed approaches using GeoCLEF as evaluation framework for GIR systems. The results obtained show that the use of proposed query expansion techniques can be a good strategy to improve the overall performance of a GIR system.

For future work, we will analyze the different types of geographic queries and then we will study in depth when is more suitable to apply these techniques in a GIR system depending on the type of the query. We will also try to analyze the performance of the expansion of the thematic part of the query, using synonyms of the keywords, for example.

Acknowledgments. This work has been partially funded by the European Commission under the Seventh (FP7-2007-2013) Framework Programme for Research and Technological Development through the FIRST project (FP7-287607). This publication reflects the views only of the author, and the Commission cannot be held responsible for any use which may be made of the information contained therein. It has been also partially supported by a grant from the Fondo Europeo de Desarrollo Regional (FEDER) through the TEXT-COOL 2.0 project (TIN2009-13391-C04-02) from the Spanish Government, and a grant from the Instituto de Estudios Giennenses through the Geocaching Urbano project (RFC/IEG2010).

References

1. Amati, G.: Probabilistic Models for Information Retrieval based on Divergence from Randomness. Ph.D. thesis, School of Computing Science, University of Glasgow (2003)
2. Andrade, L., Silva, M.J.: Relevance ranking for geographic ir. In: Purves, R., Jones, C. (eds.) GIR. Department of Geography, University of Zurich
3. Anick, P.: Using terminological feedback for web search refinement: a log-based study. In: SIGIR 2003: Proceedings of the 26th Annual International ACM SIGIR Conference on Research and Development in Informaion Retrieval, pp. 88–95. ACM, New York (2003)
4. Buscaldi, D., Rosso, P., Arnal, E.S.: Using the WordNet Ontology in the GeoCLEF Geographical Information Retrieval Task. In: Peters, C., Gey, F.C., Gonzalo, J., Müller, H., Jones, G.J.F., Kluck, M., Magnini, B., de Rijke, M., Giampiccolo, D. (eds.) CLEF 2005. LNCS, vol. 4022, pp. 939–946. Springer, Heidelberg (2006)
5. Cardoso, N., Silva, M.J.: Query expansion through geographical feature types. In: Purves, R., Jones, C. (eds.) GIR, pp. 55–60. ACM (2007)
6. Fu, G., Jones, C.B., Abdelmoty, A.I.: Ontology-Based Spatial Query Expansion in Information Retrieval. In: Meersman, R., Tari, Z. (eds.) OTM 2005, Part II. LNCS, vol. 3761, pp. 1466–1482. Springer, Heidelberg (2005)
7. Gan, Q., Attenberg, J., Markowetz, A., Suel, T.: Analysis of geographic queries in a search engine log. In: Proceedings of the First International Workshop on Location and the Web, pp. 49–56. ACM, Beijing (2008)
8. Gey, F.C., Larson, R.R., Sanderson, M., Joho, H., Clough, P., Petras, V.: GeoCLEF: The CLEF 2005 Cross-Language Geographic Information Retrieval Track Overview. In: Peters, C., Gey, F.C., Gonzalo, J., Müller, H., Jones, G.J.F., Kluck, M., Magnini, B., de Rijke, M., Giampiccolo, D. (eds.) CLEF 2005. LNCS, vol. 4022, pp. 908–919. Springer, Heidelberg (2006)
9. Gravano, L., Hatzivassiloglou, V., Lichtenstein, R.: Categorizing web queries according to geographical locality. In: Proceedings of the 12th International Conference on Information and Knowledge Management, pp. 325–333 (2003)

10. Jansen, B.J., Booth, D.L., Spink, A.: Patterns of query reformulation during web searching. JASIST 60(7), 1358–1371 (2009)
11. Jones, C.B., Purves, R.S.: Geographical information retrieval. International Journal of Geographical Information Science 22(3), 219–228 (2008)
12. Jones, R., Zhang, W.V., Rey, B., Jhala, P., Stipp, E.: Geographic intention and modification in web search. International Journal of Geographical Information Science 22(3), 229–246 (2008)
13. Kohler, J.: Analysing search engine queries for the use of geographic terms. Master's thesis, University of Sheffield - United King (2003)
14. Larson, R.: Geographic information retrieval and spatial browsing. In: Smith, Gluck, M. (eds.) Geographic Information Systems and Libraries: Patronsand Mapsand and Spatial Information, pp. 81–124 (1996)
15. Li, Y., Moffat, A., Stokes, N., Cavedon, L.: Exploring probabilistic toponym resolution for geographical information retrieval. In: Purves, R., Jones, C. (eds.) GIR. Department of Geography, University of Zurich (2006)
16. Li, Z., Wang, C., Xie, X., Wang, X., Ma, W.Y.: Indexing implicit locations for geographical information retrieval. In: Purves, R., Jones, C. (eds.) GIR. Department of Geography, University of Zurich (2006)
17. Mandl, T., Carvalho, P., Di Nunzio, G.M., Gey, F., Larson, R.R., Santos, D., Womser-Hacker, C.: GeoCLEF 2008: The CLEF 2008 Cross-Language Geographic Information Retrieval Track Overview. In: Peters, C., Deselaers, T., Ferro, N., Gonzalo, J., Jones, G.J.F., Kurimo, M., Mandl, T., Peñas, A., Petras, V. (eds.) CLEF 2008. LNCS, vol. 5706, pp. 808–821. Springer, Heidelberg (2009)
18. Perea-Ortega, J.M., García-Cumbreras, M.Á., García-Vega, M., Ureña-López, L.A.: Comparing Several Textual Information Retrieval Systems for the Geographical Information Retrieval Task. In: Kapetanios, E., Sugumaran, V., Spiliopoulou, M. (eds.) NLDB 2008. LNCS, vol. 5039, pp. 142–147. Springer, Heidelberg (2008)
19. Perea-Ortega, J.M., Martínez-Santiago, F., Montejo-Ráez, A., Ureña-López, L.A.: Geo-NER: un reconocedor de entidades geográficas para inglés basado en GeoNames y Wikipedia. Sociedad Española para el Procesamiento del Lenguaje Natural (SEPLN) 43, 33–40 (2009)
20. Perea-Ortega, J.M., Ureña-López, L.A., García-Vega, M., García-Cumbreras, M.A.: Using Query Reformulation and Keywords in the Geographic Information Retrieval Task. In: Peters, C., Deselaers, T., Ferro, N., Gonzalo, J., Jones, G.J.F., Kurimo, M., Mandl, T., Peñas, A., Petras, V. (eds.) CLEF 2008. LNCS, vol. 5706, pp. 855–862. Springer, Heidelberg (2009)
21. Sanderson, M., Kohler, J.: Analyzing geographic queries. In: Proceedings Workshop on Geographical Information Retrieval SIGIR (2004)
22. Spink, A., Jansen, B.J., Ozmultu, C.H.: Use of query reformulation and relevance feedback by excite users. Internet Research: Electronic Networking Applications and Policy 10(4), 317–328 (2000)
23. Stokes, N., Li, Y., Moffat, A., Rong, J.: An empirical study of the effects of nlp components on geographic ir performance. International Journal of Geographical Information Science 22(3), 247–264 (2008)

Learning Good Decompositions
of Complex Questions

Yllias Chali, Sadid A. Hasan, and Kaisar Imam

University of Lethbridge
Lethbridge, AB, Canada
{chali,hasan,imam}@cs.uleth.ca

Abstract. This paper proposes a supervised approach for automatically learning good decompositions of complex questions. The training data generation phase mainly builds on three steps to produce a list of simple questions corresponding to a complex question: i) the extraction of the most important sentences from a given set of relevant documents (which contains the answer to the complex question), ii) the simplification of the extracted sentences, and iii) their transformation into questions containing candidate answer terms. Such questions, considered as candidate decompositions, are manually annotated (as good or bad candidates) and used to train a Support Vector Machine (SVM) classifier. Experiments on the DUC data sets prove the effectiveness of our approach.

1 Introduction

Complex questions address an issue that relates to multiple entities, events and complex relations between them while asking about events, biographies, definitions, descriptions, reasons etc. [1]. However, it is not always understandable, to which direction one should move to search for the complete answer. For example, a complex question like *"Describe the tsunami disaster in Japan."* has a wider focus without a single or well-defined information need. To narrow down the focus, this question can be decomposed into a series of simple questions such as *"How many people had died by the tsunami?"*, *"How many people became homeless?"*, *"Which cities were mostly damaged?"* etc. Decomposing a complex question automatically into simpler questions in this manner such that each of them can be answered individually by using the state-of-the-art Question Answering (QA) systems, and then combining the individual answers to form a single answer to the original complex question has been proved effective to deal with the complex question answering problem [2,3]. However, issues like judging the significance of the decomposed questions remained beyond the scope of all the researches done so far.

It is understandable that enhancing the quality of the decomposed questions can certainly provide more accurate answers to the complex questions. So, it is important to judge the quality of the decomposed questions and then, investigate more methods to enhance their quality. In this research, we address this challenging task and come up with a supervised model to automatically learn

G. Bouma et al. (Eds.): NLDB 2012, LNCS 7337, pp. 104–115, 2012.
© Springer-Verlag Berlin Heidelberg 2012

good decompositions of complex questions. For training data generation, at first, we determine the most important sentences from a given set of relevant documents (that contains the answer to the considered complex question), and then, simplify these sentences in the second step. In the third step, questions are generated from the simplified sentences. These questions, considered as candidate decompositions of the complex question, are manually annotated (as good or bad candidates) and used to train a SVM classifier. Experiments on the DUC (Document Understanding Conference[1]) data sets prove the effectiveness of our approach. The remainder of the paper is organized as follows: Section 2 presents a brief survey of related works. Section 3 illustrates the overview of our approach. Section 4 describes the training data generation phase. Section 5 discusses the supervised model used to accomplish the considered task. Section 6 presents the evaluation results and finally, we conclude the paper in Section 7.

2   Related Work

Question decomposition has been proved effective to deal with the complex question answering problem in many studies. For example, in [2], they introduce a new paradigm for processing complex questions that relies on a combination of (a) question decompositions; (b) factoid QA techniques; and (c) Multi-Document Summarization (MDS) techniques. Necessity and positive impact of question decomposition have been also shown in many studies in the complex question answering domain [4,3]. However, there is no direct research on the problem of automatically learning good decompositions of complex questions rather several paraphrasing and textual entailment methods can be considered the most relevant tasks that have been researched. For example, in QA systems for document collections a question could be phrased differently than in a document that contains its answer [5]. Studies have shown that the system performance can be improved significantly by taking such variations into account [6]. Techniques to measure similarities between texts can be an effective way of judging the importance of a sentence. Semantic roles typically offer a significant first step towards deeper text understanding [7]. There are approaches in "Recognizing Textual Entailment", "Sentence Alignment" and "Question Answering" that use semantic information in order to measure the similarity between two textual units [8]. This indeed motivates us to find semantic similarity between a document sentence and a complex question while selecting the most important sentences. Simplifying a sentence can lead to more accurate question generation [9], so we simplify the complex sentences in this research. Different methods have been proposed so far to accomplish the task of Question Generation (QG) [10,11]. Some QG models utilize question generation as an intermediate step in the question answering process [12]. Generated questions can be ranked using different approaches such as statistical ranking methods, dependency parsing, identifying the presence of pronouns and named entities, and topic scoring [9,13].

---

[1] http://duc.nist.gov/

3   Overview of Our Approach

The main contribution of this research is to develop a classification system that given a complex question and a list of simple questions can decide whether questions in the latter are to be considered a decomposition of the former, viz. they ask about part of the information asked by the complex question. To accomplish this task, simple questions are generated from the documents containing possible answers to the complex question and used to train the classifier[2]. There is a pipeline starting with a complex question and ending with a set of (presumably) one or more simple questions. We assume that a set of relevant documents is given along with each complex question that certainly possesses potential answers to the complex question. However, it is still necessary to identify the most important sentences due to the presence of a huge number of sentences in the data set. We conduct a shallow and a deep semantic analysis between the complex question and the given document sentences to perform this task. Sentences that are found during this process can be long and complex to deal with. Hence, we pass the selected sentences through a sentence simplification module. Once we find the simple sentences, we use a sentence-to-question generation approach in order to generate corresponding questions. We claim that these questions are the potential candidate decompositions of the complex question. We judge the quality of the generated decompositions manually and annotate them into two classes: good candidate and bad candidate, considering their correctness at the question level and verifying whether they can actually satisfy the information need stated in the original complex question partially. We employ a well-known supervised learning technique: SVM, that is trained on this annotated data set and then, the learned model is used to identify good decompositions of the unseen complex questions automatically. Learning good decompositions of complex questions is unique and to the best of our knowledge, no other study has investigated this challenge before in our setting.

4   Training Data Generation

4.1   Filtering Important Sentences

We use a shallow and a deep semantic analysis to extract the most important sentences related to the complex question from the given document collection. We parse the document sentences (and the question) semantically using the Semantic Role Labeling (SRL) system, ASSERT[3]. We use the Shallow Semantic Tree Kernel (SSTK) [7] to get a similarity score between a document sentence and the complex question based on their underlying semantic structures. To do a deeper semantic analysis of the text, after getting stemmed words by using

---

[2] The same procedure is applied to generate the test data set.

[3] Available at http://cemantix.org/assert

the *OAK* system [14] for each sentence and the complex question, we perform keyword expansion using WordNet[4] [15]. For example, the word "happen" being a keyword in the question "What happened?" returns the words: *occur, pass, fall out, come about, take place* from WordNet. Thus, we find out the similar words between the sentence-question pair that gives us a similarity score. We combine this score with the score obtained from the shallow semantic analysis and select the top-scored sentences from the documents. For example[5], *"With economic opportunities on reservations lagging behind those available in big cities, and with the unemployment rate among Native Americans at three times the national average, thousands of poor, often unskilled Native Americans are rushing off their reservations."* is selected as an important sentence.

4.2 Simplifying the Sentences

Sentences that we select in the earlier stage may have complex grammatical structure with multiple embedded clauses. Therefore, we simplify the complex sentences with the intention to generate more accurate questions. We call the generated simple sentences, *elementary sentences* since they are the individual constituents that combinedly possess the overall meaning of the complex sentence. We use the simplified factual statement extractor model[6] [9]. Their model extracts the simpler forms of the complex source sentence by altering lexical items, syntactic structure, and semantics and by removing phrase types such as leading conjunctions, sentence-level modifying phrases, and appositives. For example, for the complex sentence that was selected as important in Section 4.1, we get one of the possible elementary sentences as, *"Thousands of poor, often unskilled Native Americans are rushing off their reservations"*.

4.3 Sentence-to-Question Generation

Once we get the elementary sentences, our next task is to produce a set of possible questions from them. We claim these questions to be the candidate decompositions of the original complex question. In this research, we work on generating six simple types of questions: *who, what, where, when, whom* and *how much*. We use the OAK system [14] to produce Part-Of-Speech (POS) tagged and Named Entity (NE) tagged sentences. We use these information to classify the sentences by using a sequence of two simple classifiers. The first classifies the sentences into fine classes (Fine Classifier) and the second into coarse classes (Coarse Classifier). This is a similar but opposite approach to the one described in [16]. We define the five coarse classes as: 1) Human, 2) Entity, 3) Location, 4) Time, and 5) Count. Based on the coarse classification, we consider the

---

[4] http://wordnet.princeton.edu/

[5] We look into a running example through out this paper considering the complex question: "Discuss conditions on American Indian reservations or among Native American communities."

[6] Available at http://www.ark.cs.cmu.edu/mheilman/

relationship between the words in the sentence. For example, if the sentence has the structure "Human Verb Human", it will be classified as "whom and who" question types. We define a set of ninety basic word-to-word interaction rules to check the coarse classes. We use the POS information to decompose the main verb and perform necessary subject-auxiliary inversion and finally, insert the question word (with a question mark at the end) to generate suitable questions from the given elementary sentence. For example, for the elementary sentence generated in Section 4.2, we get a simple question as: *"Who are rushing off their reservations?"*.

5   Supervised Model

For supervised learning techniques, annotated or labeled data is required as a precondition. We manually annotate the generated simple questions (i.e. candidate decomposed questions) into two classes[7]: good candidate and bad candidate, employ the Support Vector Machines (SVM) that is trained on this annotated data set and then, use the learned model to predict good decompositions from the unlabeled candidate set of decompositions (i.e. test data set) automatically. We describe our feature space, learning and testing modules in the following subsections.

5.1   Feature Space

For our SVM classifier, we use a total of thirteen features that are divided into two major categories: one that considers the correctness at the question level and other is a coverage component that measures whether a decomposed question can actually satisfy the information need stated in the original complex question partially. We automatically extract these features from the questions (for both training and testing data) in order to feed them to the supervised models for learning and then, for prediction. These features are related to the original important sentence (that is selected in Section 4.1), the input sentence (elementary sentence), the generated simple question, and the original complex question. Correctness of the questions can be measured using a composition of the following features:

Grammaticality: We count the number of proper nouns, pronouns, adjectives, adverbs, conjunctions, numbers, noun phrases, prepositional phrases, and subordinate clauses in the syntactic structures of the question and the input sentence. We set a certain threshold[8] to denote the limit up to which a candidate can

---

[7] We inspect each question to measure whether they are lexically, syntactically and semantically correct or not. We also judge each decomposed question against the original complex question and analyze further to find out whether it can ask for any information that can be found in the given data. This analysis guides us to label each question as $+1$ (good candidate) or -1 (bad candidate).

[8] The thresholds are set after inspecting the questions and the input sentences manually.

be termed as good. We also include some boolean features to encode the tense information of the main verb.

Length: We calculate the number of tokens in the question, the original source sentence, the input elementary sentence, and the answer term (that is replaced by a question type). We set a threshold on this value, too.

Presence of Question Word: We consider some boolean features to identify the presence or absence of a certain question type: *who, what, where, when, whom* and *how much*.

Presence of Pronouns: If a question has one or more pronouns, we understand that the question is asking about something that has limited reference and hence, we consider the question as *vague*. To identify whether a question includes pronouns or not, we employ a boolean feature.

The coverage component of our feature extraction module tells whether a decomposed question can satisfy the requested information need partially. To automatically encode this feature for each question, we conduct an extensive linguistic analysis and a deep semantic analysis between the decomposed question and the original complex question.

Linguistic Analysis: We use ROUGE (Recall-Oriented Understudy for Gisting Evaluation) to automatically determine the quality of a question by comparing it to the original complex question using a collection of measures [17].

Deep Semantic Analysis: We conduct a deep semantic analysis between the decomposed question and the original complex question (according to the procedure discussed in Section 4.1) that outputs a similarity score as the feature value.

5.2   Learning and Testing

Once we get the feature values for all the decomposed questions along with the associated annotation (good or bad candidate), we feed this data to the supervised learner so that a learned model is established. We use a set of 523 questions for the training purpose. Later, this model is used to predict the labels for the new set of simple questions automatically during the testing phase. Our test data set includes 350 questions. In this work, we use SVM [18] as the classifier. To allow some flexibility in separating the classes, SVM models have a cost parameter, C, that controls the trade off between allowing training errors and forcing rigid margins. We use the SVM^{light} [19] package[9] for training and testing in this work. SVM^{light} consists of a learning module and a classification module. The learning module takes an input file containing the feature values with corresponding labels and produces a model file. The classification module is used to apply the learned model to new samples. We use $g(x)$, the normalized distance from the hyperplane to each sample point, x to rank the questions.

---

[9] http://svmlight.joachims.org/

6 Evaluation and Analysis

6.1 Corpus

The recent 2005, 2006, and 2007 Document Understanding Conference (DUC) series, run by the National Institute of Standards and Technology (NIST), have modeled a real-world complex question answering task, in which a question cannot be answered by simply stating a name, date, quantity, etc. They provide several topics, a complex question related to each topic, and a set of 25 relevant documents (that contains the answer to the complex question). The task is to synthesize a fluent, well-organized 250-word summary of the documents that answers the question(s) in the topic statement. We use a subset of 10 topics[10] from the DUC-2006 data (that came from the AQUAINT corpus, comprising newswire articles from the Associated Press and New York Times (1998-2000) and Xinhua News Agency (1996-2000)) to run our experiments.

6.2 Cross-Validation

Cross-validation is an effective technique for estimating the performance of a predictive model. We apply a 3-fold cross-validation on the annotated questions (i.e. training data set) to estimate how accurately our SVM model would perform on the unlabeled set of candidate decompositions (test data set). We use a randomized local-grid search [20] for estimating the value of the trade-off parameter C. We try the value of C in 2^i following heuristics, where $i \in \{-5, -4, \cdots, 4, 5\}$ and set C as the best performed value of 0.0625 for the linear kernel[11]. In Figure 1, we show the effect of C on the testing accuracy of the SVM classifier. We can see that the accuracy is largely dependent on the value of the parameter C. The accuracy rises with the increase of the C value and reaches the peak when $C = 0.0625$. However, after that, an opposite trend is visible with the increase of the C value.

6.3 Intrinsic Evaluation

Using the learned model, the supervised SVM classifier automatically predicts the labels (good or bad candidate) of the new set of decomposed questions that are generated for each new complex question. To evaluate the performance of our approach, we manually assess the quality of these questions. Two university graduate students judged the questions for linguistic quality and overall responsiveness following a similar setting to the DUC-2007 evaluation guidelines[12]. The given score is an integer between 1 (very poor) and 5 (very good) and is guided by consideration of the following factors: 1. Grammaticality, 2. Correct question type, 3. Referential clarity (Presence of pronoun), and 4. Meaningfulness.

---

[10] We use 6 topics for training and 4 topics for testing.

[11] We found linear kernel as the best performer among all kernels.

[12] http://www-nlpir.nist.gov/projects/duc/duc2007/quality-questions.txt

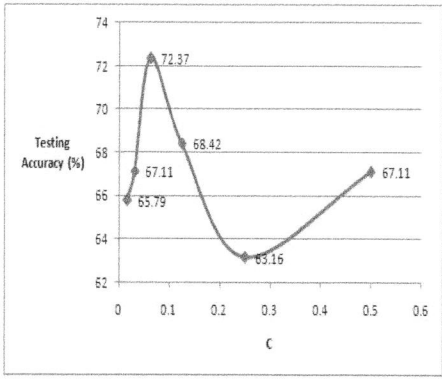

Fig. 1. Effect of C on SVM's testing accuracy

They also assigned a content responsiveness score to each question. This score is also an integer between 1 (very poor) and 5 (very good) and is based on the factor whether the question helps to satisfy the information need expressed in the original complex question. This task was performed intuitively by investigating the original complex question, the corresponding simple question at hand, and the given document collection. We compare the top-ranked good questions with the performance of a set of randomly picked (good/bad mixed) questions. For each topic, we also judge the performance of the bad questions alone[13]. Table 1 shows the evaluation results for SVM. Analyzing Table 1, we see that SVM predicted *Good Questions* improve the linguistic quality and responsiveness scores over *Mixed (Random) Questions* by 74.06%, and 100.00%, respectively whereas they outperform the Linguistic Quality and Responsiveness scores over *Bad Questions* by 20.58%, and 12.50%, respectively. These results suggest that the SVM classifier performed well to rank the decomposed questions accurately. We also see that *Bad Questions* outperform the *Mixed (Random) Questions* in terms of linguistic quality because they were small in length and had good grammatical structure. However, they could not beat the responsiveness scores meaning that small questions have limited coverage over the requested information need.

Table 1. Linguistic quality and responsiveness scores (average) for SVM

Categories	Linguistic Quality	Responsiveness
Good Questions	4.10	3.60
Mixed (Random)	2.35	1.80
Bad Questions	3.40	1.60

---

[13] *Good questions* refer to the top 50% questions with high scores predicted by the SVM classifier, whereas *bad questions* refer to the bottom 50% questions.

In another evaluation setting, the two annotators judge the questions for their overall acceptability as a good or a bad candidate decomposition and assign a score (an integer between 1 (very poor) and 5 (very good)) to each of them. The outcome of this evaluation scale is then converted into a binary (+1/-1) rating scale, deciding that a score above 3 is positive. Then, the accuracy of the SVM's binary prediction is computed with respect to the manual annotation using the following formula:

$$Accuracy = \frac{number\, of\, Correctly\, Classified\, Questions}{Total\, number\, of\, Test\, Questions} \tag{1}$$

We experimented with a total of 226 questions[14] for this evaluation and found 159 of them to be correctly classified by the SVM classifier showing an accuracy of 70.35%. An inter-annotator agreement of Cohen's $\kappa = 0.2835$ [21] was computed that denotes a fair agreement [22] between the raters.

6.4   Extrinsic Evaluation

To determine the quality of the decomposed questions based on some other task such as summarization can be another effective means of evaluation. We call it as extrinsic evaluation. We pass the top-ranked decomposed questions to the Indri search engine (i.e. a component of the Lemur toolkit[15]). For all decomposed questions, Indri returns ranked sentences from the given data set that are used to generate a 250-word summary according to the DUC guidelines. We evaluate these summaries against four human-generated "reference summaries" (given in DUC-2006) using ROUGE [17] which has been widely adopted by DUC. We consider the widely used evaluation measures Precision (P), Recall (R) and F-measure for our evaluation task. Table 2 shows the ROUGE-2 and ROUGE-SU scores of our proposed SVM system as they are used as the official ROUGE metrics in the recent DUC evaluations.

Table 2. ROUGE measures for SVM

Measures	Recall	Precision	F-score
ROUGE-2	0.0675	0.0570	0.0618
ROUGE-SU	0.1027	0.0734	0.0856

Statistical Significance: To show a meaningful comparison, in Table 3, we present the average ROUGE-2 scores (Recall) of our SVM system and the two state-of-the-art systems, *Trimmer* and *HMM Hedge* that use a Multi-Candidate Reduction (MCR) framework for multi-document summarization [23]. These well-known systems do not perform question decomposition, so, we treat them as

---

[14] We consider only the questions for which both annotators agreed on rating as positive or negative.

[15] Available at http://www.lemurproject.org/

the *non-decomposed baselines* to perform system comparisons. An approximate result to identify which differences in the competing systems' scores are significant can be achieved by comparing the 95% confidence intervals for each mean. So, we include the 95% confidence intervals to report the statistical significance of our system. ROUGE uses a randomized method named bootstrap resampling to compute the confidence intervals. Table 3 yields that the SVM system outperforms the two baseline systems. We can also see that the confidence intervals of all the systems overlap with each other meaning the fact that there is no significant difference between our system and the baseline systems.

Table 3. Comparison of different systems

Systems	ROUGE-2
Trimmer	0.0671 [0.06332–0.07111]
HMM Hedge	0.0625 [0.05873–0.06620]
SVM	0.0675 [0.06548–0.07221]

6.5   Feature Engineering

To analyze the impact of different features, we run another experiment considering the SVM classifier generated top-ranked questions. In Figure 2, we plot a graph to show the performance of different systems considering several variants of the feature space during experiments. In the figure, **Grammaticality**, **Length**, **Pronoun**, and **Coverage** indicate that the corresponding feature is not considered during experiments, whereas **OnlyGrammaticality**, **OnlyLength**, **OnlyPronoun**, and **OnlyCoverage** denote the presence of that particular feature only. **All** denotes the inclusion of all features. From the figure, we understand that if we exclude the *Grammatically* feature, the responsiveness score improves quite a lot, whereas exclusion of the *Length* feature produces good scores for both linguistic quality and responsiveness. On the other hand, if we do not consider the *Pronoun* feature, the scores have a negative impact. Again, omitting the *Coverage* feature decreases the responsiveness score. On the other hand, if we consider *OnlyGrammaticality* feature, we have better linguistic quality and worse responsiveness score. *OnlyLength* feature yields good scores for both linguistic quality and responsiveness scores whereas *OnlyPronoun* feature provides bad scores for both denoting its lower impact on the SVM learner. *OnlyCoverage* feature shows a higher responsiveness score with a moderate linguistic quality score and finally, considering *All* features yields a good linguistic quality while showing a decent performance in terms of responsiveness score. From this comparison of feature combinations, we can conclude that the inclusion of the *Pronoun*, and *Coverage* features helps to achieve the best performance for the considered task. This comparison also suggests that the chosen features are appropriate by themselves, but in combination are not able to produce a qualitative jump in performance.

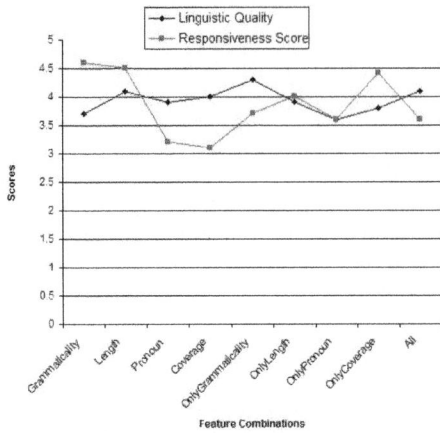

Fig. 2. Graph for different feature combinations

7   Conclusion and Future Works

We propose a supervised model for learning good decompositions of complex questions. We perform a rigorous evaluation and analysis to show the effectiveness of our approach. Furthermore, we analyze the impact of different features on the performance of the SVM classifier and conclude that the combination of the *Pronoun*, and *Coverage* features yields the best performance. In future, we plan to use more sophisticated features while investigating the potentials of other supervised approaches such as Conditional Random Fields (CRF), and Hidden Markov Models (HMM) for the learning task.

Acknowledgments. The research reported in this paper was supported by the Natural Sciences and Engineering Research Council (NSERC) of Canada-discovery grant and the University of Lethbridge.

References

1. Chali, Y., Joty, S.R., Hasan, S.A.: Complex Question Answering: Unsupervised Learning Approaches and Experiments. Journal of Artificial Intelligence Research 35, 1–47 (2009)
2. Harabagiu, S., Lacatusu, F., Hickl, A.: Answering complex questions with random walk models. In: Proceedings of the 29th Annual International ACM SIGIR Conference on Research and Development in Information Retrieval, pp. 220–227. ACM (2006)
3. Hickl, A., Wang, P., Lehmann, J., Harabagiu, S.: Ferret: Interactive question-answering for real-world environments. In: Proceedings of the COLING/ACL on Interactive Presentation Sessions, pp. 25–28 (2006)
4. Lacatusu, F., Hickl, A., Harabagiu, S.: Impact of question decomposition on the quality of answer summaries. In: Proceedings of the Fifth International Conference on Language Resources and Evaluation, LREC 2006 (2006)

5. Pasca, M.: Open-domain question answering from large text collections, 2nd edn. Center for the Study of Language and Information (2003)
6. Harabagiu, S., Hickl, A.: Methods for using textual entailment in open-domain question answering. In: Proceedings of the 21st International Conference on Computational Linguistics and the 44th Annual Meeting of the Association for Computational Linguistics, ACL, pp. 905–912 (2006)
7. Moschitti, A., Quarteroni, S., Basili, R., Manandhar, S.: Exploiting Syntactic and Shallow Semantic Kernels for Question/Answer Classificaion. In: Proceedings of the 45th Annual Meeting of the Association of Computational Linguistics, pp. 776–783. ACL, Prague (2007)
8. MacCartney, B., Grenager, T., de Marneffe, M., Cer, D., Manning, C.D.: Learning to Recognize Features of Valid Textual Entailments. In: Proceedings of the Human Language Technology Conference of the North American Chapter of the ACL, New York, USA, vol. 4148 (2006)
9. Heilman, M., Smith, N.A.: Extracting simplified statements for factual question generation. In: Proceedings of the Third Workshop on Question Generation (2010)
10. Wang, W., Hao, T., Liu, W.: Automatic Question Generation for Learning Evaluation in Medicine. In: Leung, H., Li, F., Lau, R., Li, Q. (eds.) ICWL 2007. LNCS, vol. 4823, pp. 242–251. Springer, Heidelberg (2008)
11. Chen, W., Aist, G., Mostow, J.: Generating Questions Automatically from Informational Text. In: Proceedings of the 2nd Workshop on Question Generation (AIED 2009), pp. 17–24 (2009)
12. Hickl, A., Lehmann, J., Williams, J., Harabagiu, A.: Experiments with interac- tive question-answering. In: Proceedings of the 43rd Annual Meeting of the Association for Computational Linguistics ACL 2005, pp. 60–69 (2005)
13. McConnell, C.C., Mannem, P., Prasad, R., Joshi, A.: A New Approach to Ranking Over-Generated Questions. In: Proceedings of the AAAI Fall Symposium on Question Generation (2011)
14. Sekine, S.: Proteus Project OAK System (English Sentence Analyzer) (2002), http://nlp.nyu.edu/oak
15. Fellbaum, C.: WordNet - An Electronic Lexical Database. MIT Press, Cambridge (1998)
16. Li, X., Roth, D.: Learning Question Classifiers. In: Proceedings of the 19th International Conference on Computational Linguistics, Taipei, Taiwan, pp. 1–7 (2002)
17. Lin, C.Y.: ROUGE: A Package for Automatic Evaluation of Summaries. In: Proceedings of Workshop on Text Summarization Branches Out, Post-Conference Workshop of Association for Computational Linguistics, Barcelona, Spain, pp. 74–81 (2004)
18. Cortes, C., Vapnik, V.N.: Support Vector Networks. Machine Learning 20, 273–297 (1995)
19. Joachims, T.: Making large-Scale SVM Learning Practical. In: Advances in Kernel Methods - Support Vector Learning (1999)
20. Hsu, C., Chang, C., Lin, C.: A Practical Guide to Support Vector Classification, National Taiwan University, Taipei 106, Taiwan (2008), http://www.csie.ntu.edu.tw/cjlin
21. Cohen, J.: A Coefficient of Agreement for Nominal Scales. Educational and Psychological Measurement 20(1), 37–46 (1960)
22. Landis, J.R., Koch, G.G.: The Measurement of Observer Agreement for Categorical Data. Biometrics 33(1), 159–174 (1977)
23. Zajic, D., Dorr, B.J., Lin, J., Schwartz, R.: Multi-candidate reduction: Sentence compression as a tool for document summarization tasks. Information Processing and Management 43, 1549–1570 (2007)

A Semi Supervised Learning Model for Mapping Sentences to Logical form with Ambiguous Supervision

Le Minh Nguyen and Akira Shimazu

Japan Advanced Institute of Science and Technology
School of Information Science
{nguyenml,shimazu}@jaist.ac.jp

Abstract. Semantic parsing is the task of mapping a natural sentence to a meaning representation. The limitation of semantic parsing is that it is very difficult to obtain annotated training data in which a sentence is paired with a semantic representation. To deal with this problem, we introduce a semi supervised learning model for semantic parsing with ambiguous supervision. The main idea of our method is to utilize a large amount of data, to enrich feature space with the maximum entropy model using our semantic learner. We evaluate the proposed models on standard corpora to show that our methods are suitable for semantic parsing problem. Experimental results show that the proposed methods work efficiently and well on ambiguous data and it is comparable to the state of the art method.

Keywords: Semantic Parsing, Support Vector Machines.

1 Introduction

Semantic parsing is the task of mapping a natural language sentence into a complete and formal meaning representation. This task is an interesting problem in Natural Language Processing (NLP) as it would very likely be part of any interesting NLP applications [2]. For example, the necessity of semantic parsing for most NLP applications and the ability to map natural language to a formal query or command language, are critical for developing more user-friendly interfaces.

There has been a significant amount of previous works on learning to map sentences to semantic representations. Zelle and Mooney [24] and Tang [18] proposed the empirically-based methods using a corpus of natural language(NL) sentences and their formal representation for learning by inductive logic programming (ILP). The disadvantage of the ILP approach is that it is quite complex, and slow to acquire parsers for mapping long sentences to logical forms such as the Robocup corpus.

To overcome this problem, Kate and Mooney [10] proposed a method that used transformation rules estimated from the corpus of NL and logical forms, to transform NL sentences to logical forms.

In order to improve semantic parsing accuracy, Ge and Mooney (2005) presented a statistical method [8] merging syntactic and semantic information using a semantic augmented parsing tree (SAPT).[1]

---

[1] SAPT is a syntactic tree with semantic augmented at each non-terminal node.

G. Bouma et al. (Eds.): NLDB 2012, LNCS 7337, pp. 116–127, 2012.

Similar to Ge and Mooney (2005) [8], the approach by Nguyen, Shimazu, and Phan [17] also uses the corpus of SAPT tree to estimate their semantic parsing model. Their approaches use structured SVMs [21] and learns ensemble learning of semantic parsers.

Unlike those methods using SAPT, the works proposed by [25][26] [14] map an NL sentence to its logical form using a Combinatory Categorical Grammar (CCG) with structured learning models. They have indicated that using online structured prediction along with CCG could lead to state of the art results in several domains (i.e. ATIS and Query language). Wong and Mooney (2006)[22] proposed a synchronous context free grammar frame work (SCFG) [1] to transform NL a language sentence a to semantic representation. The system was extended to work with the formal language in λ-calculus (Wong and Mooney, 2007) to deal with logical variables.

All of the above methods require training data which consists of a set of pairs of sentences and their logical representations. In fact, making the training data takes expensive and time consuming. To cope with this problem, Kate and Mooney [13] initially propose a semi-supervised learning model for semantic parsing using transductive SVMs. Along with the research line, there are some other methods for semantic parsing using unlabeled data [3][9] [20][19]. However, their approaches are designed for un-ambiguous supervision, while in real application such as Robocup casting as well as weather-focasting, we do not know exactly which semantic representation is paired with sentences. Kate and Mooney (2007a) extended their semantic parser based on string kernel for ambiguous training data in which each sentence is only annotated with an ambiguous set of multiple, alternative potential interpretations. The merit of this work is that using string kernel SVM one can deal with ambiguous and noisy data.

According to my knowledge, there is no research about using external unlabeled data for semantic parsing with ambiguous supervision. In this paper, we would like to explore whether or not unlabeled data is useful for semantic parsing on ambiguous training data. To achieve this goal, we extend the method of [12] by proposing semi-supervised learning models using unlabeled-data with word-cluster model. The contributions of our proposed method are described as follows.

- Unlike previous works, we propose a semi-supervised learning method which allows incorporating unlabeled data to improve the performance of semantic parsing using ambiguous training data. The large amount of unlabeled data can be utilized to generate a cluster model to enrich the feature space of the learning model.
- Our goal is to illustrate that unlabeled data is helpful for the task of mapping sentences to logical forms. In order to do that, we propose two semantic-parsing models, using unlabeled data to enrich the feature space of learning models. The first model is extended from the SVM-based sematic parsing model by incoporating unlabeled data to its string kernel. The second model is based on maximum entropy model (MEM) in which unlabeled data is used to enrich the feature space of MEM. In addition, we also investigate the impact of using syntactic information (i.e. part of speech tagging, chunking labels) for parsing sentences to logical forms.

The rest of this paper is organized as follows: Section 2 describes a semi-supervised method for semantic parsing in both unambiguous training data and ambiguous training data. Section 3 shows experimental results and Section 4 discusses the advantage of our method and describes future work.

SENT1: *"Which rivers run through the states bordering Texas ?"*
MR1: answer(traverse(next to(stateid('texas'))))

SENT2: *"Which is the highest point in Alaska ?"*
MR2: answer(highest(place(loc(stateid('Alaska')))))

SENT3: *"What are the major rivers in Alaska ?"*
MR3: answer(major(river(loc(stateid('Alaska')))))

Fig. 1. An example of NL sentences unambiguously paired with their MRs

2   Semantic Parsing for Ambiguous Supervision

The semantic parsing process maps sentences to their computer-executable meaning representation (MRs). The MRs are expressed as in formal languages which are defined as meaning representation languages (MLRs). We assume that all MLRS have deterministic context free grammar which ensures that every MR will have a unique parse tree.

In this section we briefly review works on semantic parsing for ambiguous supervision.

2.1   Semantic Parsing for Unambiguous Supervision Using SVMs

Figure 1 shows examples of NL sentences, in which each sentence is paired with the respective correct MR.

KRISP [11](kernel-based robust interpretation for Semantic parsing) is a system that learns a semantic parser from unambiguous training data. The main idea of the method is to find a derivation which is used to construct the logical form. A semantic derivation D, of a NL sentence s, as a parse tree of an MR such that each node of the parse tree also contains a substring of the sentence in addition to a production. According to [11], nodes of the derivation tree are defined as tuple $(\pi, [i..j])$, where π is its production and $[i..j]$ stands for a substring $s[i..j]$ of s. We say that the node covers the substring $s[i..j]$. The substrings covered by a child node are not allowed to overlap, and the substring covered by the parent must be a concatenation of the substrings covered by its children. Let $P_\pi(u)$ denote the probability that a production π of the MRL grammar covers the NL substring u. Assuming these probabilities are independent of each other, the probability of a semantic derivation D of a sentence s can be computed as:

$$P(D) = \arg\max_{(\pi, [i,j]) \in D} \prod P_\pi(s[i..j]) \tag{1}$$

The goal of semantic parsing is to find the most probable derivation of a sentence s. To achieve the goal, we apply the dynamic programming algorithm as described in [11], which is modified from the Earley algorithm [7]. In addition, we also need to estimate the probability $P_\pi(s[i..j])$ for each rule π. For each production π in the MRL grammar, KRISP collects positive and negative examples. The positive examples can be obtained for the production π if the rule π is found in the derivation for generating MR. Otherwise, we obtained negative examples. After collecting positive and negative examples,

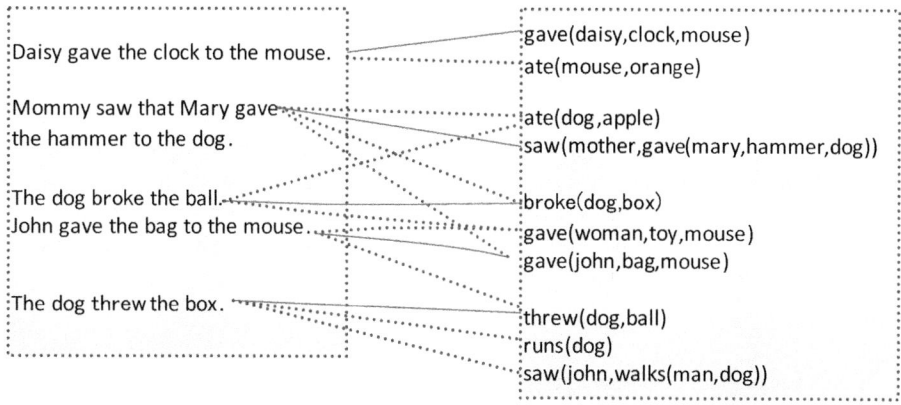

Fig. 2. Sample of ambiguous Training Data (solid: correct meaning)

a string SVM kernel model is trained for each production π using positive and negative examples. When all training processes for the productions within MRL are complete, we reparse all training examples to collect positive and negative examples corresponding to each production π. The iteration process is finished when the classifiers converge. Experimental results show that KRISP compares favorably to other semantic parsing systems [11].

2.2 Semantic Parsing with Ambiguous Supervision

In this section, we show how to adapt semantic parsing for unambiguous data to ambiguous data. As mentioned earlier, in ambiguous training data, for a given sentence we have a set of MRs corresponding to it, but there is only one for the semantic representation of this sentence. The question is how to determine which meaning representation among the set of MRs is paired with the sentence. There are two main approaches to deal with this problem. The first one applies generative alignment model[15] to align which meaning representation corresponds to the sentence. After that, the training data is obtained by using the alignment model on the whole ambiguous corpus. Then, a semantic parsing model can be leant using the corpus easily. The second approach is a joining process of alignment and training, which is named learning with ambiguous supervision [12].

Algorithm 1 shows the KRISPER algorithm [12] which is extended from KRISP algorithm [11]. Keep in mind that an ambiguous corpus forms a bipartite graph with the sentences and the MRs as two disjoint set of vertices and the associations between them as connecting edges. The set of correct NL-MR pairs from a matching on this bipartite graph which is defined as a subset of the edges with at most one edge incident on every vertex. Algorithm 1 starts by considering each pair of NL-MR as a training example with a corresponding weight. A learning model with weights for data instances is used (see line 8). Next, for each NL sentence in the training data, KRISPER estimates the confidence of generating each semantic representation (MRs) in the possible semantic forms by using the ParseEstimate function. The highest one among the possible MRs is selected to pair with the sentence for inclusion into the training data in next iterations. The new NL-MR pairs should be consistently chosen so that an MR does not get paired with more than one sentence. Kate and Mooney (2007) [12] formulated the problem of consistently selecting the best pairs in an instance of the maximum weight assignment problem on bipartite graph which was solved using the Hungarian Algorithm. Line 13 shows the BestExample procedure for selecting examples. Algorithm 1 terminates when the difference of NL-MR pairs between the current iteration and the previous iteration is less than 5%. The training model estimated on these NL-MR pairs is selected as the final learned parser.

---

1 Input ambiguous corpus: $S = (s_i; M_i), i = 1, 2, ..., N$ in which s_i is a sentence and M_i is a set of semantic representations
2 MLR grammar G
3 Output: the learning model
4 $U = \varphi$
5 **for** $i = 1, 2...N$ **do**
6     $U = U \cup \{(s_i, mr, w)|mr \in M_i, w = \frac{1}{|M_i|}\}$
7 **end**
8 $C \equiv \{C_\pi|\pi \in G\}$ = KRISP-LearningWeight(U, G)
9 **while** *not converged* **do**
10    $V = \varphi$
11   **for** $i = 1, 2, ..., N$ **do**
12        $V = V \cup \{(s_i, mr, w)|mr \in M_i, w = ParseEstimate(s_i, mr, C)\}$
13   U=BestExample(V)
14   C=KRIPSTrainSemantic(U, G)
15   **end**
16 **end**

Algorithm 1. The KRISPER Semantic Parsing [12]

2.3 Semantic Parsing with Ambiguous Supervision Using MEM

We modified the method described in [12] by using a maximum entropy model with the uses of linguistic information of sentences. One of the reason why we choose MEM [4] is that this method can easily incorporate features to the learning model. In addition, regarding the estimation of conditional probability between each sample and label, maximum entropy model serves as a direct learning method. This is more advantageous than SVM in terms of obtaining probabilistic scores. The maximum entropy model will be able to exploit those features which are beneficial, and effectively ignore those that are irrelevant. We model the probability of a class c given a vector of features x according to the MEM formulation:

$$p(c|x) = \frac{\exp[\sum_{i=0}^{n} \lambda_i f_i(c, x)]}{Z_x} \tag{2}$$

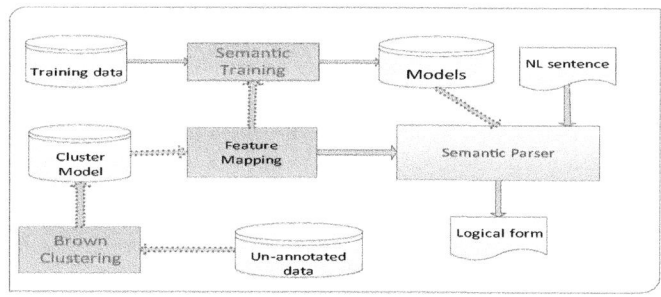

Fig. 3. A Word-Cluster Model for Semantic Parsing

Here Z_x is the normalization constant, $f_i(c, x)$ is a feature function which maps each class and vector element to a binary feature, n is the total number of features, and λ_i is a weight for a given feature function. The weight λ_i for feature $f_i(c, x)$ are estimated by using an optimization technique such as the L-BFGs algorithm [16]

One of the important components is the design of a feature space for MEM. We describe the feature set used in MEM for semantic parsing as follows. For each pair of a substring and a rule, we extract the following feature sets.

– All words with the substring are considered as features.
– Part of speech of each word with the substring is considered as a feature.
– Chunking information of each word with the substring is considered as a feature.
– Named Entity categories of words in the substring

To obtain part of speeches, chunkings and named entity categories, we used the SENNA toolkit[2] Note that we have tried with other features such as conjunctions between two consequent words, but the accuracy of the semantic parser did not improved.

3 Semi-supervised Semantic Parsing with Ambiguous Supervision

In this section we present the proposed semi-supervised semantic parsing models using word-clusters. Figure 3 shows a general framework for incorporating word-cluster modes in discriminative learning such as Maximum Entropy Models and SVM. Larger un-annotated text documents are clustered using the brown-clustering method to obtain word-cluster models. A word-cluster model is then used to enrich features space for discriminate learning models in both training and semantic parsing (testing) process. For convenience, we briefly present a summary of the Brown Algorithm as follows.

3.1 The Brown Algorithm

We can describe the Brown algorithm for generating word-cluster models as follows. The algorithm is considered as a hierarchical agglomerative word clustering algorithm

---

[2] Which is published available on http://ml.nec-labs.com/senna/

[5], in which the input is a large sequence of words $w_1, w_2, \ldots, w_n$, which are extracted from raw texts. The output is a hierarchical clustering of words—a binary tree—wherein a leaf represents a word, and an internal node represents a cluster containing the words in the sub-tree, whose root is that internal node.

The algorithm uses contextual information—the next word information—to represent properties of a word. More formally, (w) denotes the vector of properties of w (or w's context). We can think of our vector for w_i as counts, for each word w_j, of how often w_j followed w_i in the corpus:

$$(w_i) = (|w_1|, |w_2|, \ldots, |w_n|)$$

(w_i) is normalized by the count of w_i, and then we would have a vector of conditional properties $P(w_j|w_i)$. As mentioned earlier, the clustering algorithm used here is HAC-based, therefore, at each iteration, it must determine which two clusters are combined into one cluster. The metric used for that purpose is the minimal loss of average mutual information.

To use word cluster information effectively in our model at several levels of abstraction, we encode each word cluster by a bit string that describes the path from the root to the chosen internal node. The path can be encoded as follows: we start from the root of the hierarchical clustering, "0" is appended to the binary string if we go up, and "1" is appended if we go down. Table 1 shows some examples of using word-cluster for representing the word in semantic parsing data.

Table 1. Example of bit repsentations for semantic parsing data

Bit representations	Words
0101011111011110	gaining
0101011111011110	launching
1110110110111	inbound
1110110110111	outbound
101100010001	passions
101100100	forwards
101100100	steals
101100100	plays

3.2 Word-Cluster Based Features for MEM

We used the word-cluster model to enrich the feature space by using a simple method as follows. Each cluster of words is represented by a bit string as described in Sect. 3.1. The bit string of the cluster containing a word also represents the cluster information of the word. We then create an indicator function for each cluster, and use it as a selection feature:

$$f_{110101}(w) = \begin{cases} 1 \text{ if } w \text{ has bit string } 110101, \\ 0 \text{ otherwise.} \end{cases}$$

We now consider a new feature space by taking into account the word-cluster for mapping as follows.

- We keep all features as described in the previous section for MEM based-semantic parsing. It includes all lexical words, parts of speech (pos), chunking(chk), and named entities information (ner).
- Each word in the training data and test data is mapped to a bit string. We add a bit representation of each word within the substring.
- We then add a conjunction of each word with its bit representation as well as a conjunction of the bit representation of two consecutive words.

3.3   Word Cluster for SVM

Let the string kernel using in the SVM-based semantic parsing model be $K(x, y)$ where x and y are two sub strings. We now present how to use word cluster for string kernel SVM as follows. First, each word within the string x and y is replaced by a word-cluster. Then, the connecation of all word-clusters within x is considered as the bit string of x. This procedure is also applied for the string y. For convenience, let $bit(x)$ be the bit string of x and $bit(y)$ the bit string of y, respectively. The new string kernel function is the linear combination between the two kernels, and it can be computed as follows.

$$K(x, y) = a \cdot K(x, y) + b \cdot K(bit(x), bit(y)) \qquad (3)$$

The values of a and b would satisfy the condition that $a + b = 1$, and they will be selected based on experiments.

---

1 Input: an ambiguous corpus: $S = (s_i; M_i), i = 1, 2, ..., N$ in which s_i is a sentence and M_i is a set of semantic representations
2 MLR grammar G
3 Output: the learning model
4 $U = \varphi$
5 **for** $i=1, 2... N$ **do**
6      $U = U \cup \{(s_i, mr, w) | mr \in M_i, w = \frac{1}{|M_i|}\}$
7 **end**
8 $C \equiv \{C_\pi | \pi \in G\}$ =MEWeightLearn(U, G)
9 **while** *not converged* **do**
10     $V = \varphi$
11   **for** $i =1, 2,..,N$ **do**
12         $V = V \cup \{(s_i, mr, w) | mr \in M_i, w = ParseEstimate(s_i, mr, C)\}$
13   U=BestExample(V)
14   C =METrainSemantic(U, G)
15   **end**
16 **end**

Algorithm 2. The semi-supervised learning algorithm for Semantic Parsing with ambiguous supervision

Algorithm 2 shows the novel method with ambiguous supervision. The main difference of this algorithm from Algorithm 1 is that SVM with string kernel method is replaced by a maximum entropy model using word-cluster features. Line 14 shows the METrainSemantic procedure which applies maximum entropy models to train a semantic parsing model using the training data U and the grammar G.

4   Experimental Results

This section presents an experimental evaluation of the proposed method. We conduct an investigation to check whether or not word-cluster features are useful for semantic parsing. In addition, we argue whether the proposed semi-supervised framework is effective for ambiguous training data. To achieve the goal, we consider two experiments on the Robocup sports-casting corpus and AMBICHILDWORLD corpus.

In the first experiment, we consider the Robocup sports-casting corpus as described in [6] for our experiments. Table 2 shows statistics on the Robocup corpus used in the experiment.

Table 2. Statistics on the Robocup corpus

Statistics	Sports-casting data
No.of. Comments	2036
No.of extracted MR events	10452
No. of NLs w/matching MRs	1868
No. of MRs w/matching NLs	4670
Avg # of MRs per NL	2.50

We used 4 fold cross validation test in which three games among the games from 2001 to 2004 are selected as the training data and the remaining is used as the test data. The final results are obtained by averaging the results of four folds. We noticed that the use of generative alignment could help to improve the result of semantic parsing [6], so we used the alignment model of [15] to generate input for both KRISPER and the proposed semantic parsing model. Table 3 depicts the comparison of the proposed framework (S-MEM) with the baseline (MEM) and the KRISPER method [12]. It clearly indicates that the proposed framework improved over the baseline in terms of accuracy. This can be expressed that the use of word cluster features is effective for semantic parsing in the domain. It is also easy to recognize that the proposed framework is also comparable to the state-of-the-art system on semantic parsing for ambiguous data, and the best feature set outperforms the KRISPER with respect to the F-measure score. Table 3 also depicted the result of SVM-cluster which is an extension of KRIPSER by incorporating word-cluster models to string kernel using the linear combination method. As we can see, it sufficiently outperforms KRISPER and attained the best result.

Table 3 also reported the performance of our semi-supervised model with various kinds of features. It is easy to see that the combination of lexical words, parts of speech, and word-clusters attained the best result. Table 3 shows that the S-MEM significantly

Table 3. Experimental results on the Robocup corpus with feature investigation

Methods	Sports-casting data
KRISPER	75.94
SVM-cluster	**80.23**
MEM	68.63
S(semi)-MEM+wc	72.71
S-MEM+wc+pos	**77.21**
S-MEM+wc+pos+chk	74.92
S-MEM+wc+chk+chk+ner	75.39

improved over the MEM in terms of F-measure score. It is also recognized that the proposed method is favorably comparable to the state of the art system [12].

In the second experiment, we consider the best feature set which is used when exploiting our models on the first experiment. The AMBIG-CHILDWORLD corpus described in[12] is used for evaluating the proposed semantic parsing model. As described in [12], the AMBIG-CHILDWORLD corpus was constructed completely artificially but it attempts to more accurately model real-world ambiguities. We conducted 10 fold cross validation test to evaluate the proposed systems, and the F-Measure score was used to evaluate the performance of semantic parsing. We also evaluated the proposed system in various noise levels[3]. Similar to [12], we also take into account four noise levels from zero to four for evaluation. Table 4 indicated that S-MEM significantly

Table 4. Experimental results on the AMBIG-CHILDWORLD[4]

Methods	Level0	Level1	Level2	Level3
KRISPER	93.6	88.4	75.1	44.44
SVM-cluster	94	89.40	75.4	45.1
MEM	84.47	78.20	71.32	40.2
S(semi)-MEM	94.5	89.96	76.1	49.6

improves over MEM on all levels of noise (Level 0 to Level 3). As can be seen, S-MEM significantly outperforms the performance of MEM in terms of F-measure. It is comparable to the results of KRISPER. We also tested the use of word-cluster with SVM string kernel. The results show that SVM-cluster improved KRISPERs and comparable to S-MEM. This reflects that word-cluster features are useful for semantic parsing with ambiguous supervision. We believe that word-cluster features also contributed to the performance of other semantic parsing systems such as [26][23].

4.1   Error Analysis

When carefully checking the outputs of the proposed systems we found some cases (Table 4) in which the systems generate semantic representation wrongly, as follows.

---

[3] Level i means a sentence is associated with $i + 1$ meaning representations.

[4] The first replicate data is used.

– The system predicts wrong predicates (E.g: Sentence Id 1 and Id 2). For example, sentence(Id 1) shows that our system predict "badPass" meanwhile the correct one is "turnover".
– The system predicts correct predicates but wrong in arguments. Sentence (Id 3) shows this case.
– The system predicts correct predicates and arguments but the order of arguments is wrong. Sentence Id 4 and 5 show this case.

Table 5. Some examples of the proposed models

Id	NL sentences	Gold data	Semantic Parsing
1	Pink 9 loses the ball and turns it over to Purple3	turnover(purple3,pink5)	badPass(purple3, pink5)
2	Purple5 steals the ball from Pink11	steal(purple5)	turnover (purple5, pink11)
3	Purple1 kicjs the ball out to Purple4	pass(purple4, purple4)	pass(purple1, purple1)
4	donnal believes that jill walks with mine	believe(donal,walks(jill,mine))	believe(donal,walks(mine,jill))
5	father gave the bag to john	gave(father,bag,john)	gave(john,bag,father)

The errors in the first and second case are difficult to handle. One of the possibility with respect with these errors is that the ambiguousness of training data leads to the mis-corresponding between NL sentence and its logical forms. The errors in the third case is more easier to handle. We can simply use a post-processing step that generates a correct order for each predicate of the semantic representation.

5   Conclusions

This paper proposes a novel method for dealing with the problem of learning semantic parsing with ambiguous supervision. The main contribution of this paper is to show that word-clusters are useful for the semantic parsing problem. To achieve the goal, we intro-duce a semi-supervised learning framework by incorporating word-cluster models with MEM and SVM string-kernel. Experimental results on the standard corpora showed that the proposed semi-supervised learning model is effective and favorably compara-ble to the best previous systems. We also found that the semi-supervised model could significantly improve the semantic parsing based on MEM and SVM string kernel. We plan to compare our semi-supervised framework with other semi-supervised learning methods for the semantic parsing problem.

Acknowledgments. The work on this paper was partly supported by the grants for Grants-in-Aid for Young Scientific Research 22700139.

References

1. The theory of parsing. Prentice Hall, Englewood Cliffs, NJ (1972)
2. Natural language understanding, 2nd edn. Benjaming/Cumming, Mento Park (1995)
3. Artzi, Y., Zettlemoyer, L.: Bootstrapping semantic parsers from conversations. In: Proceed-ings of the EMNLP 2011, pp. 421–432 (July 2011)
4. Berger, A.L., Della Pietra, S.A., Della Pietra, V.J.: A maximum entropy approach to natural language processing. In: Computational Linguistics (1996)

5. Brown, P., Pietra, V.D., de Souza, P., Lai, J., Mercer, R.: Class-based n-gram models of natural language, pp. 467–479 (1982)
6. Chen, D.L., Kim, J., Mooney, R.J.: Training a multilingual sportscaster: Using perceptual context to learn language. Journal of Artificial Intelligence Research, 397–435 (2010)
7. Earley, J.: An efficient context-free parsing algorithm, pp. 451–455 (1970)
8. Ge, R., Mooney, R.: A statistical semantic parser that integrates syntax and semantics. In: Proceedings of CONLL (2005)
9. James Clarke, M.-W.C., Goldwasser, D., Roth, D.: Driving semantic parsing from the world's response. In: Proceedings, CoNLL (2010)
10. Kate, R.J., Mooney, R.J.: Learning to transform natural to formal languages. In: Proceedings of AAAI, pp. 825–830 (2005)
11. Kate, R.J., Mooney, R.J.: Using string-kernels for learning semantic parsers. In: Proceedings of ACL-COLING 2006, pp. 913–920 (2006)
12. Kate, R.J., Mooney, R.J.: Learning language semantics from ambiguous supervision. In: Proceedings of AAAI 2007, pp. 895–900 (2007a)
13. Kate, R.J., Mooney, R.J.: Semi-supervised learning for semantic parsing using support vector machines. In: Proceedings of NAACL/HLT 2007, pp. 81–84 (2007b)
14. Kwiatkowksi, T., Zettlemoyer, L., Goldwater, S., Steedman, M.: Inducing probabilistic CCG grammars from logical form with higher-order unification. In: Proceedings of EMNLP 2010, pp. 1223–1233 (October 2010)
15. Liang, P., Jordan, M.I., Klein, D.: Learning semantic correspondences with less supervision. In: Proceedings of ACL-IJCNLP (2009)
16. Liu, D., Nocedal, J.: On the limited memory bfgs method for large-scale optimization. Mathematical Programming, 503–528 (1989)
17. Nguyen, L.M., Shimazu, A., Phan, X.H.: Semantic parsing with structured svm ensemble classification models. In: Proceedings of the COLING/ACL 2006 (2006)
18. Tang, L.: Integrating top-down and bottom-up approaches in inductive logic programming: Applications in natural language processing and relation data mining. Ph.D. Dissertation, University of Texas, Austin, TX (2003)
19. Titov, I., Klementiev, A.: A bayesian model for unsupervised semantic parsing. In: Proceedings of ACL 2011 (2011)
20. Titov, I., Kozhevnikov, M.: Boostrapping semantic analyzers from non-contradictory texts. In: Proceedings of ACL 2010, Uppsala, Sweden (2010)
21. Tsochantaridis, T.J.I., Hofmann, T., Altun, Y.: Support vector machine learning for interdependent and structured output spaces. In: Proceedings ICML (2004)
22. Wong, W., Mooney, R.: Learning for semantic parsing with statistical machine translation. In: Proceedings of HLT/NAACL 2006 (2006)
23. Wong, W., Mooney, R.: Learning synchronous grammars for semantic parsing with lambda calculus. In: Proceedings of ACL 2007 (2007)
24. Zelle, J., Mooney, R.: Learning to parse database queries using inductive logic programming. In: Proceedings AAAI 1996, pp. 1050–1055 (1996)
25. Zettlemoyer, L., Collins, M.: Learning to map sentences to logical form: Structured classification with probabilistic categorical grammars. In: Proceedings of UAI 2005, pp. 825–830 (2005)
26. Zettlemoyer, L., Collins, M.: Online learning of relaxed ccg grammars for parsing to logical form. In: Proceedings of the EMNLP-CoNLL 2007 (2007)

On the Effect of Stopword Removal
for SMS-Based FAQ Retrieval

Johannes Leveling

Centre for Next Generation Localisation (CNGL)
Dublin City University
Dublin 9, Ireland
jleveling@computing.dcu.ie

Abstract. This paper investigates the effects of stopword removal in different stages of a system for SMS-based FAQ retrieval. Experiments are performed on the FIRE 2011 monolingual English data. The FAQ system comprises several stages, including normalization and correction of SMS, retrieval of FAQs potentially containing answers using the BM25 retrieval model, and detection of out-of-domain queries based on a k nearest-neighbor classifier. Both retrieval and OOD detection are tested with different stopword lists. Results indicate that i) retrieval performance is highest when stopwords are not removed and decreases when longer stopword lists are employed, ii) OOD detection accuracy decreases when trained on features collected during retrieval using no stopwords, iii) a combination of retrieval using no stopwords and OOD detection trained using the SMART stopwords yields the best results: 75.1% in-domain queries are answered correctly and 85.6% OOD queries are detected correctly.

1 Introduction

With the increasing popularity of mobile devices, information access *on-the-go* is becoming more popular. While modern devices such as smartphones support graphical user interfaces, both feature phones and smartphones support short message service (SMS) for texting messages. These messages were traditionally restricted to a size of 128 bytes, which inspired the use of novel methods to abbreviate the text and make it fit into the limit. Thus, compared to well-formed English, SMS contain many phenomena aiming at shortening text, including clipped word forms (e.g. *"laboratory"* → *"lab"*), replacing characters with numbers or special characters (e.g. *"forget"* → *"4get"*, *"great"* → *"gr@"*), spelling simplifications (e.g. *"right"* → *"rite"*), contractions (*"got to"* → *"gotta"*), etc. phonological approximations (e.g. *"see you"* → *"CU"*), vowel omissions (e.g. *"resident"* → *"rsdnt"*), misspellings, and omitted punctuation. This so-called *text-speak* or *textese* can be considered as a language variant and requires adaptation of natural language processing (NLP) tools to avoid performance loss.

In this paper we investigate SMS-based retrieval of frequently asked questions (FAQ), i.e. finding a correct answer to an incoming SMS question from an English

G. Bouma et al. (Eds.): NLDB 2012, LNCS 7337, pp. 128–139, 2012.

collection of FAQs on diverse topics ranging from career advise to popular Indian recipes. In particular, we are interested in the effect of stopword removal on FAQ retrieval. Tagg [11] performed a linguistic analysis of SMS and observed that *"a"* and *"the"* occur only rarely in SMS compared to other texts. This would imply that the treatment of these words (and other stopwords) has little effect for SMS-based FAQ retrieval. According to [7], stopwords were the most frequent errors in the SMS normalization (e.g. *"r"* is often incorrectly replaced with *"are"* instead of *"or"*). Effects of these errors on retrieval performance are expected to be negligible, as stopwords are removed from the IR query.

This paper aims at making the following contributions: a) investigating the effect of using different stopword lists on retrieval performance, i.e. the number of top ranked correct answers, b) investigating the effect of stopword removal on generating training instances for a classifier for out-of-domain (OOD) query detection, and c) exploring the combination of different *best settings* corresponding to the use of different stopword lists in different processing stages. Typically, separate stages in the processing pipeline of a system make use of the same resources and settings, e.g. either using no stopwords or using the same stopword list. We hypothesize that individual processing stages should be optimized separately to achieve the best performance.

The experiments described in this paper are based on data from the FIRE 2011 evaluation of the SMS-based FAQ Retrieval Task. Results are compared to the best reported results [7]. Note that experiments on SMS normalization and correction are outside the scope of this paper and that we focus exclusively on the effect of stopword removal on retrieval and OOD detection.

The rest of the paper is organized as follows: Section 2 describes the FAQ retrieval task and data. Section 3 presents related work before the experimental setup is introduced in Section 4. Section 6 contains an analysis and discussion of the experimental results described in Section 5. Conclusions and plans for future work are presented in Section 7.

2   Task Description and Data

The SMS-based FAQ retrieval task was introduced in 2011 as an evaluation benchmark at the Forum for Information Retrieval Evaluation[1] (FIRE). The objective of the FAQ retrieval task[2] was to find correct answers in a collection of frequently asked questions, given a query in *text-speak* or *textese*. Figure 1 shows an excerpt from a sample FAQ. Two types of queries can be distinguished (the type is not known in advance): in-domain (ID) queries have a corresponding correct answer in the FAQ collection; out-of-domain (OOD) queries do not have an answer in the collection. Thus, the task can be divided into two parts. The first part of the task is to find answers for ID queries, which can be facilitated by retrieving a ranked list of candidate documents. The second part of the task is to detect OOD queries and indicate that they do not have an answer (by providing

---

[1] http://www.isical.ac.in/~clia/
[2] http://www.isical.ac.in/~clia/faq-retrieval/faq-retrieval.html

"NONE" in the TREC-style results) instead of returning a list of documents. Because of the noise in SMS queries, typically a normalization and correction phase is called for. Thus, the task is similar to question answering based on noisy natural language questions. Figure 2 shows an SMS question in its original form and after normalization.

The experiments described in this paper refer to the English monolingual subtask of FAQ retrieval. Three main datasets have been provided by the organizers of the task: the FIRE FAQ collection, which contains 7251 documents, the FIRE SMS training data, which consists of 1071 questions (701 ID and 370 OOD), and the FIRE SMS test data, which consists of 3405 questions (704 ID and 2701 OOD). In addition, we employed the normalized SMS questions used by [7] for the experiments described in this paper.

```
<FAQ>
<FAQID>ENG_RECIPES_773</FAQID>
<DOMAIN>RECIPES</DOMAIN>
<QUESTION>how to preapare Popeye's Dirty Rice</QUESTION>
<ANSWER>
 1 lb spicy bulk breakfast sausage
 14 oz can clear chicken broth
 1/2 c long-grain rice
 1 t dry onion, minced

 Brown sausage in skillet untiil pink color disappears,   crumbling
 with fork. Stir in broth, rice and minced onion. Simmer gently, covered,
 18 to 20 minutes or until rice is tender and most of broth is absorbed.
 ...
</ANSWER>
</FAQ>
```

Fig. 1. Sample FAQ in XML format. Note that spelling errors are part of the original document.

3   Related Work

Stopword removal is typically applied in information retrieval (IR) to increase precision and reduce the index size. Harter [6] proposed a small list of stopwords for IR, consisting of only nine words (i.e. *"an"*, *"and"*, *"by"*, *"for"*, *"from"*, *"of"*, *"the"*, *"to"*, and *"with"*). Fox [5] describes a general method to extract a list of stopwords. In his work, the underlying corpus was the Brown corpus, from which the most frequent words were extracted. The stopword candidates were then examined manually to either remove them from the list or add missing inflectional forms. This resulted in a list of 421 stopwords. Lo, He and Ounis [9] propose using different methods to extract stopword lists adapted to the specific

```
<SMS>
 <SMS_QUERY_ID>ENG_581</SMS_QUERY_ID>
 <SMS_TEXT>gimme receepi 4 popeye styl dirty rice</SMS_TEXT>
</SMS>

<SMS>
 <SMS_QUERY_ID>ENG_581</SMS_QUERY_ID>
 <SMS_TEXT>give me recipe for popeye style dirty rice</SMS_TEXT>
</SMS>
```

Fig. 2. Sample SMS in original *textese* (top) and normalized format (bottom)

IR task instead of using a single fixed list. They evaluate a term-based random sampling approach which tries to compute how informative a term is. In their experiments on TREC data, performance similar to using the Fox stopword list is achieved and they propose that merging task-specific lists with classic lists may be beneficial. Dolamic and Savoy [2] investigate the use of two stopword lists for Okapi BM25 and variants of the Divergence From Randomness (DFR) paradigm. They report that significantly lower mean average precision can be observed when using short or no stopword lists for ad-hoc retrieval on CLEF data.

More recently, stopword removal in IR has also been investigated for non-English IR. Zou, Wang et al. [12] develop a stopword list for Chinese based on experiments on TREC 5 and 6 Chinese newspaper data. They show that a high overlap exists between the Chinese stopword list and a comparable English stopword list. El-Khair [3] examines the effect of stopword removal for Arabic IR on the LDC Arabic newswire dataset using Lemur. He applies a general stopword list, a corpus-based stopword list, and a combined one. He finds that for BM25, there is a significant performance improvement when using a stopword list. Kothari et al. [8] describe an SMS-based FAQ retrieval system. They propose SMS normalization with synonym lookup and compare performance against Lucene's Fuzzy match model on the original SMS queries. In their experiments, only single character tokens are ignored during SMS normalization as they are likely to be stopwords.

Hogan et al. [7] employ a similar approach to the experiments described in this paper. However, they combine different retrieval approaches and different techniques to OOD detection which results in a more complex system. They also use a fixed list of stopwords (the SMART stopword list) for retrieval. Their submission achieved the best results among 13 participants in the English monolingual SMS-based FAQ retrieval task in the official FIRE 2011 evaluation. In contrast to this system, the approach used for the experiments described in this paper is much simpler. Instead of generating three retrieval streams from different retrieval approaches and combining retrieval results into a single list by a linear combination of weighted scores, a single approach is used. Also, instead of using three methods for OOD detection and combining results by majority voting, a single classifier is employed for the system proposed in this paper.

Previous work on stopword removal in IR has mostly focused on different domains such as web search and recently on languages such as Chinese or Arabic, and has typically presented results for either using a stopword list or not using it. While there are many approaches to extracting a new stopword list for a particular domain, we focus mostly on applying existing stopword lists to FAQ retrieval. In addition, we also generate stopword lists containing the most frequent words in the document collection, but do not manually modify them.

To the best of our knowledge, previous research has not investigated the use of stopword lists of different sizes exhaustively and has not investigated the different effects on FAQ retrieval and out-of-domain query detection.

4   Experimental Setup

4.1   SMS Normalization and Correction

As the focus of this paper lies on experimentation with stopwords in FAQ retrieval and OOD detection, we employ the preprocessed (i.e. normalized and corrected) set of SMS queries used for the experiments described by Hogan et al. [7]. The approach to SMS normalization in [7] is token-based and includes generating a list of candidate token corrections. Token corrections include clipped forms, words extracted from manually annotated data, and words with a corresponding consonant skeleton. The candidates are scored based on n-gram frequencies and the top ranked candidate is selected for correction. This approach achieved 74% accuracy for tokens in the FAQ training data.

4.2   Retrieving Answers from FAQs

For FAQ retrieval, we used the following setup: The preprocessed and normalized SMS questions (or the SMS questions in their original form) are used a queries. FAQ question text, answer text and the combination of both are indexed in different indexes (Q, A, and Q+A, respectively). SMS questions are matched against one of the three corresponding FAQ indexes. Stopwords are removed, using one of the standard stopword lists (including the empty list). The BM25 retrieval model [10] is applied, using default parameters ($b = 0.75$, $k_1 = 1.2$, and $k_3 = 7$).

This approach yields a list of top ranked FAQs potentially containing a correct ID answer. The next step in our setup is to determine whether a query actually has an answer in the collection and return the top ranked FAQs or indicate the query as OOD, otherwise.

4.3   Filtering Out-of-Domain Queries

We view OOD detection as a classification problem, where we want to determine whether one of the top ranked answers is correct (in which case the top ranked documents are returned) or not (in which case we assume the query is OOD and *"NONE"* is returned). TiMBL [1] implements a memory-based learning approach and supports different machine learning algorithms which have been successfully applied to natural language processing problems such as part-of-speech tagging, chunking, and metonym recognition.

For the experiments described in this paper, the IB1 approach was employed to train a robust classifier distinguishing between OOD queries and ID queries. IB1 is similar to a k-nearest-neighbors approach. [3] For the experiments described in this paper, we used the overlap metric for numeric features (see Equation 1).

$$d(x_i, y_i) = \text{abs}(\frac{x_i - y_i}{max_i - min_i}) \tag{1}$$

Features for the training instances are generated during the retrieval phase on the training data, examining the top five documents. The features comprise:

- The result set size, i.e. the number of documents containing at least one query term (1 feature).
- The raw BM25 [10] document scores for the top five documents (5 features).
- The percentual BM25 score difference between consecutive top five documents (4 features).
- Normalized BM25 scores, computed by calculating the maximum possible score for the query q as the sum of IDF scores (see Equation 2) for all query terms (5 features). This method has been proposed by Ferguson et al. [4] for retrieval of Blogs.

$$score(q, d) = \sum_{t \in q} \log(\frac{N - df_t + 0.5}{df_t + 0.5}) \tag{2}$$

- The term overlap scores for the SMS query and the top five documents (5 features) as shown in Equation 1. The overlap is computed as the number of matches $m(q, d)$ between query q and document d, normalized by the query length $|q|$.

$$overlap(q, d) = \frac{m(q, d)}{|q|} \tag{3}$$

The OOD detection is then trained on the extracted instances from the training data and evaluated using *leave-one-out* validation, which is natively supported by TiMBL. The result is essentially a binary classifier which determines whether a query is OOD or if the top five results contain a correct answer. Compared to the OOD detection approach described by [7], the method proposed here is simpler because it relies on a single classifier instead of combining different classifiers by majority voting for OOD detection. The first classifier also uses TiMBL on similar features. The two additional OOD classifiers in [7] were based on the

---

[3] In contrast to the original nearest neighbor algorithm, the k in TiMBL refers to k nearest distances rather than neighbors.

normalized BM25 score and term overlap, respectively. If the corresponding value exceeds a given threshold, the query was marked as OOD. The classifier used for the experiments described in this paper integrates the features for all three classification methods into a single instance vector.

Table 1. Results on the FIRE SMS training set using the SMART stopword list

Run	FAQ fields	SMS queries	ID Correct (%)	ID Correct in Top 20
1	Q	original	400/701 (0.571)	569/701 (0.812)
2	A	original	108/701 (0.154)	282/701 (0.402)
3	Q+A	original	302/701 (0.431)	510/701 (0.728)
4	Q	normalized	**527/701 (0.752)**	660/701 (0.942)
5	A	normalized	161/701 (0.230)	411/701 (0.586)
6	Q+A	normalized	507/701 (0.723)	672/701 (0.959)

5   Retrieval Experiments and Results

5.1   Experiments on Training Data

We performed different experiments on the SMS test data to investigate the effect of stopword removal in the retrieval stage. As a first experiment, we varied which fields of the FAQs the (corrected) SMS query should be matched against: FAQ questions only (Q), FAQ answers (A), or both questions and answers (Q+A).

Performance results for the original (unprocessed) and normalized SMS queries are shown in Table 1. The results include the number of correct results at top rank (ID Correct), and the number of correct results in the top 20 results (ID Correct in Top 20). Unsurprisingly, the normalized SMS queries achieve a much higher overall performance. However, given that the answer text of FAQs is much longer than their question field, we had expected that matching against longer text would produce more correct results due to fewer vocabulary mismatches between SMS query and FAQ text. More surprisingly, matching SMS queries with FAQ questions yields the best results. This result is in accordance to the official results reported in [7], where retrieval on FAQ questions (Q) using the BM25 model achieved the best results, and using only the answer part or using it in combination with the questions consistently adds noise to the results. The same order of results is observed when stopwords are not removed. We maintain this setting (i.e. from Run 4) for the rest of our experiments.

The next set of experiments was concerned with how stopword list size affects retrieval performance for this task, which is primarily measured as the number of correct answers in the top rank. We used English stopword lists from various online resources and tools and conducted experiments on the FAQ training data based on these stopword lists. The stopword lists have been extracted from various standard IR tools and natural language processing resources. The lists have been selected to cover a various sizes to investigate the effects ranging from no stopword removal (i.e. using the empty stopword list) to aggressive stopword removal (i.e. using a large stopword list). The stopword lists include:

- empty - The empty list containing no stopwords.
- TREC HARD - The English stopword list used in the TREC HARD track which contains the three words *"a" "the"*, and *"and"*.
- Harter - The Harter stopword list consisting of nine words [6].
- Cheshire - The default English stopword list for title field from Cheshire[4] .
- Lucene - The default English stopword list from Lucene[5].
- Snowball - The default English stopword list from Snowball stemmer[6].
- Okapi - The default English stopword list from Okapi[7].
- Swish-E - The English stopword list from Swish[8].
- Lemur - The default English stopword list from Lemur[9].
- Zettair - The default English stopword list from Zettair[10].
- SMART - The default English stopword list from SMART[11].
- Brown - A list of words from closed word categories (e.g. prepositions, determiners), extracted from the part-of-speech annotated Brown corpus which is distributed as part of NLTK[12]
- Terrier - The default English stopword list from Terrier[13]. As Terrier is based on DFR, which can be sensitive to outliers, the list is quite long [2].
- Top N - The N most frequent terms extracted from the FIRE FAQ corpus.

Table 2 shows the name of the stopword list (SW), its size ($|SW|$), the number of correct ID results in the top and top 20 retrieved results, and the number of correct OOD results for classification using the same stopword list.

5.2   Out-of-Domain Query Detection

Training data for the OOD classifier is generated while running retrieval experiments on the FAQ training data, because feature values can be easily computed during retrieval. Results for OOD detection are also shown in Table 2. We found that surprisingly, OOD detection based on training data for the best retrieval run (i.e. without stopwords) decreases performance compared to using a larger list of stopwords (i.e. the SMART stopword list). A plausible explanation is that some of the classification features are sensitive to the presence of stopwords,

---

[4] http://cheshire.berkeley.edu/
[5] http://lucene.apache.org/
[6] http://snowball.tartarus.org/
[7] http://www.soi.city.ac.uk/~andym/OKAPI-PACK/
[8] http://swish-e.org/
[9] http://www.lemurproject.org/
[10] http://www.seg.rmit.edu.au/zettair/
[11] ftp://ftp.cs.cornell.edu/pub/smart/
[12] http://www.nltk.org/data
[13] http://terrier.org/

Table 2. Results on the FIRE SMS training set using different stopword lists on FAQ questions and the normalized questions

SW	$|SW|$	ID Correct (%)	ID Correct in Top 20	OOD Correct (%)
empty	(0)	**567/701 (0.809)**	671 (0.957)	809/1071 (0.755)
TREC HARD	(3)	564/701 (0.804)	671 (0.957)	759/1071 (0.742)
Harter	(9)	566/701 (0.807)	671 (0.957)	848/1071 (0.792)
Cheshire	(29)	561/701 (0.800)	670 (0.956)	847/1071 (0.791)
Lucene	(33)	559/701 (0.797)	671 (0.957)	853/1071 (0.796)
Snowball	(120)	547/701 (0.780)	671 (0.957)	870/1071 (0.812)
Okapi	(222)	539/701 (0.769)	667 (0.951)	879/1071 (0.821)
Swish-E	(337)	539/701 (0.769)	666 (0.950)	**883/1071 (0.824)**
Lemur	(418)	539/701 (0.769)	666 (0.950)	874/1071 (0.816)
Zettair	(469)	538/701 (0.768)	664 (0.947)	869/1071 (0.811)
SMART	(571)	527/701 (0.752)	660 (0.942)	877/1071 (0.819)
Brown	(679)	542/701 (0.773)	669 (0.954)	866/1071 (0.809)
Terrier	(733)	542/701 (0.773)	661 (0.943)	866/1071 (0.809)
Top 10	(10)	562/701 (0.802)	667 (0.951)	839/1071 (0.783)
Top 20	(20)	559/701 (0.797)	668 (0.953)	866/1071 (0.809)
Top 30	(30)	558/701 (0.796)	668 (0.953)	868/1071 (0.810)
Top 40	(40)	553/701 (0.789)	666 (0.950)	866/1071 (0.809)
Top 50	(50)	550/701 (0.785)	666 (0.950)	872/1071 (0.814)

which means that the corresponding features contribute differently to OOD detection depending on whether and how many stopwords are used. For example, the result set size will change drastically when stopwords are included and the term overlap scores will also depend on whether stopwords are employed or not.

5.3   Experiments on Test Data

There are different optimal settings for the retrieval stage and OOD detection with regard to the use of a stopword list: retrieval performs best when no stopwords are used, and OOD detection performs best when the Swish-E stopword list is used. We perform experiments on the FIRE test data exploring different settings to optimize performance. We expected that using individual best settings for each stage as obtained from experiments on the training data and combining these results will improve overall performance. Results for the test data are shown in Table 3.

The first four results are as expected, where using removing no or only a few stopwords yields the highest number of top ranked ID results (i.e. 0.649 using the TREC HARD stopword list) and the worst for OOD detection. Adding stopwords from the Swish-E or SMART stopword lists produces better OOD query identification (i.e. 0.863 for the SMART stopword list), but fewer correct ID results. The use of different stopword lists for different processing stages achieves the best results so far (correct ID: 529, correct OOD: 2313), outperforming the best reported result on this data set [7] (ID: 494, OOD: 2311).

Table 3. Results on the SMS test set using different stopword lists for FAQ retrieval (IR *SW*) and OOD detection (OOD *SW*)

IR *SW*	|*SW*|	OOD *SW*	|*SW*|	ID Correct (%)	OOD Correct (%)
empty	(0)	empty	(0)	455/704 (0.646)	2154/2701 (0.797)
TREC HARD	(3)	TREC HARD	(3)	457/704 (0.649)	2200/2701 (0.815)
Swish-E	(337)	Swish-E	(337)	422/704 (0.599)	2251/2701 (0.833)
SMART	(571)	SMART	(571)	394/704 (0.560)	**2333/2701 (0.863)**
empty	(0)	Swish-E	(337)	524/704 (0.744)	2245/2701 (0.831)
empty	(0)	SMART	(571)	**529/704 (0.751)**	2313/2701 (0.856)
Hogan et al. [7]				494/704 (0.701)	2311/2701 (0.856)

6   Analysis and Discussion

We observed a general trend for experiments on the training and test data that using a larger stopword lists decreases performance in terms of correctness of the top ranked document. However, the correctness of top ranked documents decreases only slightly from 671 in the best case to 660 in the worst tested case. The best retrieval performance is achieved when using no stopwords at all. One explanation is that for small queries and for this task, even stopwords can add important context beneficial for ranking documents.

However, a larger impact can be observed for results of the OOD classifier using different stopword lists. There is a difference of up to 124 correctly classified cases when comparing the worst and best results for the training data. The trend for OOD detection seems to be reversed in comparison to the retrieval stage, i.e. better results are achieved when using a stopword list, and medium or large stopword lists improve the correctness of top ranked answers. Similar effects can be observed for experiments on the test data. Thus, a combination of different stopword lists yields the highest number of top ranked correct answers.

Intuitively, one would have expected that stopwords do not contribute to retrieving correct answers (or relevant documents) and that stopwords also do not help in discriminating between ID and OOD queries. However, the role of stopwords seems to be different for IR and classification tasks. Figure 3 shows the effect of using stopword lists of different sizes on the number of correct ID and OOD queries. It can be easily seen that the number of correct ID queries is highest when using no stopwords and drops when OOD performance increases and vice versa.

7   Conclusions and Future Work

This paper investigated the effect of using different stopword lists on FAQ retrieval performance. The number of correct ID answers depends on what kind of stopword removal is applied. In contrast to most results for ad-hoc IR, stopword removal decreases IR performance for SMS-based FAQ retrieval. The effect

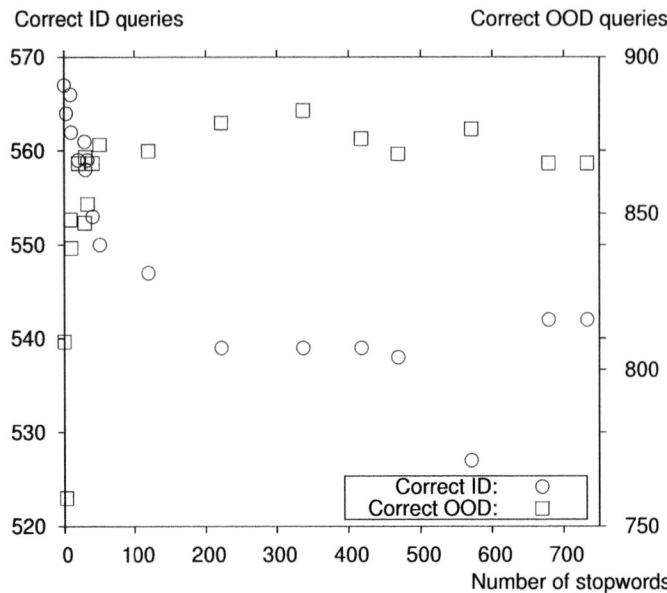

Fig. 3. Effect of stopword list length versus number of correctly answered ID queries (left) and correctly identified OOD queries (right)

of stopword removal on generating training instances for a classifier for out-of-domain query detection is almost the reverse, i.e. using a medium-size stopword list is best for training an OOD classifier. Thus, the strategy to include stopwords or not can be different in different stages of the processing pipeline. As we hypothesized, using individual best settings for each stage and combining these results improves overall performance. The best performance is obtained when combining retrieval without stopword removal with OOD classification trained on features using the SMART stopword list.

Even applying the same setting for both processing stages, i.e. not removing stopwords, would result in reasonable (but not the best) performance. However, the performance difference could even be bigger for other areas of natural language processing. For example, one may decide to train a machine translation system on training data after stopwords have been removed. A cross-language IR system could use this system to translate queries and rely on a different stopword list, e.g. to identify suitable query expansion terms.

In summary, the optimum setting – whether to remove stopwords or not, and which stopword list is used – should be carefully chosen individually for different processing stages of a retrieval system and requires fine-tuning of subsystems. This confirms our initial expectation that separate stages in the processing pipeline of a FAQ system should be optimized separately, which includes the mundane task of choosing the right stopword list. Our best results exceed the best reported result for the FAQ task [7], which was achieved by a combination of different retrieval and OOD detection approaches.

Future work will address normalization of SMS messages and other form of Microtext (e.g. Tweets) to alleviate the performance loss of NLP tools.

Acknowledgments. This research is supported by the Science Foundation Ireland (Grant 07/CE/I1142) as part of the Centre for Next Generation Localisation (CNGL) project.

References

1. Daelemans, W., Zavrel, J., van der Sloot, K., van den Bosch, A.: TiMBL: Tilburg memory based learner, version 6.2, reference guide. Technical Report 09-01, ILK (2004)
2. Dolamic, L., Savoy, J.: When stopword lists make the difference. JASIST 61(1), 200–203 (2010)
3. El-Khair, I.A.: Effects of stop words elimination for Arabic information retrieval: A comparative study. International Journal of Computing & Information Sciences 4(3), 119–133 (2006)
4. Ferguson, P., Hare, N.O., Lanagan, J., Smeaton, A., Phelan, O., McCarthy, K., Smyth, B.: CLARITY at the TREC 2011 Microblog Track. In: Proceedings of the 20th TREC Conference. National Institute of Standards and Technology (NIST), Gaithersburg, MD, USA (2011)
5. Fox, C.J.: A stop list for general text. SIGIR Forum 24(1-2), 19–35 (1990)
6. Harter, S.P.: Online information retrieval. Concepts, principles, and techniques. Academic Press (1986)
7. Hogan, D., Leveling, J., Wang, H., Ferguson, P., Gurrin, C.: DCU@FIRE 2011: SMS-based FAQ retrieval. In: 3rd Workshop of the Forum for Information Retrieval Evaluation, FIRE 2011, IIT Bombay, December 2-4, pp. 34–42 (2011)
8. Kothari, G., Negi, S., Faruquie, T.A., Chakaravarthy, V.T., Subramaniam, L.V.: SMS based interface for FAQ retrieval. In: ACL/IJNLP 2009, pp. 852–860 (2009)
9. Lo, R.T.W., He, B., Ounis, I.: Automatically building a stopword list for an information retrieval system. JDIM 3(1), 3–8 (2005)
10. Robertson, S.E., Walker, S., Jones, S., Beaulieu, M.M.H., Gatford, M.: Okapi at TREC-3. In: Harman, D.K. (ed.) Overview of the Third Text Retrieval Conference (TREC-3), pp. 109–126. National Institute of Standards and Technology (NIST), Gaithersburg (1995)
11. Tagg, C.: A corpus linguistics study of SMS text messaging. Ph.D. thesis, University of Birmingham (2009)
12. Zou, F., Wang, F.L., Deng, X., Han, S., Wang, L.S.: Automatic construction of Chinese stop word list. In: 5th WSEAS International Conference on Applied Computer Science (2006)

Wikimantic: Disambiguation for Short Queries

Christopher Boston[1], Sandra Carberry[1], and Hui Fang[2]

[1] Department of Computer Science, University of Delaware,
Newark, DE, 19716
[2] Department of Electrical and Computer Engineering, University of Delaware,
Newark, DE, 19716

Abstract. This paper presents an implemented and evaluated methodology for disambiguating terms in search queries. By exploiting Wikipedia articles and their reference relations, our method is able to disambiguate terms in particularly short queries with few context words. This work is part of a larger project to retrieve information graphics in response to user queries.

Keywords: disambiguation, short queries, context, wikipedia.

1   Introduction

Disambiguation is the fundamental, yet tantalizingly difficult problem of annotating text so that each ambiguous term is linked to some unambiguous representation of its sense. Wikipedia, the online encyclopedia hosted by the Wikimedia Foundation, has garnered a lot of interest as a tool for facilitating disambiguation by providing a semantic web of hyperlinks and "disambiguation pages" that associate ambiguous terms with unambiguous articles [3–5, 7]. For an encyclopedia, English Wikipedia is monolithic. It contains over 3.5 million articles which are connected by hundreds of millions of user-generated links. Although errors do exist in articles and link structure, Wikipedia's strong editing community does a good job of keeping them to a minimum.[1]

This paper describes the methods and performance of Wikimantic, a system designed to disambiguate short queries. Wikimantic is part of a larger digital library project to retrieve information graphics (bar charts, line graphs, etc.) that appear in popular media such as magazines and newspapers. Such graphics typically have a high-level message that they are intended to convey, such as that *Visa ranks first among credit cards in circulation*. We have developed a system for identifying this high-level message [8, 9]. We anticipate retrieving graphics relevant to a user query by relating the query to a combination of the graphic's intended message, any text in the graphic, and the context of the associated article. To do this, we must first disambiguate the words in the query.

---

[1] Here, we refer primarily to clear technical problems such as duplicate articles or dead links. We actually consider Wikipedia's alleged susceptibility to bias and error a boon since we expect to disambiguate correspondingly fallible user queries.

G. Bouma et al. (Eds.): NLDB 2012, LNCS 7337, pp. 140–151, 2012.

Although there are many existing methods that extract semantic information via disambiguation, most require large amounts of context terms or focus exclusively on named entities [1, 2, 5, 7]. Our experiments have shown that most queries are very short and that nouns and named entities convey only a portion of the query's full meaning. Thus a more robust method of disambiguation is required.

Our work has several novel contributions to disambiguation which are important for information systems. First, we disambiguate text strings that to our knowledge are the shortest yet. Second, our method is robust with respect to the terms that can be disambiguated, rather than being limited to nouns or even named entities. And third, our method can determine when a sequence of words should be disambiguated as a single entity rather than as a sequence of individual disambiguations. Furthermore, our method does not rely on capitalization since users are notoriously poor at correct capitalization of terms in their queries; this is in contrast to the text of formal documents where correct capitalization can be used to identify sequences of words that represent a named entity.

The rest of the paper is organized as follows. Section 2 discusses related research. Section 3 presents an overview of the proposed method. Section 4 describes the models that select possible concepts for a query, and Section 5 discusses how the models are used to disambiguate the individual terms in the query. Section 6 presents an evaluation of the Wikimantic system and Section 7 concludes with a general summary and suggestions for future work.

2   Related Work

Bunescu and Pasca are generally credited with being the first to use Wikipedia as a resource for disambiguation [1]. They formulated the disambiguation task to be a two step procedure where a system must (1) identify the salient terms in the text and (2) link them accurately. Though Bunescu and Pasca's work was initialy limited to named entity disambiguation, Mihalcea later developed a more general system that linked all "interesting" terms [4].

Mihalcea's keyword extractor and disambiguator relied heavily on anchor text extracted from Wikipedia's inter-article links. When evaluating the disambiguator, Mihalcea gave it 85 random Wikipedia articles with the linked terms identified but the link data removed, and scored it based on its ability to guess the original target of each link. Mihalcea achieved an impressive F-measure of 87.73[4], albeit with one caveat. Regenerating link targets is significantly easier than creating them from scratch, since the correct target must necessarily exist in Wikipedia and be particularly important to the context. Wikimantic is tasked with the more difficult problem of disambiguating all salient terms in the query indiscriminately.

Many Wikipedia based disambiguation systems use variants of Mihalcea's method which attempt to match terms in the text with anchor text from Wikipedia links [5, 10]. When a match is found, the term is annotated with a copy of the link. Sometimes, a term will match anchor text from multiple conflicting links, in which case the system must choose between them. Milne and Witten's contribution was to look for terms that matched only non-conflicting links, and use those easy disambiguations to provide a better context for the more difficult ones [5]. Given a large text string, it's always possible to find at least one trivial term to start the process. However, short strings do not reliably contain trivial terms.

Ferragina and Scaiella [3] addressed this problem by employing a voting system that resolved all ambiguous terms simultaneously. They found that good results were attainable with text fragments as short as 30 words each, which would allow for the disambiguation of brief snippets from search engine results or tweets. Although their results are very good, 30 words is still too large for the short queries we wish to process. In our evaluation, we limit our full sentence queries to a maximum of 15 terms in length. The average query length in our test set is just 8.9 words, including stop words.

Ratinov et al. define a local disambiguation method to be one that disambiguates each term independently, and a global disambiguation method to be one that searches for the best set of coherent disambiguations. Their recent work has shown that the best ranking performance can usually be obtained by combining local and global approaches [7]. Although their system was limited to named entities, our performance seems to be best when combining our own local and global approaches as well.

3   Method Overview

Given a sequence of terms, we seek to find a mapping from each salient term to the Wikipedia article that best represents the term in context. More formally, let $s = (t_1, t_2, \ldots, t_{|s|})$ be a sequence of $|s|$ terms. For every term t_j, if t_j is salient (not a function word), we wish to generate a mapping $t_j \rightarrow C_i$ where C_i is the Wikipedia article that best defines the concept t_j referenced. For example, given the sentence "Steve Jobs resigns from Apple", an acceptable mapping would link "steve" and "jobs" to the Wikipedia article about Steve Jobs, the former CEO of Apple. "resigns" would be mapped to the article Resignation, and "apple" would be mapped to the article for Apple Inc. Mapping "apple" to the article for the actual apple fruit would be unacceptable in that context.

Our general strategy is to first summarize the meaning of s, and then use that summary to choose the most probable mapping of terms to concepts. To summarize s, we construct a "Concept"[2] object that represents the general topic of s. The way we define and use Concept objects is based on our generative model as described in Section 4. Specifically, we begin by naively building a set of all Wikipedia articles that could possibly be referenced by terms in s. We weight each article with the product of its prior probability of being relevant and the degree to which terms in s match terms in the article. The weighted set is packaged up in a data structure we call a MixtureConcept. Once we have

---

[2] In this paper, we follow the convention that object types from our model are written in upper camel case. When we write "Concept", we refer specifically to the class of object from our model. When we write "concept", we are simply using the term as one would in every day speech.

the weighted set, we begin searching for the best mapping from terms in s to Wikipedia articles. We score each mapping according to the weights of the articles in the set as described in Section 5.

4   Concept Selection

4.1   Generative Model

An author encodes ideas into words and puts the words on paper. A reader may later take these words and decode them back into ideas. Our generative model is based on the premise that every idea has certain associated words that are used to talk about the idea. A person writing about the Apple Corporation may use terms like "computer", "iPhone", "Steve", or "Jobs". A reader can use a priori knowledge about these term-Concept associations to know that the writer means Apple Corporation and not the fruit when they just say "apple".

Our generative model makes the simplifying assumption that it is the Concepts themselves that generate terms in a text. When a writer wishes to write a document or formulate a query about the Apple Corporation, we say that the Concept of *Apple_Corporation* is actually generating the terms in the text directly. Therefore, a query about *Apple_Corporation* is likely to contain terms like "computer" and "iPhone" due to the Concept *Apple_Corporation*'s propensity to generate such terms.

To be more precise, our generative model states that texts are generated term by term from some topic Concept. The a priori probability of a given Concept C being the topic Concept of our text is denoted $P(C)$. For every term t, there is a probability $P(t|C)$ that C will generate t as the next term in the text. Documents, queries, and other forms of text are all considered to be of one type, TermSequence. By incorporating knowledge of all possible Concepts and their probability of generating each term, it is possible to take a TermSequence and work backwards to find the Concepts that generated it. In order to get this knowledge, we extract a set of fundamental AtomicConcepts from Wikipedia.

AtomicConcept: An AtomicConcept is a type of Concept. Like all Concepts, an AtomicConcept has an a priori probability of being a topic Concept (denoted $P(A)$ by convention when the Concept is atomic), and probabilities $P(t|A)$ of generating term t. We view each article in Wikipedia as a long TermSequence that was generated by some AtomicConcept. Since every article is unique, there is a one to one mapping between articles and the AtomicConcepts that generate them.

In order to estimate the a priori probability P(A), we look at the relative number of inter-article links that point to A's article.

$$P(A) = \frac{number\ of\ incoming\ links}{number\ of\ links\ in\ Wikipedia}$$

Since Wikipedia articles link to the other articles they discuss, the fraction of incoming links is a good estimate of the likelihood that an article's subject (its AtomicConcept) will be talked about.

To estimate $P(t|A)$, we view the article body text as a sample of terms generated by A. The probability of A generating a term t is:

$$P(t|A) = \frac{count(t, A)}{number\ of\ words\ in\ A}$$

Because articles have finite length, some terms relevant to A won't actually show up in the body text of the article. For each term not present in the article, we smooth the distribution by estimating the probability of A generating t to be the probability of t occurring in the English language.[3]

MixtureConcept: Although Wikipedia covers a wide range of topics, it would be overly simplistic to assume that each and every real world text can be accurately summarized by just one AtomicConcept. A query about Apple Inc.'s profits on the iPod Touch might better be summarized with a mixture of the AtomicConcepts *Apple_Inc.*, *iPod_Touch*, and *Profit_(accounting)*. Thus we instead model the topic using a MixtureConcept which is a set of weighted AtomicConcepts. When a MixtureConcept generates a term, it randomly selects one of its AtomicConcepts to generate in its stead. The weight of an Atomic-Concept tells us the probability that it will be the one selected to generate, and all weights necessarily sum to 1. Like all Concepts, a MixtureConcept M has an a priori probability $P(M)$ of being the topic, and probabilities $P(t|M)$ of generating a given term t.

Let MixtureConcept $M = \{(w_i, A_i)|\ i = 1...n\}$

where w_i = the weight of A_i in M

$$P(M) = \sum_{i=1}^{n} w_i * P(A_i)$$

$$P(t|M) = \sum_{i=1}^{n} w_i * P(t|A_i)$$

If a term sequence s discusses Apple Inc., summarizes its profits and briefly mentions the iPod Touch, the MixtureConcept for s may look something like:

$$M_s = \{(0.5, Apple\_Inc.), (0.3, Profit\_(accounting)), (0.2, iPod\_Touch)\}$$

$$P(M_s) = 0.5*P(Apple\_Inc.)+0.3*P(Profit\_(accounting))+0.2*P(iPod\_Touch)$$

---

[3] In Wikimantic, we use Microsoft n-Grams to give us P(t). Because probabilities from Wikipedia and Microsoft n-Grams each sum to 1, the sum of $P(t \mid A)$ over all t equals than 2. In practice, estimated probability values for AtomicConcepts are always stored as elements of normalized collections, which ensures that no probability value falls outside the range [0,1].

To find the likelihood of the term "iPhone" in any term sequence with M_s as it's topic, one would calculate

$$P(\text{``iPhone''}|M_s) = 0.5 * P(\text{``iPhone''}|Apple\_Inc.)$$
$$+0.3 * P(\text{``iPhone''}|Profit\_(accounting))$$
$$+0.2 * P(\text{``iPhone''}|iPod\_Touch)$$

The key problem is how to estimate the weight w_i of each AtomicConcept. In the following, we first present a method that uses the content of a concept's article to estimate w_i and then a second method that uses references between concepts.

4.2   Content-Based Topic Modeling:

To build a MixtureConcept M that represents the meaning of a TermSequence s constituting a query, we first populate the set with AtomicConcepts and then weight the AtomicConcepts. Our base method uses the content of a concept's article to estimate the concept's weight in the MixtureConcept.

To construct the elements of the set, we look at every subsequence of terms in s and attempt a direct lookup in Wikipedia. Any article that has a title that matches a subsequence of terms in s is added to the set. Any article that is disambiguated by a page whose title matches a subsequence of terms in s is also added to the set. Finally, all articles that share a disambiguation page with an article already in the set are added. For example, if $s = $ "Steve Jobs resigns":

Steve → Matches title of disambiguation page *Steve*. Add all articles disambiguated by that page.

Jobs → Matches the title of a redirect page that points to *Jobs_(Role)*. Add *Jobs_(Role)* and all other articles that *Jobs_(disambiguation)* link to.

Resigns → Matches the title of a redirect page that points to *Resignation*. Add *Resignation* and all other articles that *Resignation_(disambiguation)* link to.

Steve Jobs → Matches the title of the article *Steve_Jobs*.

Jobs Resigns → Matches nothing, so no articles added.

Steve Jobs Resigns → Matches nothing, so no articles added.

Once our unweighted set M is populated, it will contain a large number of candidate AtomicConcepts of varying degrees of relevance, and we rely on weights to mitigate the impact of spurious concepts. We weight each AtomicConcept according to the probability that every term in s was generated by that AtomicConcept, ignoring stopwords.

$$w_i = P(A_i|s) = \prod_{j=1}^{|s|} P(A_i|t_j) \tag{1}$$

$$P(A_i|t_j) = \frac{P(t_j|A_i) * P(A_i)}{P(t_j)}$$

This weighting schema ensures that an AtomicConcept will only get a high score if it is likely to generate all terms in the sequence. A Concept like $Jobs\_(Role)$ may have a high probability of generating "jobs", but its low probability of generating "steve" will penalize it significantly. We can expect the Concept $Steve\_Jobs$ to generate "steve", "jobs", and "resigns" relatively often, which would give it a larger weight than $Jobs\_(Role)$ would get.

Once M, the estimated topic concept, has been populated and weighted, it can be used to guide disambiguation. Since AtomicConcepts in M are weighted by their propensity to generate terms in s, our first method makes the somewhat strong assumption that the probability of an AtomicConcept generating the string s is roughly equal to its probability of being the correct mapping for a term in s.

4.3   ReferenceRank: M_R

Our second method, ReferenceRank, uses references between AtomicConcepts to estimate the weights w_i. In M_R, AtomicConcepts are weighted according to the probability that they describe the sense of a given term in s, rather than the probability that they will generate a given term in s. Consider the following example where M might be misleading if it is weighted by probability of generating.

$$M_1 = \{(0.5, \text{Apple\_Inc.}),(0.5,\text{Whole\_Foods})\}$$
$$M_2 = \{(0.5, \text{Apple\_Inc.}),(0.2,\text{iPhone}), (0.2,\text{Apple\_Safari}), (0.1,\text{iPad})\}$$

In the text described by M_1, the topic is 50% about Apple Inc. and 50% about the grocery store Whole Foods. In the document described by M_2, the topic is 50% about Apple Inc. and 50% about various Apple products. Since $iPhone$, $Apple\_Safari$, and $iPad$ are all concepts that are likely to generate the term "apple" (referring to the company), one would expect Apple Inc. to be referenced more often in M_2 than M_1. However, Apple Inc. is weighted equally in M_1 and M_2. It's subtle, but there is a very real difference between the probability that a Concept will generate terms in our query and the probability that a Concept will be referred to by concepts associated with other terms in our query. To account for this, we extend our generative model by making the claim that Concepts generate references to other concepts as well as terms.

Given that AtomicConcept A_1 generated a reference to another concept, the probability that the referenced concept is A_2 is estimated as the probability that clicking a random link in A_1's article will lead directly to the article for A_2.

$$P(R_{A_2}|A_1) = \frac{number\ of\ links\ from\ A_1\ to\ A_2}{total\ number\ of\ links\ originating\ at\ A_1}$$

The probability that a MixtureConcept M will generate a reference to a Concept C (denoted R_C) is just a mixture of the probabilities of the AtomicConcepts in M generating a reference to C:

$$P(R_C|M) = \sum_{i=1}^{n} w_i * P(R_C|A_i)$$

where w_i = the weight of A_i in M (Equation 1)

For a TermSequence s, we compute the special MixtureConcept M_R that contains all relevant AtomicConcepts weighted by their probability of being referenced.

$$\text{Let MixtureConcept } M_R = \{(w_i, A_i) \mid i = 1...n\}$$

$$\text{where } w_i = P(R_{A_i} \mid M)$$

This reweighting in M_R is very similar to one iteration of the PageRank algorithm, where nodes in a graph vote for other nodes to which they link. In our case, AtomicConcepts in M vote for other AtomicConcepts in M, and the power of each vote is proportional to the weighting of that AtomicConcept in M.

5 Disambiguation

The weights of M and M_R tell us a lot about the probability that a term in s will reference a given AtomicConcept A. Since AtomicConcepts in M_R are weighted by their chances of being referenced in s, and each term we are disambiguating comes from s, one could simply use A's weight in M_R as an estimate of the probability that A is the correct sense of the term. As it turns out, M_R's reliance on Wikipedia's relatively sparse link structure causes some problems. "Do Life Savers cause tooth decay?" is a perfectly reasonable query, but there are no direct links between the articles for Life Savers and tooth decay. This means that neither will receive a vote and therefore their weights in M_R will be zero. Although M_R is useful when links are found, it must be supplemented with information from M. For this reason, the mixture $P(A|s) = (1-d)*M+d*(M_R)$ is used. The optimal value of d is determined experimentally.

In many disambiguation papers[1, 2, 4, 7], the important term strings are assumed to be marked ahead of time and the system must simply choose the single best Wikipedia article for the marked string. For queries, the number of mappings are not known a priori, which makes disambiguation considerably more difficult. Does "life saver" refer to the brand of candy or a person who saved a life? If we are talking about junk food, then "life saver" should entail a single mapping to the AtomicConcept $Life\_Saver$, otherwise it entails two separate mappings to $Life$ and $Saver$. Although these kinds of conflicts seem like they should be rare, the vast coverage of Wikipedia actually makes them common. Company names, book titles, and music album titles are particularly troublesome since they are often common phrases; moreover, they are often the topics of graphs in popular media and thus occur in user queries for these graphs.

If it weren't for these conflicts, disambiguation would be simple. For each term t, one could simply choose the AtomicConcept A that maximizes $P(A|s)$ from the list of all AtomicConcepts that were added to the MixtureConcept by t in Section 4.2.

If a problematic sequence of terms like "life saver" or "new york city" is found, every possible breakdown of the sequence is disambiguated. Each breakdown yields a unique candidate set of disambiguations that is scored according to its new york city

$P(Concept_{New} \mid s) * P(Concept_{York} \mid s) * P(Concept_{City} \mid s)$

| New | York | City |

$P(Concept_{New\_York} \mid s)^2 * P(Concept_{City} \mid s)$

| New_York | City |

$P(Concept_{New} \mid s) * P(Concept_{York\_City} \mid s)^2$

| New | York_City |

$P(Concept_{New\_York\_City} \mid s)^3$

| New_York_City |

Fig. 1. Product method's scoring of four possible disambiguations of "new york city"

probability of being the correct one. The scoring is calculated using either the Mixture method or the Product method.

5.1 Product Method

The Product method scores a candidate set as the product of the probabilities of each mapping of term to AtomicConcept. Figure 1 depicts the four candidate sets that are considered when the string "new york city" is broken down. We use italics to refer to AtomicConcepts by name, so $P(Concept_{New}|s)$ refers to the probability of the Concept *New* being the disambiguation of the term "new". The first set contains the three AtomicConcepts *New*, *York*, and *City*. The score of the set is simply the product of their probabilities multiplied together. When n adjacent terms should be disambiguated as a single entity, the Product method scores it as n disambiguations of the entity, as shown by the fourth row in Figure 1, where the score for the sequence "new york city" is $P(Concept_{New\_York\_City})$ to the third power.

5.2 Mixture Method

The Mixture method treats a set of possible disambiguations as a mixture of the AtomicConcepts that disambiguate terms in the set. The AtomicConcepts in M_{set} are given equal weight. Under the Mixture method, a set's score is simply equal to the average of the probability values of all Atomic Concepts in the set; once again, n adjacent terms that were disambiguated as referring to a single entity are counted as n disambiguations. For example, the fourth line of Figure 2 shows the sequence "new york city" being disambiguated as a single entity but the score in this case is just the probability of the concept *New_York_City* (ie., the average of the scores for three disambiguations of the sequence).

6 Evaluation

Our system, named Wikimantic, includes four alternative methods for disambiguation: the Product and Mixture methods with M as the topic model and the Product and Mixture methods with $(1-d)*M+d*M_R$ as the topic model.

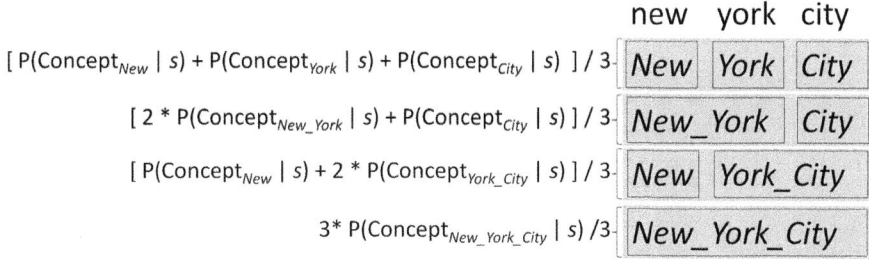

Fig. 2. Mixture method's scoring of four possible disambiguations of "new york city"

Each method was evaluated using 70 queries from the Trec 2007 QA track and 26 queries collected for our Information Graphic Retrieval project. The QA track was chosen because we intend to eventually incorporate Wikimantic into a larger system that operates on short grammatically correct full sentence questions, but it is worth noting that Wikimantic is in fact entirely agnostic to the grammatical structure of its input. The queries acquired from the Information Graphic Retrieval Project were collected from human subjects who were given information graphics and told to write queries they might have used to find them. All queries contain at least one (but usually more) salient word that must be disambiguated. The word count of each query is no less than 4 and no greater than 15. Out of the 850 words in the set, evaluators identified about 349 nouns (they disagreed on a couple due to ambiguous phrasing of the queries). About 110 words were content words that were not nouns. We present results for disambiguating just nouns and for disambiguating all non-function words.

To measure correctness, we gave the system results to two evaluators and instructed them to decide for each term whether the linked page correctly described the meaning of the word as it was used in the query. The general rule was that a disambiguation was wrong if a better page could be found for the term. For non-nouns, it was considered correct if a verb or adjective was linked to its noun-equivalent article. For example, it would be acceptable to annotate the term "defect" (to betray) with the page "Defection". If a term appeared in the query with a sense that has no equivalent article in Wikipedia, the evaluators were instructed to mercilessly mark the output wrong.

Tables 1 and 2 present statistics on precision and recall for the four methods. Precision is equal to the number of terms correctly mapped to concepts divided by the number of terms mapped by the system. Recall is equal to the number of correct mappings divided by the number of terms fed to the system.

Overall, the Product method fared better than the Mixture method, and performance was better on nouns than on non-nouns. With the Mixture method, it's possible for an obviously incorrect mapping to be offset by a high scoring one. With the Product method, a mapping with a near-zero probability will cause the score for the entire set to be near-zero. The Product method is therefore a more conservative scoring method that favors well rounded sets over sets with some likely and some unlikely references. The exceptional performance on nouns

Table 1. Performance on Nouns Only

Performance (Nouns Only)			
	Topic Model: M		Model: (1-d)*M+d*M_R
	Mixture	Product	Mixture	Product
Precision	78.71	80.51	81.38	82.76
Recall	75.21	76.93	77.65	79.08
F-Measure	76.92	78.68	79.47	80.88

Table 2. Performance on all Non-function Words

Performance (All Terms)			
	Topic Model: M		Model: (1-d)*M+d*M_R
	Mixture	Product	Mixture	Product
Precision	66.82	68.28	69.52	70.57
Recall	61.47	62.45	63.96	64.61
F-Measure	64.04	65.23	66.62	67.46

seems to be partly due to Wikipedia's greater coverage of nouns. Additionally, Wikimantic did not incorporate a stemmer, which occasionally prevented it from recognizing matches between alternate conjugations of the same verb.

For each of the two methods described in Section 5.1, we evaluated Wikimantic using the mixture $(1 - d) * M + d * M_R$ with varying values of d. Improved performance occurred for small values of d (d<.2). Although the optimal value of d was found to be very small (d = 0.0001), the effects of ReferenceRank were still surprisingly significant. MixtureConcepts often get weighted in such a way that one AtomicConcept has virtually all the weight, which gives it extremely high voting power. The top AtomicConcept's votes are then so powerful that they have disproportionate sway over the lesser AtomicConcepts. The small value of d works to correct for this.

Our results show that our system Wikimantic has very good success at disambiguating terms in short queries, even without capitalization or a priori identification of multi-word strings that should be mapped to a single concept.

7   Conclusion

In this paper, we presented a robust disambiguation method that performs well on short text fragments in which context words are scarce, that is not limited to nouns, that does not rely on correct capitalization, and that can determine when a sequence of words should be disambiguated as a single entity. Thus the approach will be useful in retrieval systems that must handle short user queries. Our disambiguation method used a two step process in which topic concepts were hypothesized via a local approach and refined with a global approach. Our experimental results show the success of the methodology and that a combination of M and M_R (ReferenceRank) has the potential to improve results, but that the disproportionate weighting of MixtureConcepts causes the top AtomicConcept to have too much voting power. Future work will explore a smoother method of weighting MixtureConcepts to overcome this problem.

Acknowledgments. This work uses Microsoft Web N-gram Services and was supported by the National Science Foundation under Grants III-1016916 and IIS-1017026.

References

1. Bunescu, R., Pasca, M.: Using Encyclopedic Knowledge for Named Entity Disambiguation. In: Proceedings of the 11th Conference of the European Chapter of the Association for Computational Linguistics, pp. 9–16. EACL, Trento (2006)
2. Fader, A., Soderland, S., Etzioni, O.: Scaling Wikipedia-based Named Entity Disambiguation to Arbitrary Web Text. In: WikiAI 2009 Workshop at IJCAI 2009 (2009)
3. Ferragina, P., Scaiella, U.: TAGME: On-the-fly annotation of short text fragments (by Wikipedia entities). In: Proceedings of the 19th ACM International Conference on Information and Knowledge Management, pp. 1625–1628. ACM, New York (2010)
4. Mihalcea, R.: Csomai. A.: Wikify!: Linking documents to encyclopedic knowledge. In: Proceedings of the Sixteenth ACM Conference on Conference on Information and Knowledge Management, pp. 233–242. ACM, New York (2007)
5. Milne, D., Witten, I.H.: Learning to Link with Wikipedia. In: Proceedings of the 17th ACM Conference on Information and Knowledge Management, pp. 509–518. ACM, New York (2008)
6. Page, L., Brin, S., Motwani, R., Winograd, T.: The PageRank Citation Ranking: Bringing Order to the Web. Technical report, Stanford InfoLab (1999)
7. Ratinov, L., Roth, D., Downey, D., Anderson, M.: Local and Global Algorithms for Disambiguation to Wikipedia. In: Proceedings of the 49th Annual Meeting of the Association for Computational Linguistics: Human Language Technologies, pp. 1375–1384. Association for Computational Linguistics (2011)
8. Elzer, S., Carberry, S., Zukerman, I.: The Automated Understanding of Simple Bar Charts. Artificial Intelligence 175(2), 526–555 (2011)
9. Wu, P., Carberry, S., Elzer, S., Chester, D.: Recognizing the Intended Message of Line Graphs. In: Goel, A.K., Jamnik, M., Narayanan, N.H. (eds.) Diagrams 2010. LNCS, vol. 6170, pp. 220–234. Springer, Heidelberg (2010)
10. Li, C., Sun, A., Datta, A.: A Generalized Method for Word Sense Disambiguation Based on Wikipedia. In: Clough, P., Foley, C., Gurrin, C., Jones, G.J.F., Kraaij, W., Lee, H., Mudoch, V. (eds.) ECIR 2011. LNCS, vol. 6611, pp. 653–664. Springer, Heidelberg (2011)

Polish Language Processing Chains
for Multilingual Information Systems[*]

Maciej Ogrodniczuk and Adam Przepiórkowski

Institute of Computer Science
Polish Academy of Sciences
ul. Jana Kazimierza 5, Warsaw, Poland
{maciej.ogrodniczuk,adam.przepiorkowski}@ipipan.waw.pl

Abstract. The ATLAS project, started in March 2010, intends to create
a multilingual language processing framework integrating the common set
of linguistic tools for a group of European languages, among them Polish. The chained tools producing multi-level UIMA-encoded annotation
of texts can be used by NLP applications for complex language-intensive
operations such as automated categorization, information extraction, machine translation or summarization.

This paper concentrates on applications of ATLAS language processing chains to multilingual information systems, with particular interest in
processing Polish. Inflectional characteristics of this language offers the
possibility to comment on a few more advanced functions such as multiword unit lemmatisation, vital for real-life presentation of extracted
phrases. Several sample applications using the NLP chain are also presented.

1 Introduction

The ATLAS project[1] [8] offered the possibility to test interoperability of several
NLP tools for Polish together with other project languages — Bulgarian, English,
German, Greek and Romanian — in multilingual information systems. After selecting the common integration and annotation framework, the language tools
capable of providing necessary linguistic information have been evaluated and
adapted to form ready-to-use processing chains. Currently three linguistically-
aware online services make use of this functionality: i-Publisher (Web-based content management system), i-Librarian (a digital library of scientific works) and
EUDocLib/PLDocLib sites (for browsing and searching through EUR-LEX documents and, respectively, acts of Polish Parliament).

---

[*] The work reported here was carried out within the Applied Technology for Language-
Aided CMS project co-funded by the European Commission under the Information
and Communications Technologies (ICT) Policy Support Programme (Grant Agreement No 250467).
[1] See http://www.atlasproject.eu/ for a detailed information about the project.

G. Bouma et al. (Eds.): NLDB 2012, LNCS 7337, pp. 152–157, 2012.

2   Language Processing Chains

Combining diverse tools into a single framework required selection of an integration platform for linguistic annotation. Unstructured Information Management Architecture (UIMA) had been chosen from among other reputable architectures (such as General Architecture for Text Engineering), mainly because its potential for scalability achievable by decomposition of language processing applications into components replicable over a cluster of network nodes.

The NLP tools have been integrated into UIMA annotation framework by means of wrapping them into UIMA-compatible primitive engines, chained into aggregate engines. In certain cases it required their technical adaptation to continuous use (without frequent loading of models, processing rules, etc.) and enforcing their thread-safety.

Besides, wrappers were designed to maintain a uniform type system developed for ATLAS, with document-level and text-level properties, the latter comprising annotations of paragraphs, tokens (POS tags and morphosyntactic categories), noun phrases (with semantic heads) and named entities.

3   Polish Language Tools

A number of existing tools for the processing of Polish were used within the project. According to the agreed annotation model, supporting the target application of linguistic data, particularly their display to the user, the tools were in certain cases reconfigured and extended to provide additional information such as ready-to-display normalized versions of base forms of identified multiword noun phrases.

3.1   Segmentation and Morphosyntactic Tagging

Sentence- and token-level segmentation information is provided by Morfeusz [16], also a lemmatizer and morphological analyzer for Polish. It uses positional tags starting with POS information followed by values of morphosyntactic categories corresponding to the given part of speech [12]. Current version of the tool, Morfeusz SGJP, is based on linguistic data coming from The Grammatical Dictionary of Polish [13].

Morphosyntactic disambiguation is performed by Pantera [1], a rule-based tagger of Polish using an optimized version of Brill's algorithm adapted for specifics of inflectional languages. The tagging is performed in two steps, with a smaller set of morphosyntactic categories disambiguated in the first run (part of speech, case, person) and the remaining ones in the second run. Due to the free word order nature of Polish, the original set of rule templates as proposed by Brill has been extended to cover larger contexts.

3.2   Noun Phrase Extraction and Multi-word Expression Lemmatisation

Noun phrases are identifier by Spejd [2], a shallow parsing engine using cascade grammars, able to co-operate with various taggers of Polish, including Pantera (see above). In Spejd parsing rules are defined using a cascade of regular grammars which match against orthographic forms, base forms or morphological interpretations of particular words. Spejd's specification language is used, which supports a variety of actions to perform on the matching fragments: accepting and rejecting morphological interpretations, agreement of entire tags or particular grammatical categories, grouping (syntactic and semantic head may be specified independently), etc. Users may provide custom rules or may use one of the provided sample rule sets.

Apart from identifying noun phrases, the Spejd grammar of Polish [6] created within the National Corpus of Polish [11] is the basis for the nominal groups lemmatiser [4] developed throughout the project which combines the lemmatisation task with shallow parsing. The parsing structures are used in lemmatisation schemata written separately for each grammar rule and operating on the matched strings and structure.

3.3   Named Entity Recognition

Identification of named entities is supported by NERF tool [14] – a statistical CRF-based named entity recognizer trained over 1-million manually annotated subcorpus of the National Corpus of Polish and successfully used in the process of automated annotation of its total 1,5 billion segments.

NERF annotation model is consistent with general requirements of the ATLAS framework, defined to cover dates, money, percentage and time expressions, names of organizations, locations and persons. Normalized versions of entities are provided to facilitate extraction and comparisons (e.g. values conforming to xsd:date and xsd:time types for date/time expressions and ISO currency codes for money expressions).

4   Current Work

The project targets at practical approach to numerous advanced linguistic issues such as coreference resolution, summarization, machine translation or categorization, often creating synergies with other ongoing initiatives. Initial versions of tools providing the above-mentioned operations are scheduled to be integrated by the time of NLDB 2012.

4.1   Text Summarization and Coreference Resolution

Despite intensive worldwide research in the field, just a few general summarization systems for Polish have been implemented so far. The first of them was PolSumm2 [3], a modular, text extraction-based system monitoring inter-sentence relations, anaphors and ellipses. In 2007 Lakon [5], a heuristic summarizer used for testing various sentence selection methods was implemented and in 2010 – the extraction-based machine learning system of Świetlicka [15]. They are currently being evaluated against shallow language-independent clause- and marker-based extractive summarizer implemented by Romanian partners and adapted to ATLAS.

To achieve better results, a coreference resolution module is currently being integrated into the summarization engine with two approaches being evaluated. The first of them is a language-neutral coreference resolver RARE [10], the second one is the results of the first attempts of coreference resolution for Polish carried out within the *Computer-based methods for coreference resolution in Polish texts* project financed by the Polish National Science Centre. Currently used end-to-end implementation [9] adopts a rich rule-based approach, integrating syntactic constraints (elimination of nested nominal groups), syntactic filters (elimination of syntactic incompatible heads), semantic filters (wordnet-derived compatibility) and selection (weighted scoring).

4.2 Categorization and Machine Translation

General language-independent categorization tools for heterogeneous domains have been integrated into ATLAS. The engine employs different categorization algorithms, such as Naïve Bayesian, relative entropy, Class-Feature Centroid, Support Vector Machines and Latent Dirichlet Allocation, with results consolidated by a voting system.

The machine translation engine currently being evaluated combines an example-based component with a statistical, domain-factored approach powered by Moses [7]. After the most appropriate translation model and example-based sub-component are selected based on the results from categorisation engine, the translation database and the statistical component are used to provide the translation output. Before the engine reaches its maturity, a third-party solution using Microsoft Bing is being used by the demo interfaces described below.

5 Polish Linguistic Chain-Based NLP Demo Applications

The UIMA linguistic annotation chains for Polish have been tested together with higher-level linguistic functions such as summarization, categorization and machine translation in a several sample Web site interfaces implemented as demonstrations of the technology.

5.1 i-Librarian and EUDocLib

i-Librarian (http://www.i-librarian.eu/) is a free online library that assists authors, students, young researchers, scholars, librarians and executives to easily create, organise and publish various types of documents; EUDocLib (http://eudoclib.atlasproject.eu/) is a publicly accessible repository of EU documents from the EUR-LEX collection which provides easier access to relevant documents in the user's language.

Both sites are capable of processing documents in supported languages in order to automatically categorize, summarize and annotate content with important noun phrases and named entities. They also provide annotation-based content navigation (such as list of similar documents) and machine-translated excerpts of documents used for document categorization and clustering.

5.2   PLDocLib

PLDocLib (http://www.atlasproject.eu/pl/) is a language processing chain-powered Web site offering full-text search and category-based browsing of around 1000 acts of Polish Sejm. For each document a set of recognized named entities, automatically clustered important noun phrases (with their weights) and a list of similar documents is produced. For presentation, base forms of multiword units are generated and manually assigned categories (retrieved from the document source) are used.

6   Conclusions

While the number of resources and tools for European languages grows rapidly, their interoperability leaves much to be wished for, which hinders development of multilingual information systems. In this paper we reported on a practical exercise in making Polish language processing tools interoperable.

At the technical level, the interoperability is ensured by integrating the tools within the UIMA platform. In order to do that, the tools themselves did not need to be substantially modified, but appropriate wrappers around them had to be implemented.

At the more interesting linguistic level, the tools were created with the intention of using them in sync, although they had never before been combined into a processing chain like the one described here. In particular, the Pantera tagger assumes positional tagsets of the kind employed by the morphological analyser Morfeusz, the shallow parsing system Spejd can in principle deal with any morphosyntactic tagsets, but frontends exist reading the format produced by tools like Pantera and, moreover, the grammar developed within the National Corpus of Polish assumes the same tagset (by now standard in Polish NLP). Although Polish NER tools were developed in the same project, they have not been used together with the shallow parser so far.

The deployment of these tools in ATLAS applications serves as an important proof of concept that the intended interoperability of these tools is indeed possible and relatively straightforward.

References

1. Acedański, S.: A Morphosyntactic Brill Tagger for Inflectional Languages. In: Loftsson, H., Rögnvaldsson, E., Helgadóttir, S. (eds.) IceTAL 2010. LNCS, vol. 6233, pp. 3–14. Springer, Heidelberg (2010)

 2. Buczyński, A., Przepiórkowski, A.: Spejd: A Shallow Processing and Morphological Disambiguation Tool. In: Vetulani, Z., Uszkoreit, H. (eds.) LTC 2007. LNCS (LNAI), vol. 5603, pp. 131–141. Springer, Heidelberg (2009)
 3. Ciura, M., Grund, D., Kulików, S., Suszczanska, N.: A System to Adapt Techniques of Text Summarizing to Polish. In: Okatan, A. (ed.) Proceedings of the International Conference on Computational Intelligence (ICCI 2004), pp. 117–120. International Computational Intelligence Society, Istanbul (2004)
 4. Degórski, Ł.: Towards the Lemmatisation of Polish Nominal Syntactic Groups Using a Shallow Grammar. In: Bouvry, P., Kłopotek, M.A., Leprévost, F., Marciniak, M., Mykowiecka, A., Rybiński, H. (eds.) SIIS 2011. LNCS, vol. 7053, pp. 370–378. Springer, Heidelberg (2012)
 5. Dudczak, A., Stefanowski, J., Weiss, D.: Evaluation of Sentence-Selection Text Summarization Methods on Polish News Articles. Foundations of Computing and Decision Sciences 1(35), 27–41 (2010)
 6. Głowińska, K., Przepiórkowski, A.: The design of syntactic annotation levels in the National Corpus of Polish. In: Proceedings of the 7th International Conference on Language Resources and Evaluation, LREC 2010. ELRA, Valletta (2010)
 7. Koehn, P., Hoang, H., Birch, A., Callison-Burch, C., Federico, M., Bertoldi, N., Cowan, B., Shen, W., Moran, C., Zens, R., Dyer, C., Bojar, O., Constantin, A., Herbst, E.: Moses: open source toolkit for statistical machine translation. In: Proceedings of the 45th Annual Meeting of the ACL on Interactive Poster and Demonstration Sessions, ACL 2007, pp. 177–180 (2007)
 8. Ogrodniczuk, M., Karagiozov, D.: ATLAS — The Multilingual Language Processing Platform. Procesamiento del Lenguaje Natural 47, 241–248 (2011)
 9. Ogrodniczuk, M., Kopeć, M.: End-to-end coreference resolution baseline system for Polish. In: Proceedings of the 5th Language & Technology Conference (LTC 2011), Poznań, Poland, pp. 167–171 (2011)
10. Postolache, O.: RARE: Robust Anaphora Resolution Engine. Master's thesis, University of Iasi (2004)
11. Przepiórkowski, A., Górski, R.L., Łazinski, M., Pęzik, P.: Recent developments in the National Corpus of Polish. In: Proceedings of the 7th International Conference on Language Resources and Evaluation, LREC 2010. ELRA, Valletta, Malta (2010)
12. Przepiórkowski, A., Woliński, M.: A Flexemic Tagset for Polish. In: Proceedings of Morphological Processing of Slavic Languages, EACL 2003 (2003)
13. Saloni, Z., Gruszczyński, W., Woliński, M., Wołosz, R.: Grammatical Dictionary of Polish – Presentation by the Authors. Studies in Polish Linguistics 4, 5–25 (2007)
14. Savary, A., Waszczuk, J., Przepiórkowski, A.: Towards the Annotation of Named Entities in the National Corpus of Polish. In: Proceedings of the 7th International Conference on Language Resources and Evaluation, LREC 2010, Valletta, Malta, ELRA (2010)
15. Świetlicka, J.: Machine learning methods in automatic text summarization (in Polish). Master's thesis, Warsaw University, Poland (2010)
16. Woliński, M.: Morfeusz – a practical tool for the morphological analysis of Polish. In: Kłopotek, M.A., Wierzchoń, S.T., Trojanowski, K. (eds.) Proceedings of the International Intelligent Information Systems: Intelligent Information Processing and Web Mining 2006 Conference, Wisła, Poland, pp. 511–520 (2006)

GPU-Accelerated Non-negative Matrix Factorization for Text Mining

Volodymyr Kysenko[1], Karl Rupp[2,3], Oleksandr Marchenko[1],
Siegfried Selberherr[2], and Anatoly Anisimov[1]

[1] Faculty of Cybernetics, Taras Shevchenko National University of Kyiv, Ukraine
[2] Institute for Microelectronics, TU Wien, Austria
[3] Institute for Analysis and Scientific Computing, TU Wien, Austria

Abstract. An implementation of the non-negative matrix factorization algorithm for the purpose of text mining on graphics processing units is presented. Performance gains of more than one order of magnitude are obtained.

1 Introduction

The automatic extraction of high-quality information from text, typically referred to as text mining, has received a lot of attention in many areas. Due to the vast amount of text to be processed, there is a virtually insatiable need for computational power. We address this demand by an implementation of the non-negative matrix factorization (NMF) algorithm [1] and its application to document clustering on graphics processing units (GPUs).

In contrast to traditional central processing units (CPUs), the computational power of modern GPUs already ranges into the tera-floating point operations per second (TFLOPs) regime. The higher flexibility of modern GPU architectures also allows for the use of GPU-hardware for non-graphics application. As a consequence, such general purpose computations on graphics processing units (GPGPUs) have gained a lot of popularity recently. Significant performance gains within scientific applications have been reported in many different applications, e.g. [2].

The high computational power of GPUs can only be accessed, if the underlying algorithm provides a sufficiently high degree of fine-grained parallelism in order to occupy thousands of light-weight threads simultaneously. Since this constitutes a considerable paradigm-shift compared to single-threaded implementations, existing codes often need to be rewritten in order to benefit from GPU acceleration. This problem is at least partially addressed by software libraries, which provide basic functionalities via a high-level interface. In the context of the NMF algorithm, basic linear algebra operations are required, which are well-studied and well-developed for GPUs [5]. A number of different libraries exists and most of them rely on CUDA [3] technology, for example, CUBLAS and MAGMA. In our work the C++ OpenCL-based [4] Vienna Computing Library (ViennaCL) [6,7] is used. It offers a convenient means to run custom compute

G. Bouma et al. (Eds.): NLDB 2012, LNCS 7337, pp. 158–163, 2012.

kernels not provided with the library and allows for a comparison on a broader range of GPUs from different vendors.

Aspects of document clustering and their link to matrix decompositions are discussed in Sec. 2. A formal description of the NMF algorithm is given in Sec. 3, whereas implementation details are discussed in Sec. 4. Results are discussed in Sec. 5 and a conclusion is drawn in Sec. 6.

2   Document Clustering

Clustering is a process of collecting a set of objects into subsets (clusters) such that objects inside a cluster are in a certain predefined sense more similar to each other than to objects from other clusters.

In this work we consider the application of clustering for grouping documents by topics. More formally, for a set of m text documents, the goal of clustering by topics is to find a division of documents into groups such that documents from one group share common topics. It is crucial to note that the list of topics is typically not known prior to the clustering process. Therefore, such a clusterization can also be interpreted as an automatic document categorization and topics extraction process.

Traditional methods of document clustering are often based on the vector space model (VSM) [9]. In this model every document is represented as a term-frequency vector. Therefore, VSM uses words as a measure for similarity between different documents. Various metrics can be defined for the vectors of terms, for example a cosine distance or an Euclidian distance. Due to this representation of documents in terms of linear algebra, traditional clustering techniques can be applied.

One of the major disadvantages of the VSM model is that terms are assumed to be statistically independent from each other. However, this is often not the case in real-world texts, where topics, concepts, and semantics are key features of each document. In order to extract these key features from the texts, special techniques referred to as *feature extraction* have been developed. The goal of such methods is an extraction of core concepts from the individual texts and a representation of documents as a combination of these. Singular value decomposition [10] or NMF [12] are often used as a basis for feature extraction methods. Methods of this type are also known as low-rank matrix approximation or latent semantic analysis. Further note that applications of these methods are not limited to document clustering. They has been successfully used for solving a wide range of natural language processing problems such as cross-language retrieval, information indexing [10], or selectional preference induction [11].

For the remainder of this work only NMF is considered. Generally, NMF is a decomposition of a matrix V into a product of two matrices W and H, where the additional constraint of non-negativity of the entries in the matrices V, W, H is imposed. It is important to note that such a decomposition is not necessarily unique and neither W nor H need to be orthogonal. NMF factorization became popular with the publication of Lee and Seung [1], where two algorithms for computing non-negative matrix factorizations are proposed.

Other methods such as the projected gradient method [13], alternating non-negative least squares [14], or sparse encoding [15] were proposed later. For this paper, the original Lee-Seung method, which is also known as *multiplicative update rules algorithm*, with the Frobenius norm as a cost function is considered.

In text mining NMF is usually applied to the term-document matrix (TD matrix). Every row in the TD matrix corresponds to one document, while every column of the matrix corresponds to one term. For m documents and a total number of n terms, the TD matrix V consequently is of size $m \times n$. NMF is used to decompose V into two matrices W and H with sizes $m \times k$ and $k \times n$ respectively, where typically $k \ll \min(m, n)$ is set to the expected number of clusters. In the following we will refer to the parameter k as the *number of features*. Intuitively, k can be interpreted as follows: The TD matrix V ('documents' $\times$ 'terms') is decomposed into a product of two matrices. The matrix W relates documents and features, while H relates features and terms. As a result of matrix product properties, every document is thus presented as a linear combination of the extracted features. Traditional clustering techniques[8] can therefore be applied to the rows of W.

NMF as a tool for natural language processing possesses several advantages over other feature extraction methods. First, the matrices W and H have non-negative entries only, which results in an easier interpretation in terms of text mining. Second, the columns of W do not need to be orthogonal. Hence, the extracted topics are allowed to share common senses, which seems to be quite usual for real-world documents.

3   The Non-negative Matrix Factorization Algorithm

In the following, the NMF algorithm proposed by Lee and Seung [1] using the Frobenius norm is described. The objective function is

$$\min_{W,H} ||V - WH||_F^2 , \tag{1}$$

where the entries of W and H need to be non-negative. Note that the minimum in (1) is typically non-zero. Given arbitrary initial matrices W_0 and H_0, the NMF algorithm consists of an iterative application of the following two steps:

$$(H_k)_{i,j} = (H_{k-1})_{i,j} \times \frac{(W_{k-1}^T V)_{i,j}}{(W_{k-1}^T W_{k-1} H_{k-1})_{i,j}} , \tag{2}$$

$$(W_k)_{i,j} = (W_{k-1})_{i,j} \times \frac{(V H_{k-1}^T)_{i,j}}{(W_{k-1} H_{k-1} H_{k-1}^T)_{i,j}} . \tag{3}$$

Here $(\cdot)_{i,j}$ refers to the entry in row i and column j of the matrix in parentheses.

In practise, (2) and (3) are repeated until either a stationary point or a maximum number of iterations is reached. Lee and Seung proved two main properties of this algorithm. First, the objective function (1) is non-increasing with k. Second, W and H become constant, if and only if they represent a stationary point.

4   Implementation

The iterative NMF algorithm (2) and (3) can almost directly be implemented with the features provided by ViennaCL. Two types of matrix operations are required: matrix-matrix-multiplications and element-wise manipulations. While the former are provided by ViennaCL directly, the latter requires a simple custom OpenCL kernel for setting up the matrix indicated by parentheses in (2) and (3).

In the typical case, where k is small compared to m and n, the computationally most expensive operations are the matrix-matrix products with V. This is the case, because only V is of size $m \times n$, while at least one of the dimensions of all other matrices are given by k.

In practical applications it is often observed that the TD matrix V is sparse. As a consequence, computational efficiency can be improved substantially and memory requirements can be reduced significantly by exploiting the structural information. Memory consumption is particularly a concern when using GPUs due to the typically limited amount of memory on GPU adapters compared to main memory on the host machine. Thus, by using a sparse matrix storage format such as the compressed sparse row format (CSR)[16], we are able to process a much bigger set of documents than in the case of a dense matrix type.

It is important to note that (2) and (3) in principle require products with V from the left and from the right. However, the CSR-format used for the storage of V allows for an efficient implementation of the matrix-matrix product only, if V is multiplied from the left, which is not the case for the operation $W^T V$. There are two possible remedies in this case: The first is to store V in a compressed column format in addition. Besides additional memory, this requires a dense-matrix-sparse-matrix multiplication kernel. The second option is to set up and store V^T in CSR format in addition. Rewriting $W^T V = ((W^T V)^T)^T = (V^T W)^T$ enables the reuse of the multiplication kernel, but another entry-wise manipulation kernel must be provided. Since the entry-wise manipulation kernel is simpler, we stick with the second option.

5   Benchmark Results

For our experiments we have used recent mid- to high-end consumer hardware: An Intel Core i7 960 CPU with 3.2 GHz clock freqency, an NVIDIA GeForce 470 GTX GPU, and an AMD Radeon 6970 HD GPU.

For a comparison of our GPU-accelerated implementation we developed a purely CPU-based version of the NMF algorithm. The goal of this version is not only performance numbers, but also validation of numerical accuracy. The open-source linear algebra library Eigen [17] was used for this purpose. The CPU version was compiled with maximal optimization level and with support for SIMD instructions (SSE2) and OpenMP enabled.

Our benchmarks include both real-world and artificially generated matrices. As a benchmark based on real-world data, the *Newsgroups (NG)* dataset [18] is used. Texts in this dataset are messages gathered from 20 different newsgroups

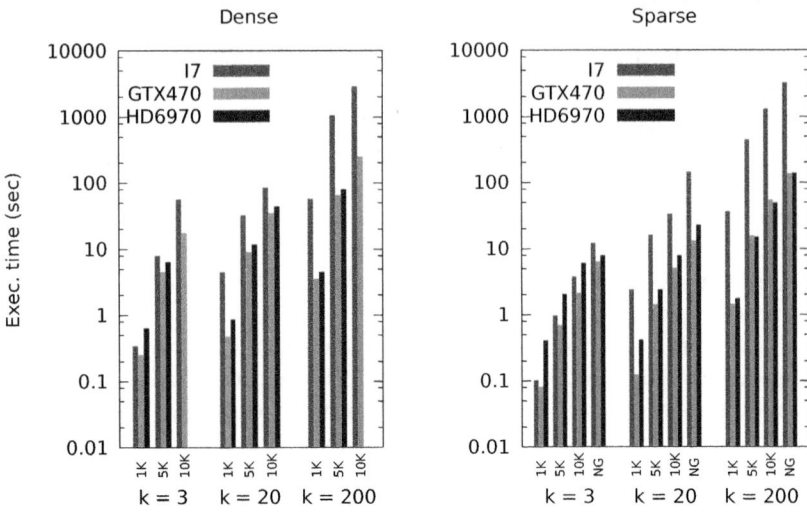

Fig. 1. Execution times for all test matrices on different platforms

with a total number of 18827 documents. Due to the nature of the newsgroup messages, which are usually quite short, the TD matrix V for this dataset is very sparse. The resulting size of the term-document matrix is $18\,827 \times 87\,014$, while the number of non-zero elements is $1\,553\,867$. Note that a dense matrix of the same dimensions would require the storage of more than one billion entries, which may already exceed the memory available in current desktop computers. As an artificial benchmark we generated a set of random matrices, with sizes varying from 1000×1000 (1K) to 10000×10000 (10K). All matrices from this set have about 1% of non-zero elements, which allows for the use in both dense and sparse tests. In both cases values of 3, 20, and 200 are used for the number of features k.

Benchmark results of our experiments with all matrices are shown in Fig. 1. It can be seen that the use of a sparse matrix type instead of a dense matrix type leads to a performance gain on both CPU and GPU. While the gain of using a sparse matrix type is about one order of magnitude for $k = 3$ on the CPU, the difference is only a factor of two for $k = 200$. On GPUs the performance difference is around five to ten for $k = 3$ and $k = 20$, and around a factor of three to five on both the NVIDIA and the AMD GPU. In the latter case, the dense matrix-matrix multiplication on GPUs achieves higher performance due to the more uniform size of the matrices involved, thus the difference is smaller. A comparison of execution times for CPU and GPUs on all test matrices reveals a difference in execution times by a factor of up to 25 for the largest matrices and $k = 200$. For $k = 3$ a performance gain from the use of GPUs of only a factor of up to two is obtained, whereas $k = 20$ leads to a gain of almost one order of magnitude.

6   Conclusion

Our investigations of the acceleration of the NMF algorithm by GPUs for the use in text mining tasks show that a performance gain of more than one order of magnitude can be obtained. Consequently, modern GPU hardware must not be ignored for such purposes, whenever performance is of importance.

Large parts of the NMF algorithm are ported to GPUs by reusing high-level functionality already provided with common GPU libraries such as ViennaCL. Therefore, the entry-barrier to GPU computing is lower than it may appear at first sight. Most linear algebra functionality is ready to be reused within other algorithms commonly employed in the field of text mining.

Our implementation of the NMF algorithm presented in this work is to be made freely available with ViennaCL 1.3.0.

References

1. Lee, D.D., Seung, H.S.: Algorithms for Non-Negative Matrix Factorization. In: NIPS (2000)
2. Hwu, W.W. (ed.): GPU Computing Gems. MKP (2011)
3. NVIDIA CUDA, http://www.nvidia.com/
4. Khronos Group. OpenCL, http://www.khronos.org/opencl/
5. Agullo, E., et al.: Numerical Linear Algebra on Emerging Architectures: The PLASMA and MAGMA projects. J. Phys.: Conf. Ser. 180(1) (2009)
6. ViennaCL, http://viennacl.sourceforge.net/
7. Rupp, K., et al.: ViennaCL - A High Level Linear Algebra Library for GPUs and Multi-Core CPUs. In: Proc. Intl. Workshop on GPUs and Scientific Applications (GPUScA 2010), pp. 51–56 (2010)
8. Xu, R., Wunsch II, D.: Survey of Clustering Algorithms. IEEE Trans. on Neural Networks 16(3), 645–678 (2005)
9. Salton, G., et al.: A Vector Space Model for Automatic Indexing. Communications of the ACM 18(11), 613–620 (1975)
10. Deerwester, S., et al.: Indexing by Latent Semantic Analysis. Journal of the American Society for Information Science 41(6), 391–407 (1990)
11. Van de Cruys, T.: A non-negative tensor factorization model for selectional preference induction. In: Proc. Workshop on Multiword Expressions, pp. 83–90 (2009)
12. Xu, W., et al.: Document Clustering Based on Non-Negative Matrix Factorization. In: Proc. 26th Intl. Conf. Research and Development in Information Retrieval, pp. 267–273 (2003)
13. Pauca, V., et al.: Text Mining Using Non-Negative Matrix Factorizations. In: Proc. 4th SIAM Intl. Conf. Data Mining (2004)
14. Amy, L., Carl, M.: ALS Algorithms Nonnegative Matrix Factorization Text Mining. SAS NMF Day (2005)
15. Hoyer, P.: Non-Negative Sparse Coding. In: Proc. IEEE Workshop on Neural Networks for Signal Processing (2002)
16. Saad, Y.: Iterative Methods for Sparse Linear Systems. SIAM (2003)
17. An Overview of Eigen. First Plafrim Scientific Day. Bordeaux (May 31, 2011), http://eigen.tuxfamily.org/
18. 20 Newsgroups, http://people.csail.mit.edu/jrennie/20Newsgroups/

Generating SQL Queries Using Natural Language Syntactic Dependencies and Metadata

Alessandra Giordani and Alessandro Moschitti

Department of Computer Science and Engineering
University of Trento
Via Sommarive 14, 38100 POVO (TN) - Italy
{agiordani,moschitti}@disi.unitn.it

Abstract. This research concerns with translating natural language questions into SQL queries by exploiting the MySQL framework for both hypothesis construction and thesis verification in the task of question answering. We use linguistic dependencies and metadata to build sets of possible SELECT and WHERE clauses. Then we exploit again the metadata to build FROM clauses enriched with meaningful joins. Finally, we combine all the clauses to get the set of all possible SQL queries, producing an answer to the question. Our algorithm can be recursively applied to deal with complex questions, requiring nested SELECT instructions. Additionally, it proposes a weighting scheme to order all the generated queries in terms of probability of correctness.

Our preliminary results are encouraging as they show that our system generates the right SQL query among the first five in the 92% of the cases. This result can be greatly improved by re-ranking the queries with a machine learning methods.

Keywords: Natural Language Processing, Question Answering, Metadata, Information Schema, SQL.

1 Introduction

NLIDB propose a large body of work based manual work for grammar specification and dataset annotation. However the task of question answering, translating natural language (NL) into something understandable by a machine, in an automatic way is rather challenging as it is not possible to hand-crafting all the needed rules.

We address this problem by generating SQL queries whose structure and components match with NL concepts (expressed as words) and grammar dependencies. Our new idea consists on how and where this matching can be found, i.e., exploiting existing knowledge that comes along with each database (metadata). The resulting matching between metadata and words allows for building sets of query components (clauses). These are combined together using a smart algorithm to generate a set of SQL queries, taking into account also the structure of the starting NL.

G. Bouma et al. (Eds.): NLDB 2012, LNCS 7337, pp. 164–170, 2012.

1.1 Motivations and Problem Definition

A database is not just a collection of data: at design time, domain experts organize entities and relationships giving proper names to tables and columns, defining constraints and specifying the type of the stored data. This additional information is known as metadata and is stored in an underlying database called INFORMATION_SCHEMA (IS, for brevity) that contains, for each database, tables containing columns referring to table names and column names. This self-reference allows for querying metadata with the same technique and technology used to query the database embedding data that answer a given question. In other words, we can execute several SQL queries over IS and a target database and combine their result sets to generate the final SQL query in a straightforward and very intuitive way.

In addition, we can perform question answering against multiple databases, since IS stores metadata of all databases that reside in a single machine. Moreover there is no need to use tailored dictionaries since metadata already embeds a rich knowledge based on the experience of the domain expert that designed the database. Another important feature that should be taken into account is the potential complexity of the NL question (subordinates, conjunction, negation) that can find its own matching in nested SQL query. The general SQL query that our system can deal with has the form:

$$\text{SELECT DISTINCT } COLUMN \text{ FROM } TABLE \text{ [WHERE } CONDITION] \quad (1)$$

We can *project* a single $COLUMN$ and eventually apply aggregation operators that summarize it by means of SUM, AVG, MIN, MAX and COUNT. Optionally, we can also *select* data for which a $CONDITION$ holds. This is represented as a logical expression where basic conditions, in the form e_L OP e_R, with OP=$\{<,>,\text{LIKE,IN}\}$, may be combined with AND, OR, NOT operators. While e_L (as well as $COLUMN$) is always in the form `table.column`, e_R could be a numerical value, a string value or a nested query. The meaning of $TABLE$ is more straightforward, since it should contain table name(s) to which the other two clauses refer. This clause could just be a single table or be a join operation, which selectively pairs tuples of two tables.

We found a mapping algorithm that matches dependencies between NL components and SQL structure that allows to build a set of possible queries that answers a given question. To represent textual relationships of the NL sentence we use typed dependency relations. The Stanford Dependencies Collapsed (SDC) representation [1] provide a simple and uniform description of binary grammar relations holding between a governor and a dependent (each dependency is written as $rel(gov, dep)$ where gov and dep are words in the sentence).

The question answering task of finding an SQL query Q that retrieves an answer for a given NL question q reduces to the following problem. Given question q represented by means of its typed dependencies collapsed SDC_q, generate the three sets of clauses $\mathcal{S}, \mathcal{F}, \mathcal{W}$ such that the set $\mathcal{A}$ =SELECT $\mathcal{S}\times$ FROM $\mathcal{F}\times$ WHERE $\mathcal{W}$ contains all possible answers to q and select the one that maximizes the probability of being the correct answering query Q. In the next section, we show how we can efficiently generate such triples.

$$(1)root(\text{ROOT, are}),$$
$$(2)nsubj(\text{are, capital}),$$
$$(3)prep\_of(\text{capital, state}),$$
$$(4)nsubj(\text{border, state}),$$
$$(5)rcmod(\text{state, border}),$$
$$(6)advmod(\text{populat, most}),$$
$$(7)amod(\text{state, populat}),$$
$$(8)dobj(\text{border, state})$$

$$\Pi = \{\text{capital, state}\}$$
$$\Sigma = \{\text{are}\} \Rightarrow \Sigma = \phi$$

$$\Pi' = \{\text{state, border}\}$$
$$\Sigma' = \{\text{border, state}\}$$

$$\Pi'' = \{\text{most, populat, state}\}$$
$$\Sigma'' = \phi$$

Fig. 1. Categorizing stems into projection and/or selection oriented sets

2 Building Clauses Sets

The first step before generating all possible queries for a question q is to create their components $\mathcal{S}, \mathcal{F}, \mathcal{W}$, i.e. SELECT, FROM and WHERE clauses, starting from a dependency list SDC_q. This list should be (a) preprocessed using pruning, stemming and adding synonyms, (b) analyzed to create the set of stems used to build $\mathcal{S}$ and $\mathcal{W}$ and (c) modified/cleaned to keep dependency used in a eventual recursive step in order to generate nested queries.

First we prune those relations that are useless for our processing and then reduce *govs* and *deps* to stems [2] to obtain the optimized list SDC_q^{opt}. An example showing $SDC_{q_1}^{opt}$ with respect to question q_1: *"What are the capitals of the states that border the most populated state?"* can be found in Figure 1.

Then for each grammatical relation in SDC_q^{opt} we apply an iterative algorithm that adds these stems respectively to Π and/or Σ categories accordingly to a set rules that for lack of space cannot be listed here. However, the key idea is to exploit projection-oriented relations (e.g. *ROOT* and *nsubj*) and selection-oriented ones (e.g. *prep* and *obj*) to categorize stems recursively.

Next, we use Π to search in metadata all fields that could match with projection-oriented stems. Based on how many matchings are found, a weight w is inferred to each projection, obtaining the SELECT clause set $\mathcal{S}$.

Instead, the selection-oriented set of stems Σ should be divided into two distinct sets of stems Σ_L and Σ_R. The set Σ_L contains stems that find their matching in IS, whereas for remaining stems $\Sigma_R = \Sigma - \Sigma_L$ we should look up in the database to find a matching. In order to build the WHERE clauses set $\mathcal{W}$, $\forall e_L \in \mathcal{W}_L, \forall e_R \in \mathcal{W}_R$ we first generate basic expressions $expr = e_L \text{ OP }_R$ and combine them by means of conjunction and negation, keeping only those expressions $expr$ such that the execution of $\pi_{count(*)}\left(\sigma_{expr}(table)\right)$[1] does not lead to an error for at least a *table* in the database.

It is worth nothing that Σ_R could be the empty set, e.g. when a WHERE condition requires nesting; in this case e_R will be the whole subquery. Moreover, also Σ_L could be empty. This is not surprising since, in a SQL query the WHERE clause is not mandatory. However the absence of selection-oriented stems doesn't necessarily mean that $\mathcal{W}$ should be empty. When this happens all tables and

---

[1] π and σ represent projection and selection operators of relational algebra.

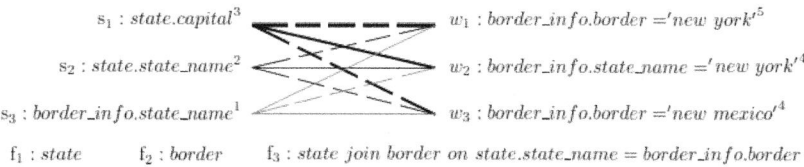

$s_1 : state.capital^3$ $w_1 : border\_info.border ='new\ york'^5$

$s_2 : state.state\_name^2$ $w_2 : border\_info.state\_name ='new\ york'^4$

$s_3 : border\_info.state\_name^1$ $w_3 : border\_info.border ='new\ mexico'^4$

$f_1 : state$ $f_2 : border$ $f_3 : state\ join\ border\ on\ state.state\_name = border\_info.border$

Fig. 2. Possible pairings for q_2: *"Capitals of states bordering New York"*

columns of the database are taken into account to find valid conditions: $W^*_R = \{t.c|t \in \pi_{table\_name}(IS.Columns)$ and $c \in \pi_{column\_name}(IS.Columns)\}$.

Last, once the two sets S and W have been constructed, the generation of the FROM clause $\mathcal{F}$ is straightforward. This set should just contain all tables to which clauses in S and W refer, enriched by pairwise joins. Again, this task could be performed running SQL queries over IS, and especially exploiting metadata stored in table KEY_COLUMN_USAGE. This table identifies all columns in current databases that are restricted by primary/foreign key constraints.

3 Generating Queries

The starting point for finding an answering query is to generate the set $\mathcal{A} = \{S \times \mathcal{F} \times W\} \cup \{S \times \mathcal{F}\}$. If such query exists there should be a pairing $\langle s, f, w \rangle \in \mathcal{A}$, such that the execution of SELECT s FROM f [WHERE w] retrieve the correct answer. Given that on average each clause set contains up to ten items, their product could result in a very huge set. Actually when generating all pairs some preliminary conditions are verified, e.g. tables appearing in SELECT and WHERE clauses should as well appear in the FROM clause, otherwise the execution of that query will fail. This avoids to generate incorrect queries and to waste time trying to execute them.

At this point the set $\mathcal{A}$ contain all valid pairings, among which there are still someone not useful, e.g. meaningless queries: those projecting the same field compared to a value in the selection. For example the pairing $\langle s_3, f_2, w_2 \rangle$ in Figure 2 answers the question *"Which state is Texas?"* that is clearly useless.

In particular there are redundant queries that once optimized can lead to a duplicate in the set; hence its cardinality is lower than in theory. For example, the pairing $\langle s_2, f_3, w_1 \rangle$ involves that the columns *state.state_name* and *border_info.border* are the same, so w_2 would select same rows of $w'_1 : state.state\_name ='new\ york'$, but this means that table *border_info* is not used and this pairing would be equivalent to $\langle s_2, f_1, w'_1 \rangle$, which is a meaningless query.

As we already noticed in previous sections, we add a weight to each clauses in S and W. This weight is a simple measure consisting in counting how many stems originated the clause. When pairing clauses the combined weight is just computed as the sum of its components, and it's used to order the obtained set $\bar{\mathcal{A}}$ of possible useful queries from the most probable to the less one.

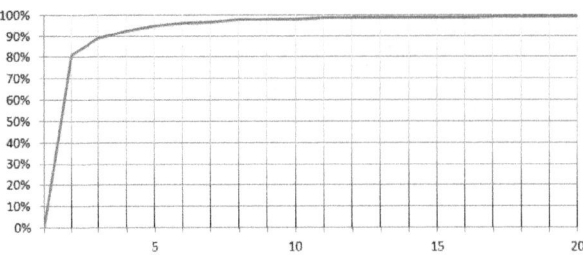

Fig. 3. Precision values when seeking for the answer in a larger subset

In Figure 2 the higher probability is highlighted by thicker connection lines (dashed lines illustrate pruned queries). The final ordered set answering q_2 is
$$\bar{\mathcal{A}} = \left\{ \langle s_1, f_3, w_2 \rangle^7, \langle s_3, f_2, w_1 \rangle^6, \langle s_2, f_3, w_2 \rangle^6, \langle s_1, f_1 \rangle^3, \langle s_2, f_1 \rangle^2, \langle s_3, f_2 \rangle^1 \right\}.$$

The right answer can be derived from the pairing with highest weight, that is: `SELECT state.capital FROM state join border on state.state_name = border_info.border WHERE border_info.state_name='new york'`.

It is worth noting that there could be more queries with the same weight. To cope with that we privilege queries that involve less joins and those that embed the most meaningful (i.e. referenced) table, e.g. `state` in the case of GEOQUERY. Note, for example, the order of the second and third pairings in $\bar{\mathcal{A}}$: they have been swapped since f_3 contains a join while f_2 doesn't.

4   Preliminary Experiment and Related Work

In a first preliminary experiment, we have studied the effectiveness of our automatic generation of SQL queries for a given NL question by testing the accuracy of selecting the one that retrieves the correct answer.

We started from the set of 800 NL questions given in the GEOQUERY corpus, trying to generate a result set equivalent to the one of associated SQL query in the corpus. Originally these questions were paired with meaning representations [3] that have been translated [4] into SQL queries. However, 23 of 800 queries are wrong or cannot be executed without leading to a MySQL error.

The first implementation of the algorithm as illustrated in this paper, failed to generate the set of possible queries for 71 questions, leading to a recall of 88%. These failures may be due to: (i) empty clauses set $\mathcal{S}$ and/or $\mathcal{W}$, e.g., "*How many square kilometers in the us?*" does not contain any useful stem; (ii) mismatching in nested queries, e.g. "*Count the states which have elevations lower than what alabama has*" contains an implicit reference to a missing piece of question; and (iii) ambiguous questions, e.g. "*Which states does the colorado?*" from which we retrieve an incomplete dependency set.

For all remaining questions from which we succeed in generating an ordered list of possible queries, we find that the query on top of this list retrieves the correct result set 81% of the times where as it can be found within the first 10 generated queries with a precision of 99%, according to the growing precision curve shown in Figure 3. As pointed out in the plot, the right query is found among the first three in 92% of the cases.

As the previous work suggests [5], similar outcome has been obtained using different approaches: relying on semantic grammar specified by an expert user [6], enriching the information contained in the pairs [4] and implementing ad-hoc rules in a semantic parser [3,7]. Our system instead, requires no intervention since the database metadata already contain all the needed information. This result is encouraging since it compares favorably with the state of the art: with respect to the Precise system [4] (100% Precision and 77% Recall) and the Krisp [3] system (94% Precision and 78% Recall). Our system can provide an answer in 88% of the cases, achieving an accuracy of 81%. If we consider valid an answer given in the top three, our accuracy increases to 95%, achieving 99% on the 10-top ranked. Note that the accuracy at top-ranked answer can be improved by learning a re-ranker, as explained in [8], which can move correct answers in the top position.

5 Conclusions and Future Work

In this paper, we approach the question answering task of implementing a NL interface to databases by automatically generating SQL queries based on grammatical relations and matching metadata. The complexity of generated queries is fairly high indeed, since we can deal with questions that require nesting, aggregation and negation in addition to basic projection, selection and joining (e.g. *"How many states have major non-capital cities excluding Texas"*). To our knowledge, the underlying idea that we propose for building and combining clauses sets is novel. Preliminary experiments on automatic question generation system show a satisfactory accuracy, i.e. 81%, although large improvement is still possible.

In the future we plan to extend this research by introducing learning approaches to classify and rank generated queries.

Acknowledgement. This research has been partially supported by the European Community's Seventh Framework Programme (FP7/2007-2013) under the grant #247758: ETERNALS – Trustworthy Eternal Systems via Evolving Software, Data and Knowledge, and the grant #288024: LiMoSINe – Linguistically Motivated Semantic aggregation engiNes.

References

1. Marie-Catherine de Marneffe, B.M., Manning, C.D.: Generating typed dependency parses from phrase structure parses. In: Proceedings LREC 2006 (2006)
2. Porter, M.: Porter stemmer, http://tartarus.org/~martin/PorterStemmer/
3. Kate, R.J., Mooney, R.J.: Using string-kernels for learning semantic parsers. In: Proceedings of the 21st ICCL and 44th Annual Meeting of the ACL, pp. 913–920. Association for Computational Linguistics, Sydney (2006)

4. Popescu, A.M., Etzioni, O., Kautz, H.: Towards a theory of natural language interfaces to databases. In: Proceedings of the 2003 International Conference on Intelligent User Interfaces. Association for Computational Linguistics, Miami (2003)
5. Giordani, A., Moschitti, A.: Corpora for automatically learning to map natural language questions into sql queries. In: Proceedings of LREC 2010. European Language
Resources Association (ELRA), Valletta (2010)
6. Minock, M., Olofsson, P., Näslund, A.: Towards Building Robust Natural Language
Interfaces to Databases. In: Kapetanios, E., Sugumaran, V., Spiliopoulou, M. (eds.)
NLDB 2008. LNCS, vol. 5039, pp. 187–198. Springer, Heidelberg (2008)
7. Ruwanpura, S.: Sq-hal: Natural language to sql translator,
http://www.csse.monash.edu.au/hons/projects/2000/Supun.Ruwanpura
8. Giordani, A., Moschitti, A.: Syntactic Structural Kernels for Natural Language Interfaces to Databases. In: Buntine, W., Grobelnik, M., Mladenić, D., Shawe-Taylor,
J. (eds.) ECML PKDD 2009. LNCS, vol. 5781, pp. 391–406. Springer, Heidelberg
(2009)

Two-Stage Named-Entity Recognition Using Averaged Perceptrons

Lars Buitinck and Maarten Marx

Information and Language Processing Systems
Informatics Institute
University of Amsterdam
{l.j.buitinck,maartenmarx}@uva.nl

Abstract. We describe a simple approach to named-entity recognition (NER), aimed initially at the Dutch language, but potentially applicable to other languages. Our NER system employs a two-stage architecture, with handcrafted but dataset-independent features for both stages, and is on a par with state-of-the-art systems described in the literature. Notably, our approach does not depend on language-specific assets such as gazetteers. The resulting system is quite fast and is implemented in less than 500 lines of code.

1 Introduction

Named-entity recognition (NER) is the problem of detecting, in running text, all names or *definite descriptors*, including the names of persons, organizations, geographical entities, etc. We describe a machine-learned approach to this problem for the Dutch language, for which no publicly available NER software exists. Our approach, however, makes no language-specific assumptions and should be portable to similar languages. We intend to use the results of NER tagging to enhance searching in parliamentary hansards; thus, in addition to a full-text query, users will get the opportunity to look for all occurrences of a certain person's or organisation's name.

1.1 Problem Description

In most machine-learned approaches to NER (Tjong Kim Sang, 2002), it is considered a sequence labeling problem,[1] where tokens are annotated with BIO-tags, indicating *beginning* of, *inside*, or *outside* a name. In addition, the names are annotated with a category. The following sentence from the SoNaR (Oostdijk et al., 2008) corpus exemplifies the NER problem.

Bij de stad [LOC Falujah] is een [LOC Amerikaanse] [PRO Black Hawk]-helikopter neergestort.

---

[1] But see Sarawagi and Cohen (2004) for an argument against this view, and for a view of NER as a *segmentation* problem.

G. Bouma et al. (Eds.): NLDB 2012, LNCS 7337, pp. 171–176, 2012.

We note that the recognition of names and their classification are really two distinct problems, and tackle the two using separate classification algorithms, as described in more detail in section 2.

1.2 Related Work

An two-stage approach to NER was tried and found effective by Wang and Patrick (2009). The difference between our approach and theirs is that Wang and Patrick use a conditional random field (CRF) to do a first attempt at solving the complete problem of detection and classification, followed by a voting procedure to re-classify the detected entity names.

Similarly, Desmet and Hoste (2010) use a combination of various classifiers (CRF, SVM, k-NN) combined into a voting ensemble with voting weights learned by a genetic algorithm; they obtain good results on a Dutch NER task, but conclude that "[t]he performance gain of the ensemble system over the best individual classifier [...] is not very large and comes at a high computational cost." We avoid this cost by using a single linear classifier in each stage of our system.

2 Architecture

We break up the NER problem into two stages and use a two-stage architecture to solve it. The first stage is the recognition stage, where we try to find the actual entity references in running text. The second stage is that of classification, where the entity references are divided into various classes; the exact classes depend on the training and evaluation sets in use (see section 3).

In both stages, we use the averaged perceptron algorithm, a simplification of the voted perceptron algorithm of Freund and Schapire (1999), to train a linear classifier for the problem at hand, using a handcrafted set of features detailed below. We train our perceptrons for 40 iterations.

Our implementation language is the Learning Based Java (LBJ) language and assorted libraries (Rizzolo and Roth, 2010). By leveraging these, we have managed to implement our full system in less than 500 lines of code.

In the rest of this section, we discuss the features used in both stages of our NER pipeline. It should be noted that none of these are particularly new (Bogers, 2004, pp. 29–31), although we are not aware of NER systems that use separate feature sets for recognition and classification of entity names.

2.1 Recognition Stage

In the recognition stage, we use the following basic features to determine the BIO tag of a single token:

- The identity of tokens in a size-five window, both with and without their relative position to the word under scrutiny. E.g., the word "Obama" in the window "all of Obama's plans are" would yield token features "all", "of", "Obama's", "plans" and "are", as well as token-at-position features "-2: all", "-1: of", "0: Obama's", "+1: plans" and "+2: are".

- The part-of-speech tags of the above (assumed part of the training and test data).
- The conjunction of the previous two features; i.e. word identity+POS as a combined feature with window size of five.
- Boolean features for several patterns (regular expressions) of the token under scrutiny: first letter capitalized, all capitals, contains digit, all digits, contains dash (–), contains dash followed by number and a limited matcher for Roman numerals, ^[IVX]+$.
- Capitalization features, as in the previous bullet point, for a token context window.
- Prefixes and suffixes of the current and previous token, up to four characters in length.

We extend this feature set with the classifier's output to the two previous tokens to turn our token-level classifier into a sequence classifier. We also add the conjunction of the previous two predictions, the conjunction of capitalization pattern and the previous prediction, and the conjunction of word window features with the previous prediction. Preliminary experiments with smaller feature sets showed a strong increase in classifier accuracy when introducing these conjunctive features.

2.2 Classification Stage

In the classification stage, we assign a class label to a complete entity name as discovered by the recognition stage. We therefore have a different type of inputs than in the previous stage: sequences of tokens in context instead of single tokens, e.g.

... de stad [Falujah] is een ...

Hence, we use a different set of features, comprising:

- Token identity inside the sequence.
- Identity of the two tokens preceding the sequence and the tokens following it.
- Affixes, up to length four, of the tokens inside the sequence.
- The length of the sequence.
- The capitalization pattern of the sequence, expressed as a string on the alphabet $(L|U|O)*$ to indicate alternations of upper and lower case initial letters (with O denoting not a letter).
- The occurrence of digits, dashes and all-capital tokens inside the sequence.

Note that we do not employ any so-called "gazetteer" features (lookup of entity names in a pre-compiled list). We chose not to use a gazetteer since no high-quality gazetteer was available for the Dutch language and available time prohibited us from compiling a gazetteer ourselves.

3 Evaluation

We evaluate our approach on two different datasets. The CoNLL'02 dataset (Tjong Kim Sang, 2002) consists of 309.686 tokens, containing 19901 names, split into 65% training, 22% validation and 12% test material (measured by the number of tokens). CoNLL'02 divides named entities into four categories: PER (person), LOC (location), ORG (organisation) and MISC (all other).

Table 1 shows our results on the CoNLL'02 test set. We note that the overall F_1 score is just below that of Wu et al. (2002), the second-best result reported in the CoNLL'02 shared task proceedings.

Table 1. Results on the CoNLL'02 test set

	Precision	Recall	F_1
LOC	80.11%	77.52%	78.79
MISC	77.09%	70.85%	73.84
ORG	73.68%	63.49%	68.21
PER	73.39%	85.15%	78.84
Overall	75.79%	74.50%	75.14

For reference, we show results for the Stanford Named Entity Recognizer (Finkel et al., 2005) on the CoNLL'02 test set (retrained on the CoNLL training set) in table 2; [2] re-using the Stanford NER package is a common choice for practical named entity recognition in the Dutch language area. We note that our system performs slightly better on this set, despite being smaller and simpler than Stanford's.

Table 2. Stanford NER results on the CoNLL'02 test set

	Precision	Recall	F_1
LOC	84.26%	74.68%	79.18
MISC	73.27%	65.12%	68.96
ORG	76.63%	66.55%	71.24
PER	73.56%	87.16%	79.78
Overall	76.02%	73.46%	74.72

Our second dataset is that used by Desmet and Hoste (2010, 2011), a selection from the SoNaR corpus (Oostdijk et al., 2008): 206421 tokens containing 13818 names, taken from Dutch newscast autocue text. SoNaR divides named entities into six categories: PER, LOC, ORG, PRO (product), EVE (event) and MISC.

---

[2] Courtesy Andrei Vishneuski, U. Amsterdam.

Table 3 shows our results on the SoNaR set, obtained by a stratified three-fold cross validation. We use the exact same division of the dataset into three folds as was used by Desmet and Hoste (2010). We note that the overall F_1 score is slightly lower than the 84.44 reported by Desmet and Hoste (2011) for their classifier ensemble trained by a genetic algorithm, but on a par with the 83.77 for their CRF classifier.

From table 3, it is obvious that our system's performance is somewhat lacking with respect to the classes PRO, EVE and MISC; we attribute this to two factors. First is the fact that our features have been optimized for the CoNLL'02 dataset, where two of these categories are largely subsumed by the MISC class. Second is the fact that these categories are relatively less well-represented in the data; each occurs less than a thousand times, while the categories LOC, PER and ORG occur 6624, 3290 and 2461 times, respectively.

Table 3. Results on the SoNaR set (macro-averaged over 3-fold cross validation)

	Precision	Recall	F_1
EVE	86.80%	60.08%	70.99
LOC	87.22%	92.14%	89.61
MISC	73.20%	53.91%	62.07
ORG	80.38%	77.89%	79.09
PER	82.43%	82.19%	82.30
PRO	70.51%	48.49%	57.46
Overall	83.95%	83.17%	83.56

4   Conclusion

We have shown that a two-stage approach with careful feature engineering, using a standard classification algorithm, can result in a named-entity recognizer that is competitive with state-of-the-art systems for the Dutch language, even without the use of gazetteers. In particular, our "poor man's approach" to sequence classification is on a par with specialized models such as conditional random fields.

Our system, however, is at present weaker than it could be when evaluated on the new SoNaR NER-tagged corpus. Further feature engineering could be used to improve its performance on this data set with its fine-grained classification.

Acknowledgments. The research for this paper was funded by Clarin-NL (http://www.clarin.nl) under project number 10-009 (WiP).

References

Bogers, T.: Dutch named entity recognition: Optimizing features, algorithms, and output. Master's thesis, Tilburg University (2004)

Desmet, B., Hoste, V.: Dutch named entity recognition using classifier ensembles. In: Proc. 20th Meeting of CLIN, pp. 29–41 (2010)

Desmet, B., Hoste, V.: Dutch named entity recognition using classifier ensembles. In: Proc. 23rd Benelux Conference on Artificial Intelligence (2011)

Finkel, J.R., Grenager, T., Manning, C.D.: Incorporating non-local information into information extraction systems by gibbs sampling. In: Proc. 43rd Annual Meeting of the ACL, pp. 363–370 (2005)

Freund, Y., Schapire, R.E.: Large margin classification using the perceptron algorithm. Machine Learning 37(3), 277–296 (1999)

Oostdijk, N., Reynaart, M., Monachesi, P., Van Noord, G., Ordelman, R., Schuurman, I., Vandeghinste, V.: From D-Coi to SoNaR: A reference corpus for Dutch. In: Proc. Int'l Conf. on Language Resources and Evaluation, LREC (2008)

Rizzolo, N., Roth, D.: Learning Based Java for rapid development of NLP systems. In: Proc. Int'l Conf. on Language Resources and Evaluation, LREC (2010)

Sarawagi, S., Cohen, W.W.: Semi-Markov conditional random fields for information extraction. Advances in Neural Information Processing Systems 17, 1185–1192 (2004)

Tjong, E.F., Sang, K.: Introduction to the CoNLL-2002 shared task: Language-independent named entity recognition. In: Roth, D., van den Bosch, A. (eds.) Proc. 6th Conf. on Computational Natural Language Learning (CoNLL), pp. 155–158 (2002)

Wang, Y., Patrick, J.: Cascading classifiers for named entity recognition in clinical notes. In: Proc. Workshop on Biomedical Information Extraction (WBIE), pp. 42–49 (2009)

Wu, D., Ngai, G., Carpuat, M., Larsen, J., Yang, Y.: Boosting for named entity recognition. In: Roth, D., van den Bosch, A. (eds.) Proc. 6th Conf. on Computational Natural Language Learning, CoNLL (2002)

Using Natural Language Processing
to Improve Document Categorization
with Associative Networks

Niels Bloom

Pagelink Interactives, Sherwood Rangers 29, 7551 KW Hengelo, The Netherlands
University of Twente, P.O. Box 217, 7500 AE Enschede, The Netherlands
n.bloom@pagelink.nl

Abstract. Associative networks are a connectionist language model with the ability to handle large sets of documents. In this research we investigated the use of natural language processing techniques (part-of-speech tagging and parsing) in combination with Associative Networks for document categorization and compare the results to a TF-IDF baseline. By filtering out unwanted observations and preselecting relevant data based on sentence structure, natural language processing can pre-filter information before it enters the associative network, thus improving results.

Keywords: Associative Networks, WordNet, Stanford Natural Language Parser, Natural Language Processing, Document Categorization.

1 Introduction

Ordering large sets of documents, such as the articles in Wikipedia, into a hierarchical structure of categories allows users to find information more easily. However, the changing nature of such data offers special challenges in creating and maintaining that hierarchy.

An associative network is a connectionist language model that is capable of automatically categorizing libraries while term frequency-inverse document frequency or TF-IDF [6] is a method to determine how important a word is to a document in a set of documents. We wish to prove that associative networks outperform a TF-IDF baseline in document categorization and that the results can be further improved by using natural language processing or NLP-techniques for disambiguation and for detection of relevant text elements.

We tested associative networks without NLP, with a part-of-speech tagger and with full language parsing and compared the results to each other and to TF-IDF baselines. For the NLP we used the Stanford Natural Language Parser [3,10]. In all cases, Princeton WordNet [2,5] was used to link words to concepts.

2 Associative Network

An associative network is a mathematical model of associative thinking used by humans to solve certain problems [4,7,8]. It contrasts with case-based reasoning

G. Bouma et al. (Eds.): NLDB 2012, LNCS 7337, pp. 177–182, 2012.

approaches based on this work by not relying on formally defined cases and ontologies, but using a connectionist model as a base instead.

Associative networks are modelled as a graph: each node represents an observation, such as the occurrence of a specific concept, and each edge a link between these observations. Edges with higher weights represent more closely related observations. For example, the concepts of *fly* and *mosquito* are more closely related than *fly* and *insect*. Using the same connectionist foundation [1] as neural networks, each node can be activated, which spreads the input signal to the connected nodes proportionally to the edge weight. Thus, high weight edges spread the signal more strongly than low weight edges.

In our case, a document activates each word concept found in the document. Concepts that have more relevance to the document are activated more than others, thus spreading further and stronger through the associative network.

2.1 Creation of an Associative Network

To construct an associative network, we used Princeton WordNet [2,5] as a foundation as it provides both a thesaurus and dictionary data in one system, is well documented and supported and is a well-accepted standard. The network was initialised by creating a graph consisting of a single node for each synset (word concept) in WordNet. These nodes where then connected according to the relationships between those synsets in WordNet. For example, the synset for *fly* has the sister-term *mosquito* and a hypernym *insect* so the nodes for these synsets were connected. All edge weights were initialised at a value of 1.

2.2 Training of an Associative Network

Not all relationships in WordNet are equal in conceptual distance and thus they should not have the same weight in an associative network. However, the number of steps to go from one synset to another does not directly determine the conceptual distance between the two. To make associative networks learn the actual conceptual distance between synsets, they can be trained using back-propagation to give more weight to closer connections and lower weight to more distant ones. As a result, when a node in that associative network is activated, activation is spread towards more closely related synsets more easily.

In our case, training was done by creating a training library composed of 30 manually selected Wikipedia articles with each article being closely related by topic to exactly one other article and not related to the other 28 articles, thus forming fifteen pairs of articles. Articles were selected from documents in the same category as the test set (see Section 3.1), but from different sub-categories.

A newly initialised associative network was activated for each of the 30 articles in random order to determine which were the most closely related. Depending on the correctness of the result, positive or negative reinforcement was applied by using back-propagation to adjust the weights in the network. This cycle was repeated until the associative network produced the correct matching article for the entire library. By training the associative network in this way, we increased the weight of edges between closer concepts such as *fly* and *mosquito* while reducing the weight of edges to more distant concepts such as *insect.*

2.3 Improving the Activation Pattern

The quality of the activation pattern, that is the nodes in an associative network that are activated for a certain input, is important for getting good results. Our assumption is that we can use NLP-techniques to improve the quality of the input for associative networks.

To test the effect of NLP techniques, we used the Stanford Part-Of-Speech Tagger [10] to help exclude synsets which do not match the part of speech from the activation pattern and the Stanford Natural Language Parser [3] to boost the activation of key words in each sentence. Simply put, these techniques modify which nodes in the associative network are activated and by how much.

3 Design of the Experiment

To prove the above assumption, we tested associative networks with three types of natural language processing: 1) a basic associative network without NLP, 2) an associative network with part-of-speech tagging, and 3) an associative network with a natural language parser.

The task was to sort 50 related articles into five predefined categories. The systems were evaluated by determining how well they matched the manual categorization made by the Wikipedia user base. Each system also calculated a baseline value using TF-IDF. In each of the cases, the natural language processing between the associative network and its TF-IDF baseline was identical.

3.1 Constructing the Test Sets

We created five test sets of 50 articles, each within a single category. The five categories were manually selected from Wikipedia, based on the criteria that they should be general topics with many articles and sub-categories. Categories selected included topics such as "Biology", "Philosophy" and "Nature".

Next, five sub-categories were randomly selected within each main category. Each sub-category had to contain at least ten qualifying articles and had to have a listed main article. An article qualified if it had a content of at least one thousand words, and if it was not marked as a stub or list article.

Finally ten qualifying articles were randomly selected from each of the five sub-categories, to give a test set of 50 articles. For example, the category "Biology" might have "Genetics", "Biochemistry", "Mycology", "Neuroscience" and "Ecology" as sub-categories. These were then mixed up randomly. For each article we wished to identify to which sub-category it belongs. To determine this, each sub-category's main article was used as a target, with which articles were compared to see how closely they matched.

3.2   Setup of the Experiment

First, the text of each Wikipedia article in each test set was converted to three sets of synsets as described below. In each case, every synset in a set was given an activation value: the more important and frequent a synset was in relationship to the text, the higher the activation value. Once each set of synsets was constructed they were given to a trained associative network (see Section 2.2) as input, spreading the activation across the network.

The five target articles, each the main article of a sub-category, were connected in turn to the associative network, and received activation spread as the sets of synsets for each of the 50 articles were activated. The target article receiving the highest activation value was selected and the associated sub-category was determined as the one in which the test article should be categorized.

As mentioned, three methods were used to construct the set of synsets. Each system was made to determine which article matched which sub-category, based only on their textual content. No other information, such as links, was used. An accuracy score was established based on how many articles were sorted correctly. For example, if 40 articles were sorted correctly, the accuracy score would be 80%.

3.3   Methods to Construct the Set of Synsets

Basic Associative Network.   A basic associative network was established using WordNet [2,5]. To determine a sub-category for each article, first the text in each article was split into words, which were matched with the corresponding synsets. If a word-form matched multiple synsets, it was not disambiguated, but all matching synsets were used. Thus, for each article a set of synsets was established, with a count of how many times each synset was matched.

These values were used as weights to connect the article to the associative network. Once all 50 test and five target articles were connected, each of the test articles was activated individually. The spread of this activation through the network activated each of the main articles to a different degree, depending on the available paths and the associated weights between the articles being activated. The amount of activation that spread to the main articles was then established, with the category of the one receiving the highest activation being selected as the category to place the test article in.

Associative Network with Part-of-Speech Tagger.   Linking words to all synsets with the word as a word-form is not very precise. For example, the word *fly* in the sentence *I like to fly* matches both the act of *flying* and the insect *fly*.

To help correct for this, the Stanford Part-Of-Speech Tagger [10] was used to get a better match between words and synsets: using the default tagger, the type of each word was established. For every word, the synsets were filtered based on this type, with non-applicable synsets being removed. In the earlier example *fly* would be tagged as a verb, so it would not match the synset for the insect. The improved set of synsets was then used in the same way as described above to determine a sub-category for each article.

Associative Network with Natural Language Parser. Not every part of a sentence is equally important. For example, in the sentence *I want to fly to Paris in a jet*, the word *fly* is more important than the word *in*. We would like to know which words are more relevant before activating the network. Though the associative network can establish which concepts are important to some extent, if words that have less relevance in the sentence are activated less, the associative network will be better able to identify relevant concepts in turn.

Using the Stanford Natural Language Parser [3], each sentence in the documents was tagged and parsed to determine the dependency relations between its words. In the earlier example sentence seven connections are made, five of which include the word *fly*. This implies that the word *fly* is crucial in the sentence.

Based on this idea that words connected to many other words in the sentence are more relevant to the sentence, the weight of each word's associated synsets was increased with a factor of the total amount of connections in the sentence to the amount of connections involving that word. Once weighted, the same techniques as above were used to sort the articles into the different sub-categories.

3.4 TF-IDF Baseline

Because TF-IDF is an unsupervised learning method, while associative networks are a supervised method, we expect the baseline to have a lower level of performance than associative networks. To establish the baseline, we calculated the TF-IDF value based on synsets (rather than raw words), as determined by the three methods described above. The resulting values were used to determine the distance between each test article and the five target articles. Each test article was assigned the sub-category of the target article to which it had the shortest distance. The distance was calculated using the following method:

$$D(a_x, m_y) = \sum_{n=1}^{s} |tfidf(a_x, S_n) - tfidf(m_y, S_n)| \tag{1}$$

where $D(a_x, m_y)$ is the distance between the article a_x and the main article m of sub-category y. S is the set of synsets 1...s in the corpus and $tfidf(a, S_j)$ is the term frequency / inverse document frequency of the synset S_j in article a.

4 Experimental Results

Table 1 lists the average results over five sets of 50 documents. The associative network approach clearly outperformed the TF-IDF baselines. Not only did the TF-IDF approach gain nothing from NLP, the results actually worsened, most likely because relevant words now activate fewer synsets: in turn fewer synsets match with the target articles. By contrast, associative networks clearly benefit from the improved match between word-forms and synsets, gaining six percent-point with part-of-speech tagging and another three using the parser.

Table 1. Results of the experiment

	TF-IDF accuracy	AN accuracy
No NLP	34 %	78 %
Part-of-speech tagger	31 %	84 %
Natural language parser	34 %	87 %

5   Conclusions and Future Work

Because of the associations made by the associative network, terms relevant to the text are brought forward even if they are not mentioned in the text or are only mentioned a few times. As expected, the associative network connects better with the underlying meaning of the text than a TF-IDF based approach.

In this work, we have compared associative networks against a TF-IDF baseline. As future work, we plan to set them up against supervised learning methods [9] to see whether associative networks can outperform these as well.

NLP-techniques help to extract the meaning from the text, filtering out synsets that share a common word-form with the synset which actually corresponds to the word, and identifying key words in each sentence. As a result, the accuracy of the network in categorising articles is increased.

We have used only basic NLP-techniques in our tests, yet these have already increased the success of the associative networks. Further research into the combination of associative networks and NLP may improve the results even further.

References

1. Bechtel, W.: Connectionism and the philosophy of mind: an overview. The Southern Journal of Philosophy 26, 17–41 (1988)
2. Fellbaum, C.: WordNet: An Electronic Lexical Database. MIT Press, Cambridge (1998)
3. Klein, D., Manning, C.: Fast Exact Inference with a Factored Model for Natural Language Parsing. Adv. in Neural Information Processing Systems 15, 3–10 (2003)
4. Marcus, G.F.: The Algebraic Mind: Integrating Connectionism and Cognitive Science. MIT Press, Cambridge (2001)
5. Miller, G.: WordNet: A Lexical Database for English. Communications of the ACM 38(11), 39–41 (1995)
6. Ramos, J.: Using TF-IDF to Determine Word Relevance in Document Queries. In: Proceedings of the First Instructional Conference on Machine Learning, iCML (2003)
7. Schank, R.C.: Dynamic Memory: A Theory of Learning in Computers and People. Cambridge University Press, New York (1982)
8. Schank, R.C., Abelson, R.P.: Scripts, Plans, Goals and Understanding. Erlbaum, Hillsdale, New Jersey (1977)
9. Sun, J., Chen, Z., Zeng, H., Lu, Y., Shi, C., Ma, W.: Supervised latent semantic indexing for document categorization. In: Proceedings for ICDM, pp. 535–538 (2004)
10. Toutanova, K., Klein, D., Manning, C., Singer, Y.: Feature-Rich Part-of-Speech Tagging with a Cyclic Dependency Network. In: Proceedings of HLT-NAACL, pp. 252–259 (2003)

MLICC: A Multi-Label and Incremental Centroid-Based Classification of Web Pages by Genre

Chaker Jebari

Ibri College of Applied Sciences, Sultanate of Oman
jebarichaker@yahoo.fr

Abstract. This paper proposes an improved centroid-based approach to classify web pages by genre using character n-grams extracted from URL, title, headings and anchors. To deal with the complexity of web pages and the rapid evolution of web genres, our approach implements a multi-label and incremental scheme in which web pages are classified one by one and can be affected to more than one genre. According to the similarity between the new page and each genre centroid, our approach either adjust the genre centroid or considers the new page as noise page and discards it. Conducted experiments show that our approach is very fast and achieves superior results over existing multi-label classifiers.

Keywords: Multi-label, Incremental, Centroid, genre classification, character n-grams.

1 Introduction

With the explosive growth of the web, genre classification of web pages become more and more important to find the desired information among thousands of web pages returned by a search engine. Generally speaking, the term "genre" describes a characterization of document with respect to its form and functional traits. A broad number of studies on genre classification of web pages have been proposed in the literature, without deeply addressing two important challenges: the complexity of web pages and the evolution of web genres. To deal with these challenges, we propose in this paper an improved centroid-based classification approach called MLICC. Our approach implements a multi-label and incremental classification scheme in which web pages are classified one by one. According to the similarity between the new web page and each genre centroid, our approach either adjusts the genre centroids by incorporating the new page in the genre profile or considers this page as noise and discards it.

The remainder of the paper is organized as follows. In Section 2 we present an overview of the previous work on genre classification of web documents by focusing only on those trying to solve the two challenges listed above. In Section 3, we present an overview about multi-label classification methods. Section 4 describes our approach MLICC. Section 5 presents the results achieved by our approach as well as a comparison with other multi-label methods. Finally, Section 6 summarizes the main points of this paper and presents some future works.

G. Bouma et al. (Eds.): NLDB 2012, LNCS 7337, pp. 183–190, 2012.

2   Genre Classification of Web Pages

Genres found in web pages (also called cybergenres) are characterized by the triple <content, form, functionality> (Shepherd & Watters, 1998). The content and form attributes are common to non-digital genres and refers to the text and the layout of the web page respectively. The functionality attribute concerns exclusively digital genres and describes the interaction between the user and the web page.

The content can be represented by vectors of terms or n-grams extracted from the text of the Web pages and therefore the analysis of content is based on the existence or absence of a term within web pages of a given genre, and the existence or absence of objects such as images, video, sound, etc. For example, many FAQ web pages either have the term 'FAQ' and/or 'frequently asked questions' near the top of the web page.

The form attribute is represented by the layout structure of the web page which is usually determined using the hypertext links. For example, a faculty homepage is a mix of course links, lists of publications, and usually an image of a person near the top of the page.

The functionality of a Web page can be represented by noting the presence or absence of executable code such as JavaScript and applets, or the number of hyperlinks on the page.

A broad number of studies on genre classification of web pages have been proposed in the literature, without deeply addressing two important challenges: the complexity of web pages and the evolution of web genres.

A web page is a complex object that is composed of different sections belonging to different genres. For example, a conference web page contain information on the conference, topics covered, important dates, contact information and a list of hypertext links to related information. This complexity need to be captured by a multi-label classification scheme in which a web page can be assigned to multiple genres. The need for multi-label classification to identify the genre of web pages is noticed by two studies (Santini, 2007), (Vedrana, Mitja, & Matjaž, 2009).

Given the dynamic nature of the web environment, web genre evolves rapidly. Usually, this evolution consists in adapting old genres or producing new genres (Shepherd & Watters, 1998). Adapted genres are genres that exist in some other medium, whereas novel genres are unique to the web and they could be evolved from existing genres or they could be completely original. In this paper we focus only on how we can adapt existing genres. To tackle this problem, a classification model that is based on incremental and adaptive learning is more appropriate.

3   Multi-label Classification Methods

Multi-label classification methods are grouped into two main categories: problem transformation and algorithm transformation. The first category of methods is algorithm independent and transforms a learning problem into one or more learning problems. The algorithm adaptation methods extend existing learning algorithms to deal with multi-label data directly (Tsoumakas & Katakis, 2007).

The most popular transformation method is called Binary relevance (BR) that learns one binary classifier for each label independently of the rest of labels and outputs the union of their predictions. This method does not consider the relationships between labels and it can do ranking if classifier outputs scores (Godbole & Sarawagi, 2004).

The most recent adapted algorithm called MLKNN has been proposed by Zhang and Zhou (Zhang & Zhou, 2005). This algorithm transforms the original data set into $|L|$ data sets D_l that contain all examples of the original data set, labeled with the label l if the example belongs to l and $\neg l$ otherwise. After that, MLKNN applies the KNN algorithm for each label λ_j and considers those that are labeled at least with the label λ_j as positive and the rest as negative. MLKNN can be extended to produce a ranking of the labels as an output.

4   MLICC Approach

Our approach is based on three main steps. The first step consists in representing a web page by selecting a relevant subset of character n-grams. The second step builds genre centroids using a set of training web pages. Finally, the third step, incrementally classifies new web pages using the centroids generated in the previous step.

4.1   Representing a Web Page

In our approach a web page is represented by a selected set of character n-grams (contiguous n characters) extracted from the following data sources: URL text (UT), title text (TT), heading text (HT) and anchor text (AT). In this paper, we varied the length n of the character grams between 2 and 7. We used the Vector Space Model (VSM) to represent the web pages (Salton, 1989). In this model, a web page p_i is represented by a vector of character n-grams as follows:

$$p_i = \left(C_{i1}, ..., C_{ij}, ..., C_{i|p_i|} \right) \tag{1}$$

Where C_{ij} is the character n-gram j found in the web page i. Each character n-gram C_{ij} is associated with a weight w_{ij} calculated using the normalized TFIDF weighting schema (Sebastiani, 2002).

4.2   Building Genre Centroids

The centroid of the genre g_i is calculated using the normalized sum formula. In this formula, we firt compute the summed genre centroid GC_i^S by summing all web page vectors belonging to genre g_i as follows:

$$GS_i^S = \sum_{p_i \in g_i} p_i \tag{2}$$

Next, we evaluate the normalized genre centroid GS_i^N as follows:

$$GS_i^N = \frac{GS_i^S}{\left\| GS_i^S \right\|} \tag{3}$$

Where $\left\| GS_i^S \right\|$ denotes the 2-norm of the centroid vector GC_i^S.

For each genre g_i, we determine the similarity threshold σ_i that allows a web page to belong to more than one genre or to no genre at all. This threshold is calculated using the sum formula as follows:

$$\sigma_i = \frac{1}{|g_i|} \cdot \sum_{p_j \in g_i} sim\left(p_j, GC_i^N\right) \tag{4}$$

Where $sim\left(p_i, GC_i^N\right)$ is the cosine similarity between the normalized genre centroid GC_i^N and a training web page p_j. Next, we recalculate for each genre g_i a new training set ng_i by removing noise pages. A web page is considered noise for a genre g_i if its similarity with its normalized centroid is below the initial threshold calculated using the original training set. After that, we recalculate for each genre its centroid and its similarity threshold based on the new training set using the same formulas (3) and (4).

4.3   Incremental Classification of New Web Pages

The classification of a new page p denoted by $\Phi(p)$, is a ranking of the genres according to their similarities with the genre centroids and is defined as follows:

$$\Phi(p) = \left\{ g_i / sim\left(p, GC_i^N\right) \geq \sigma_i \right\} \tag{5}$$

The centroid for each genre $g_i \in \Phi(p)$, is adjusted as follows:

$$GC_i^N = \frac{GC_i^N + p}{\left\| GC_i^N + p \right\|} \tag{6}$$

To prevent deterioration in prediction accuracy over time our approach has to adjust the similarity threshold of each genre $g_i \in \Phi(d)$ after the classification of each web page using the following formula:

$$\sigma_i = \frac{S_i + sim\left(p, GC_i^N\right)}{|g_i|} \tag{7}$$

Where, S_i is the sum of the similarities between the adjusted centroid and all web pages belonging to the genre g_i except the new web page p.

4.4   Time and Space Complexity

Broadly speaking, an algorithm must have a reasonable computational complexity to be of practical significance. The major advantage of centroid-based classifiers is the linear complexity. In this section we discuss theoretically the computational complexity of our approach for both training and testing phases. First and foremost, we will give some general notations: let $G=\{g_1, ..., g_i, ..., g_{|G|}\}$ the set of predefined genres, m_i the number of training web pages belonging to the genre g_i, r_i is the number of training web pages of the genre g_i after refinement ($r_i \leq m_i$) and n the number of new web pages.

The training phase of our approach consists of four steps. In the first step, we calculate the initial centroid for each genre, while the second step consists in computing the optimal similarity threshold for each genre. Each of these steps has a linear complexity of $O(m_i)$. In the third step, we build a refined set of training web pages for each genre. Since we used the similarity and the optimal threshold, which are already calculated, this step does not need computational effort. Finally, in the fourth step we recalculate the genre centroids using the refined set of training web pages. This step has a linear complexity of $O(r_i)$. Hence, we can conclude that the training and testing phases have linear complexities of $O(m_i \times |G|)$ and $O(n \times |G|)$ respectively and therefore, we can say that our approach has a linear complexity.

5   Evaluation

5.1   Corpus

In our approach we used the corpus MGC (Vidulin, Luštrek, & Gams, 2007). This corpus was gathered from internet and consists of 1539 English web pages classified into 20 genres as shown in the following table. In this corpus each web page was assigned by labelers to primary, secondary and final genres. Among 1539 web pages, 1059 are labeled with one genre, 438 with two genres, 39 with three genres and 3 with four genres. It is clear from the following table that the corpus MGC is unbalanced, meaning that the web pages are not equally distributed among the genres.

Table 1. Composition of MGC corpus

Genre	# of web pages	Genre	# of web pages
Blog	77	Gateway	77
Children's	105	Index	227
Commercial/promotional	121	Informative	225
Community	82	Journalistic	186
Content delivery	138	Official	55
Entertainment	76	Personal	113
Error message	79	Poetry	72
FAQ	70	Adult	68
Shopping	66	Prose fiction	67
User input	84	Scientific	76

In this paper we used the average precision (Precision), the ranking loss (RankLoss), One-error (OneError) and Hamming Loss (HamLoss) metrics (Read, Pfahringer, & Holmes, 2008). In our experiments, we followed the 10 × 10 cross-validation procedure which consists in randomly split the corpus into 10 equal parts. Then we used 9 parts for training and the remaining part for testing. This process is performed 10 times and the final performance is the average of the 10 individual performances.

5.2   Experiments and Results

This section discusses the results provided by two experiments. The purpose of the first experiment is to identify the data source for which our approach achieves the best performance, whereas the objective of the second experiment is to show the importance of the incremental aspect in genre classification.

Experiment1: Effect of Data Source

In this experiment we compared the classification performance using different data sources: URL text (UT), title text (TT), heading text (HT), anchor text (AT), the combination of all the previous sources (CT) and the entire text of the web page (ET). It can be observed from Table 2 that results obtained using the combination of all character n-grams extracted from all data sources outperform those achieved using only character n-grams extracted from each data source alone. Moreover, using character n-grams extracted from the entire text of the web page we achieved less results.

Table 2. Classification performance using different data sources

Data source	Precision	RankLoss	OneError	HamLoss
UT	0.36	0.23	0.21	0.16
TT	0.78	0.10	0.08	0.08
HT	0.90	0.08	0.06	0.08
AT	0.72	0.11	0.17	0.08
CT	**0.94**	**0.06**	**0.05**	**0.06**
ET	0.65	0.13	0.08	0.09

Experiment2: Effect of Incremental Classification

In this experiment we varied the percentage of training web pages from 10% to 90% by step of 10%. From Table 3, we can say that our approach requires a small set of training pages to achieve good results. This can proof that incremental classification of web pages leads to better results than batch classification. Moreover, we can see that the best results are reported with adjusted centroids rather than stable centroids.

Table 3. Classification performance for different sizes of training sets and for adjusted/stable centroids

% of training set	Precision	RankLoss	OneError	HamLoss
10%	**0.94/0.81**	**0.06/0.12**	**0.05/0.11**	**0.06/0.09**
20%	0.76/0.63	0.03/0.08	0.15/0.21	0.09/0.11
30%	0.72/0.58	0.04/0.09	0.17/0.19	0.12/0.14
40%	0.62/0.55	0.06/0.06	0.18/0.18	0.13/0.14
50%	0.54/0.44	0.07/0.10	0.19/0.22	0.15/0.16
60%	0.64/0.64	0.05/0.08	0.17/0.18	0.12/0.13
70%	0.51/0.52	0.06/0.07	0.17/0.19	0.13/0.14
80%	0.49/0.42	0.08/0.09	0.15/0.16	0.09/0.10
90%	0.39/0.27	0.08/0.07	0.15/0.18	0.09/0.11

6   Comparison with Other Multi-label Classifiers

In this section, we compare our approach with other multi-label classification methods presented previously in section 3 and implemented in the Mulan toolkit[1]. These algorithms are Rakel, BR-SVM, MLKNN and BPMLL. To provide a valid comparison, we compared these algorithms to each other using the corrected re-sampled paired t-test with significance level of 5%. From Table, we can say that our classifier outperforms all multi-label classification algorithms.

Table 4. Classification performance provided by each classifier

	Precision	RankLoss	OneError	HamLoss
MLICC	**0.94**	**0.08**	**0.05**	**0.06**
Rakel	0.92	0.11	0.09	0.09
BR-SVM	0.89	0.10	0.11	0.09
MLKNN	0.77	0.09	0.10	0.07
BPMLL	0.89	0.12	0.19	0.10

The execution speed is also another important comparison aspect. Generally, the execution speed is based on training and testing times. In this experiment, we compare our approach with all algorithms used before such as Rakel, BR-SVM, MLKNN and BPMLL in terms of training and testing time measured in second. The achieved results are presented in Table 5.

Table 5. Time spent by each classifier in training and testing phases

	Training	Testing	Total time
MLICC	**7**	**12**	**19**
Rakel	15	14	29
BR-SVM	12	14	28
MLKNN	10	16	26
BPMLL	16	18	34

---

[1] http://mulan.sourceforge.net/index.html

It is clear from Table 5 that our approach is the fastest. This result is obvious because we exploits only character n-grams extracted from specific web page elements, such as title, headings, links and URL rather than the entire text of the web page. The method that takes longer for training is BPMLL, because they are the ones that apply the most complicated transformation to the training web pages.

7   Summary and Future Work

In this paper, we proposed a new multi-label centroid-based approach for genre classification of web documents using the character n-grams extracted from the document URL, title, headings and links. To cope with the complexity of web pages and the genre evolution, our approach implements a multi-label and incremental classification schema. The experiments conducted using a known multi-labeled corpus show that our approach is very fast and provides encouraging results in comparison with other multi-label classifiers. Since our approach uses only character n-grams, we can say that is language independent, thereby it can be useful in many NLP applications such as information extraction, information retrieval, text summarization, etc. As part of the future work, our approach should be tested on another data set, preferably with more examples.

References

Han, E.-H(S.), Karypis, G.: Centroid-Based Document Classification: Analysis and Experimental Results. In: Zighed, D.A., Komorowski, J., Żytkow, J.M. (eds.) PKDD 2000. LNCS (LNAI), vol. 1910, pp. 424–431. Springer, Heidelberg (2000)

Godbole, S., Sarawagi, S.: Discriminative Methods for Multi-labeled Classification. In: Dai, H., Srikant, R., Zhang, C. (eds.) PAKDD 2004. LNCS (LNAI), vol. 3056, pp. 22–30. Springer, Heidelberg (2004)

Read, J., Pfahringer, B., Holmes, G.: Multi-label Classification Using Ensembles of Pruned Sets. In: 8th IEEE International Conference on Data Mining (2008)

Salton, G.: Automatic Text Processing: The Transformation Analysis and Retrieval of Information by Computer. Addison-Wesley (1989)

Santini, M.: Automatic Identification of Genr. Web Pages. PhD thesis, University of Brighton, UK (2007)

Sebastiani, F.: Machine Learning in Automated Text Categorization. ACM Computing Surveys 34(1), 1–47 (2002)

Shepherd, M., Watters, C.: Evolution of Cybergenre. In: 31 Hawaiian International Conference on System Sceinces (1998)

Tsoumakas, G., Katakis, I.: Multi-label classification: An overview. International Journal of Data Warehousing and Mining 3(3), 1–13 (2007)

Vedrana, V., Mitja, L., Matjaž, G.: Multi-Label Approaches to Web Genre Identification. Journal of Language and Computational Linguistics 24(1), 97–114 (2009)

Vidulin, V., Luštrek, M., Gams, M.: Using Genres to Improve Search Engines. In: 1st International Workshop: Towards Genre-Enabled Search Engines: The Impact of Natural Language Processing, Borovest, Bulgaria, pp. 45–51 (2007)

Zhang, M., Zhou, Z.: A K-Nearest Neighbor based Algorithm for Multi-label Classification. In: 1st IEEE International Conference on GrC, China (2005)

Classifying Image Galleries into a Taxonomy Using Metadata and Wikipedia

Gerwin Kramer[1], Gosse Bouma[1], Dennis Hendriksen[2], and Mathijs Homminga[2]

[1] Information Science, University of Groningen
gerwinkramer@gmail.com, g.bouma@rug.nl
[2] Kalooga, Groningen
{dennis.hendriksen,mathijs.homminga}@kalooga.com
http://www.kalooga.com

Abstract. This paper presents a method for the hierarchical classification of image galleries into a taxonomy. The proposed method links textual gallery metadata to Wikipedia pages and categories. Entity extraction from metadata, entity ranking, and selection of categories is based on Wikipedia and does not require labeled training data. The resulting system performs well above a random baseline, and achieves a (micro-averaged) F-score of 0.59 on the 9 top categories of the taxonomy and 0.40 when using all 57 categories.

Keywords: gallery, image gallery, classification, hierarchical classification, taxonomy, Wikipedia.

1 Introduction

Organizing and managing the overwhelming amount of images on the web has become an active area of research. Image classification is the task of (automatically) classifying images into semantic categories. For this purpose, either visual clues can be used or text and metadata surrounding an image.

In this paper, we concentrate on a special case of image classification, namely *image gallery*[1] classification. An image gallery is a (part of a) website that displays a collection of images. Usually, this collection has a certain topic in common like a person, group, event or place. We are especially interested in these galleries, because they are often suitable as illustration in (on-line) news items. By categorizing galleries, we hope to improve the accuracy of a method that finds suitable illustrations for news items. We only use textual metadata, such as the title and image captions for classification. Galleries are classified into a hierarchically structured set of predetermined categories, our taxonomy.

The traditional, statistical, supervised, approach to image classification requires an (up-to-date) labeled training-set, which can be hard to create and

---

[1] Example galleries can be extracted from several locations on the web, e.g. IMDB (http://www.imdb.com), Flickr (http://www.flickr.com/), and various publisher's websites.

G. Bouma et al. (Eds.): NLDB 2012, LNCS 7337, pp. 191–196, 2012.

maintain. We avoid the use of a training-set by adopting an ontology-based classification technique based on [3]. The ontology we use is extracted from the category system of Wikipedia. Entities found in the metadata are linked to Wikipedia pages, and the most important Wikipedia categories of those entities are chosen as categories for the image gallery.

2 Related Work

The idea of creating a taxonomy of image galleries is not totally new. ImageNet [1] aims to (manually) assign images to concepts in WordNet. The goal is to provide a visual data set to be used for training and benchmarking classification algorithms. Text-based approaches to image classification can be tracked back to 1970s [15]. An ontology can be beneficial for text-based classification, for instance to bridge 'lexical' or 'semantic gaps' [11,10].

For the purpose of ontology-based research, the online and open encyclopedia Wikipedia is a valuable source [4]. A Wikifier [6,7,2] is a system that links entities to Wikipedia pages. Wikipedia can be used to calculate *relatedness* between terms as well [12]. The relatedness measure can be used to rank concepts in a text by importance (topic indexing) [5,9]. Most articles in Wikipedia are assigned one or more, hierarchically structured, categories. [14] acknowledge that the Wikipedia category graph can be used for various NLP tasks, although using it as is, has several drawbacks. [8] extract a taxonomy from the Wikipedia category graph that is more suitable for NLP applications.

3 Taxonomy

Our taxonomy is a hierarchical structured set of categories for image classification. Its categories are chosen by their relevance for image gallery classification and the needs of the Kalooga application that automatically suggests galleries for news items.

The root category of the taxonomy is *Contents*. Image galleries that fit none of the descendant categories should be classified here. The first-level categories are *People* (has 11 descendant categories and 2 sublevels), *Sports* (has 28 descendant categories and 2 sublevels), *Vehicles* (has no descendant categories), *Places* (has 2 descendant categories and 2 sublevels), *Entertainment* (has 8 descendant categories and 2 sublevels), *Arts* (has no descendant categories), *Animals* (has 6 descendant categories and 1 sublevel), and *Plants* (has no descendant categories).

Each category in the taxonomy is linked to a Wikipedia category from the Wikipedia category graph. This mapping enables us to perform the entity classification process described in section 4. Categories could in theory be mapped to a category graph in another language as well to enable classification of galleries in that language.

In hierarchical classification, a gallery should be classified in the most specific category or categories. If no category is suitable, *Contents* should be assigned. The decision was made that a category should only be assigned when it applies to each of the images of the gallery.

4   Methodology

The intuition behind the presented method is that shared and/or relevant categories of important entities in gallery text will be categories of the gallery. In the following subsections, each step of this method is described.

Extract Gallery Text. Metadata, such as the URL, title, description, image captions, and keywords found in the HTML of the gallery is merged into one text (URL's are decomposed into words). The title and URL words are included twice, to boost their weighting. Stopwords (including typical 'gallery' words such as *picture* and *gallery*) are removed.

Extract Entities. For the process of recognizing entities from the text, an in-house *wikifier* is used that recognizes entities and links these to Wikipedia pages. A semantic graph is constructed in which the nodes are entities and the edges the semantic relatedness between them. The computationally inexpensive Wikipedia link-based measure (WLM) is used [12] to compute relatedness. To filter out unlikely edges in the graph, relatedness must be above a certain threshold.

Score Entities by Importance. We use a variant of Averaged PageRank weighting (APW) [9], which takes into account the centrality of the entity in the semantic graph and the relative frequency of the entity.

For computing centrality, we use a variant of *closeness centrality* [13]. In 1, we first calculate closeness centrality (CC) of an entity within its subgraph and then multiply by the size of the subgraph ($distance(a, b) = 1 - SR_{a,b}$ and g the size of the subgraph), because an entity from a large subgraph is intuitively of higher importance:

$$CC_{e_i} = g \cdot \frac{g - 1}{\sum_j distance(e_i, e_j)} \qquad (1)$$

The relative frequency of an entity is calculated using $tf\text{-}idf$, in which entities are treated as terms and Wikipedia as the document collection. The term count in the corpus is then the inlink count of the entity's article. The calculation of the total combined entity score is presented in (2). If the maximum centrality score is 0, then only the tf-idf part is used.

$$ES_{e,g} = \frac{1}{2} \left(\frac{CC_e}{CC_{max}} + \frac{tf\text{-}idf_{e,g}}{tf\text{-}idf_{max}} \right) \qquad (2)$$

After the scoring is applied to each entity, a threshold is applied to filter out noise.

Find and Score Entity Categories. In this step, we try to find taxonomy categories for each entity by using its Wikipedia article. Each article is connected to categories of the Wikipedia category graph. By searching through the ancestors of the article's categories in this graph, we try to find broader categories that are *also present in the taxonomy*. To prevent inclusion of highly general or irrelevant categories, a maximum search depth is applied of 5 steps. Every retrieved category that matches a taxonomy category receives a relevance score.

$$ECS_{c,e} = \min(0.5, \frac{1}{d_{min}}) + \min(0.5, \frac{n_{path}}{d_{avg}}^2) \tag{3}$$

This entity's category score $ECS_{c,e}$ is the combination of the minimum distance d_{min} to travel from entity e to category c and the number of paths n_{path} from e to c, normalized by the average path distance d_{avg}. We like to classify the entity as specific in the taxonomy as possible. Therefore, if a certain category from the taxonomy is found, candidate ancestor categories are deleted.

Select Gallery Categories. The final step of the process is selecting categories for the gallery. For each category, we calculate the final gallery category score (GCS) as denoted in (4).

$$GCS_{c,g} = \sum_k ES_{e_k,g} \cdot ECS_{c,e_k} \tag{4}$$

The score is based on the amount of entities belonging to a candidate category c of gallery g, the importance of these entities (ES), and how relevant the category is to each entity (ECS). After calculating GCS for each candidate, a cut-off is applied to preserve only the most relevant categories.

5   Experiments

Test Set. Six persons manually assigned categories from the taxonomy to galleries. This resulted in a total of 734 English galleries of which 223 were randomly selected and 511 were found by manually finding galleries for categories. Most galleries (90%) had only one category.

Performance Measures. Because the number of galleries per category is very unevenly distributed, we report both micro and macro averaged precision, recall, and F-score.

In classic precision and recall measures, the hierarchical structure of the taxonomy is not taken into account. For hierarchical classification, this may be too pessimistic. Therefore, we measured precision and recall at the 9 top-level categories, and for all 57 categories.

Baseline. A random baseline classifier was created with 90 percent of chance to pick one category for a gallery and 10 percent chance to pick two categories for a gallery. The average of 100 runs is taken. A most frequent baseline is not used, because the distribution in the test set does not reflect the actual distribution of categories.

Thresholds. During tests on the development set, we found the following thresholds to result in a reasonable balance between precision and recall: Minimum semantic relatedness SR: 0.20; Minimum entity score ES: 0.45; Minimum entity category score ECS: 0.45; The gallery category score $GCS_{c,g}$ must be at least 50% of the highest GCS; Maximum number of gallery categories: 3;

6   Results

In Table 1, we can see that the classifier performs significantly better than the random baseline. The poor results of the random baseline give an indication of the complexity of the classification problem. Unsurprisingly, the classifier performs better in the 9 broader categories than for all 57 categories.

Table 1. Overall performance

Categories	Measure	Method	Precision	Recall	F-score
9 top-level categories	micro-avg	random	0.20	0.20	0.20
		classifier	0.54	0.66	0.59
	macro-avg	random	0.14	0.14	0.14
		classifier	0.54	0.70	0.61
All 57 categories	micro-avg	random	0.02	0.02	0.02
		classifier	0.35	0.46	0.39
	macro-avg	random	0.02	0.02	0.02
		classifier	0.48	0.47	0.48

The higher performance of the macro-averaged measures, compared to the micro-averaged measures indicates that the classifier performs relatively well on galleries from some small categories from the test-set and/or it performs relatively poor on galleries from some large categories from the test set. The lowest performance was achieved in the *Places* categories. This is largely due to most galleries being related to a place and entities on Wikipedia are highly related to places.

7   Conclusions and Future Work

We have seen how, and to what extent, image galleries can be classified into a taxonomy using metadata and Wikipedia. For the classification in large distinct categories, the presented method looks promising. However, fine grained classification is erroneous.

In future research, existing issues can be addressed by investigating some unexplored techniques. First, a change in the taxonomy might help to address low precision of the *Places* categories by making the location of a gallery a property. Second, the entity classification errors could be addressed by investigating additional ontologies and/or knowledge sources besides the Wikipedia category graph. Third, taking gallery context information into account might increase performance.

Finally, it would be interesting to find the *inter-annotator agreement* (IAA) about categories for gallery text. With the results of this test, a realistic target precision and recall for automated classification can be established. A comparison with a state of the art statistical classifier would also be interesting.

References

1. Deng, J., Dong, W., Socher, R., Li, L.J., Li, K., Fei-Fei, L.: Imagenet: A large-scale hierarchical image database. In: CVPR 2009 (2009)
2. Hoffart, J., Yosef, M., Bordino, I., Fürstenau, H., Pinkal, M., Spaniol, M., Taneva, B., Thater, S., Weikum, G.: Robust disambiguation of named entities in text. In: Proc. of EMNLP, pp. 27–31 (2011)
3. Janik, M., Kochut, K.: Training-less ontology-based text categorization. In: ECIR Workshop on Exploiting Semantic Annotations in Information Retrieval (ESAIR 2008), pp. 3–17. Citeseer (2008)
4. Medelyan, O., Milne, D., Legg, C., Witten, I.: Mining meaning from wikipedia. International Journal of Human-Computer Studies 67(9), 716–754 (2009)
5. Medelyan, O., Witten, I., Milne, D.: Topic indexing with wikipedia. In: Proceedings of the AAAI WikiAI Workshop (2008)
6. Mihalcea, R., Csomai, A.: Wikify!: linking documents to encyclopedic knowledge. In: CIKM 2007: Proceedings of the Sixteenth ACM Conference on Conference on Information and Knowledge Management, pp. 233–242. ACM, New York (2007)
7. Milne, D., Witten, I.: Learning to link with wikipedia. In: Proceeding of the 17th ACM Conference on Information and Knowledge Management, pp. 509–518. ACM (2008)
8. Ponzetto, S., Strube, M.: Deriving a large-scale taxonomy from wikipedia. In: AAAI, pp. 1440–1445. AAAI Press (2007)
9. Tsatsaronis, G., Varlamis, I., Nørvåg, K.: Semanticrank: ranking keywords and sentences using semantic graphs. In: Proceedings of the 23rd International Conference on Computational Linguistics, pp. 1074–1082. Association for Computational Linguistics (2010)
10. Wang, H., Liu, S., Chia, L.: Does ontology help in image retrieval?: a comparison between keyword, text ontology and multi-modality ontology approaches. In: Proceedings of the 14th Annual ACM International Conference on Multimedia, pp. 109–112. ACM (2006)
11. Wang, P., Hu, J., Zeng, H., Chen, Z.: Using wikipedia knowledge to improve text classification. Knowledge and Information Systems 19(3), 265–281 (2009)
12. Witten, I., Milne, D.: An effective, low-cost measure of semantic relatedness obtained from wikipedia links. In: Proceeding of AAAI Workshop on Wikipedia and Artificial Intelligence: an Evolving Synergy, pp. 25–30. AAAI Press, Chicago (2008)
13. Wolfe, A.: Social network analysis: Methods and applications. American Ethnologist 24(1), 219–220 (1997)
14. Zesch, T., Gurevych, I.: Analysis of the wikipedia category graph for NLP applications. In: Proceedings of the Second Workshop on TextGraphs: Graph-Based Algorithms for Natural Language Processing, pp. 1–8. Association for Computational Linguistics, Rochester (2007)
15. Zhu, Q., Lin, L., Shyu, M., Liu, D.: Utilizing context information to enhance content-based image classification. International Journal of Multimedia Data Engineering and Management (IJMDEM) 2(3), 34–51 (2011)

Supervised HDP Using Prior Knowledge

Boyi Xie and Rebecca J. Passonneau

Columbia University, Center for Computational Learning Systems,
475 Riverside Drive MC 7717, 10115 New York, USA
{xie,becky}@cs.columbia.edu

Abstract. End users can find topic model results difficult to interpret and evaluate. To address user needs, we present a semi-supervised hierarchical Dirichlet process for topic modeling that incorporates user-defined prior knowledge. Applied to a large electronic dataset, the generated topics are more fine-grained, more distinct, and align better with users' assignments of topics to documents.

Keywords: topic modeling, hierarchical Dirichlet process, supervised learning.

1   Introduction

Topic modeling, a method to discover semantic themes that permeate large collections of electronic documents provides a high level view of content in each document and over the collection. It is typically unsupervised, scales well, and can be done with little text pre-processing. Typically, however, only some of the topics look meaningful to end users, and it is unclear how useful a given topic model might be for improving user access to a collection. We introduce an approach to topic modeling that sacrifices some predictive power to incorporate an independent knowledge source.

This paper proposes a method to incorporate into topic modeling semantic categories of interest to users, and allows the user to control the degree to which this prior knowledge supervises the learning. Our method incorporates *a priori* semantic categories in two ways. The categories are used to initialize the set of topics, thus corresponding to topic labels. These categories are also used to label words in the vocabulary that have an *a priori* association with the categories.

We are collaborating with university librarians on a web archive of human rights sites in the use of a controlled vocabulary from the library domain, and a digital library collection. Their goals are to facilitate research on human rights websites, and to preserve sites at risk of being taken down. For subject indexing they use Library of Congress Subject Headings (LCSH), which are an integral part of bibliographic control. We demonstrate the use of a set of human rights LCSH terms in our method, and investigate how librarians and non-librarians assign the generated topics to documents in the collection.

Sections 2 (related work) and 3 (methods) provide context for the experiments described in section 4. We contrast the number and distinctness of topics in results from standard (unsupervised) hierarchical Dirichlet process (HDP) and three levels of supervised HDP. We compare unsupervised to lightly supervised models in our user study. In the user study results, topics from the supervised topic model align more closely with librarians' assignments of topics to websites.

G. Bouma et al. (Eds.): NLDB 2012, LNCS 7337, pp. 197–202, 2012.

2   Related Work

[4] first introduce latent Dirichlet allocation (LDA) to topic modeling, building on the idea from Latent Semantic Analysis (LSA) that hidden semantic dimensions condition the distribution of words in documents. LDA replaces matrix decomposition with a probabilistic model. [10] discuss the hierarchical Dirichlet process (HDP) and use a non-parametric Bayesian model.

Previous work on topic modeling has investigated the introduction of supervision. The models (e.g. [3], [8]) are based on document labels while we supervise using word characteristics. Labeled LDA in [9] that rely on supervision on words are relevant to our study, and z-label in [2] reflects a similar idea of domain dependent modeling. However, all these are LDA models while ours is based on hierarchical Dirichlet process, which is a non-parametric Bayesian approach. As such, it is more readily extended to supervision of the number of topics and concentration of information within topics.

3   Methods

We aim for scalability, thus we assume the existence of an unknown number of mixtures in any corpus. Nonparametric Bayesian methods are appropriate, as they define a model with an infinite limit of finite mixtures. In a hierarchical Dirichlet process (HDP) model, each data grouping is associated with a mixture model, which in turn contributes to a global model. Gibbs sampling is used for inference. Hyper parameters are determined by user-defined labels and word distributions within and across documents.

Our goal is to produce a semi-supervised topic model in which the topics align more or less tightly with pre-defined user categories. As part of this process, words in the vocabulary can have a more or less strong *a priori* association with these categories. Given a set of categories defined by the user, a parameter $\lambda \in [0, 1]$ controls the degree of supervision. When $\lambda = 0$, topics emerge from the data. When $\lambda = 1$, the initial topics will coincide with the labels in a one-to-one mapping.

In our framework, we assume that different metrics or procedures can be used to define the association between a user-defined category, such as an LCSH term, and words in the vocabulary. For this experiment, we associate words in the vocabulary to LCSH terms if they appear in the official definition of the term (see section 4).

In our semi-supervised hierarchical topic model, we incorporate knowledge from user-provided labels, together with the word distribution, to infer the model. There are two sets of parameters. The first set consists of the global level topic distribution over the corpus (θ_0 in *supervisedSampling* algorithm), the local level topic distribution over documents (θ_d), and the word distribution over topics. The other is a set of hyper parameters such as the concentration parameters for global and local level topics (γ and α). We use blocked Gibbs sampling that infers the two sets of parameters in turn. More detailed explanations can be found in [10] and [5].

Our *supervisedHDP* algorithm shown below starts with an initialization using user-defined labels, and uses this initial state to estimate initial hyper parameters. Lines 8 and 9 control the degree of supervision: *supLag* sets which iterations to supervise, and λ controls the probability a word gets supervision. In our *supervisedSampling*

Algorithm 1. supervisedHDP	**Algorithm 2.** supervisedSampling

// initialization
1 **for** *each document* $d \in D$ **do**
2   **for** *each word* $w \in V$ **do**
3   supervisedSampling();
 	
4 sample hyper parameters γ and α;
   // main iteration
5 **for** *iter* < *max iteration* **do**
6   **for** *each document* $d \in D$ **do**
7   **for** *each word* $w \in V$ **do**
8   **if** $iter\%supLag = 0$
            then
9             $X \sim Bernoulli(\lambda)$;
10   **if** $X == 1$ **then**
11   supervisedSampling();
 	
12   **else**
13   do ordinary HDP
           sampling;
 	
14   **else**
15   do ordinary HDP
        sampling ;
 	
16   sample hyper parameters γ and α;
	1 **if** *wordLabelRelation contains* w **then**
	2 sample a label $l_{d,w} \sim Mult(\psi_w)$;
	3 **if** *labelToTopic has* $l_{d,w}$ **then**
	4 $z_{d,w} \longleftarrow$
	        $labelToTopic.get(l_{d,w})$;
	5 **else**
	6 $z_{d,w} \longleftarrow$ a new topic;
	7 $labelToTopic.put(l_{d,w}, z_{d,w})$;
	
	8 **else**
	9 **if** *no topic created in the corpus* **then**
	10 $t_{d,w} \longleftarrow$ a new table;
	11 $z_{d,w} \longleftarrow$ a new topic;
	12 $topicToTable_d.put(z_{d,w}, t_{d,w})$;
	13 **else**
	14 $\boldsymbol{\theta}_0 \sim Dir(\gamma)$;
	15 $z_{d,w} \sim Mult(\boldsymbol{\theta}_0)$;
	
	16 **if** $topicToTable_d$ *has* $z_{d,w}$ **then**
	17 $t_{d,w} \longleftarrow topicToTable_d.get(z_{d,w})$;
	18 **else**
	19 $\boldsymbol{\theta}_d \sim Dir(\alpha)$;
	20 $t_{d,w} \sim Mult(\boldsymbol{\theta}_d)$;
	21 $topicToTable_d.put(z_{d,w}, t_{d,w})$;

algorithm, words first sample a label based on $wordLabelRelation$ (ψ_w), then local and global topics are inferred. $labelToTopic$ keeps track of the mapping between topics and labels, due to our assumption that a topic can be viewed as a mixture of labels (including none), and a label as a mixture of topics. They are different projections on different coordinates of interest.

4 Experiment

Data. The dataset contains 423 websites, addressing human rights issues all over the world. They are maintained by official organizations or individuals, and are in more than ten languages including English, Spanish, French, Chinese and Arabic. Each website home page and the depth one pages are crawled, concatenated and treated as a single document. We smooth our dataset to facilitate topic modeling. HTML headings and tags are removed. After we remove non-English websites, Flash sites and non-content ones, we end with a collection of 201 sites. We applied the Stanford CoreNLP named entity recognizer for organization, place and person names. The total size of the dataset is nearly 6 million words (30,000 per site). Removal of stop words and rare words, and concatenation of named entity words yields 29,392 unique terms.

Four Topic Models. We applied the following algorithms to the dataset: hierarchical Dirichlet process (HDP); supervision at initialization (InitialSup); supervision at initialization and every 20 iterations, with a 10% probability for words to be generated under supervision (LightlySup); supervision at every iteration, with a 50% probability for words to be generated under supervision (HeavilySup). For the three semi-supervised methods, librarians provided a set of labels (229 LCSH terms) related to human rights.

LCSH is a dynamic resource whose authoritative definitions are available on the web.[1] To assign word-level labels, we rely on the descriptive text that defines a term along with other textual fields (e.g., *General Notes*) and lists of variant terms, related terms and narrower terms. For each word w in our vocabulary, we associate w with each LCSH term l using the normalized term frequency of w in the descriptive text of the authority record for l. The word *police*, for example, is associated with four labels (LCSH terms) with the following strengths: *Bodyguards* (0.80), *Training* (0.09), *Peace-keeping forces* (0.07), *Extraordinary rendition* (0.04). About 10% of the vocabulary is associated with LCSH terms (2.9 per word on average).

User Study. We conducted a user study to measure how well users' assignments of topics to websites aligns with the relevant topic model under two conditions: the HDP and LightlySup topic models. For each condition, we randomly selected 15 sites, and from the topics for these sites, we selected 16 topics. Three librarians and three graduate students were recruited to participate in two one-hour sessions on different days to match topic word clouds to web sites. All did LightlySup on the first day, HDP on the second. Participants were instructed to browse each website and assign zero to three topics, along with percentages to reflect coverage.

5   Results

Two intuitive criteria that have been proposed for good topics are that they should be more fine-grained, and have fewer words in common [7]. Table 1 presents descriptive statistics for the results of the four methods at 200 iterations, illustrating that with greater supervision, topics are more numerous, and have fewer words in common. HeavilySup has 66.9% more topics than LightlySup, and 87.4% more than the average of HDP and InitialSup (col. 1). Columns two and three give the total number of distinct words among the top 1000 words across all topics for each method, and the distinct words per topic, followed by the ratio of topics per word. The first three methods have about the same ratio, while HeavilySup has half again as many.

To measure topic distinctness, we used the Jaccard coefficient [6], the ratio of the size of the intersection of two sets to the size of their union. Values range from 0 for disjoint sets to 1 for identical sets. Column 5 of Table 1 gives the average of the Jaccard coefficient for the sets of top twenty words from all pairs of topics for each method. The HDP topics have the most words in common, and LighlySup has the fewest.

Another contrast between the supervised and unsupervised methods pertains to coverage, in the sense of how many websites a topic gets assigned to (S/T, col. 6), and how much of a website it represents (T/S, col. 7). Supervised topics have higher S/T and

---

[1] http://id.loc.gov/authorities/subjects.html

lower Jaccard scores, which suggests they are relatively more distinct and less likely to be *vacuous* [1]. More supervised topics are assigned to each site (T/S) because each topic is more specific.

All the supervised methods are initialized by a set of topics corresponding directly to the human rights LCSH terms. Any new topic necessarily has no *a priori* association with an LCSH term, thus at any iteration after the first, some topics will link to an LCSH term and

Table 1. Descriptive Statistics (200 Iterations). Jacc is Jaccord score of 10^{-3}; S/T refers to sites per topic and T/S refers to topics per site.

Method	Topics	Vocab.	Ratio	Jacc	S/T	T/S
HDP	131	22,233	0.0059	13.11	4.11	2.68
InitialSup	138	22,652	0.0061	10.78	6.07	4.16
LightlySup	151	24,259	0.0062	8.34	6.21	4.67
HeavilySup	252	26,740	0.0094	9.14	10.47	13.12

others might not. At the 200th iteration, 89% of LightlySup topics and 94% of Heavily-Sup topics are associated with LCSH terms. Figure 1 illustrates topics from HeavilySup with and without an associated LCSH term. Font size in the word cloud represents the probability of the word in the topic.

Figure 1a, associated with the LCSH term *Government and the press*, reflects characteristics specific to this dataset: websites related to this term are often about Tibet and freedom of the press in China. Here, supervision yields topics that relate the data to the terminology in ways that an unsupervised method would not. Figure 1b is a contrasting example of a topic that is not associated with an LCSH term. It accounts for a signifi-

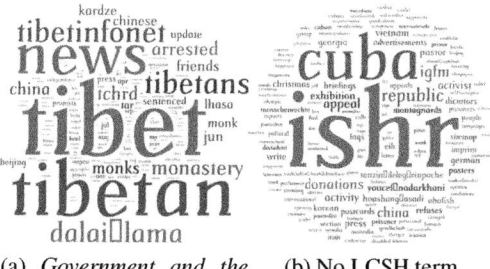

(a) *Government and the press.*

(b) No LCSH term

Fig. 1. Two HeavilySup Topics

cant proportion of the content of two websites. Thus supervised topic modeling can still find hidden relations among sites.

5.1 User Study Results

On average, the LightlySup and HDP models assign 2.00 and 1.63 topics per site, respectively. As shown in col. 2 of Table 2, librarians assign topics at a more consistent rate across models (1.88 for HDP vs. 2.11 for Lightly-Sup) while students' rates are less so (1.38 vs. 1.89). Columns 3-5 report precision/recall and f-measure (F) of the user assignments compared with the topic model assignments.

Precision on the user task is more critical to the librarians' ultimate goals than the other mea-

Table 2. User results

	Topic	Prec	Rec	F
Librarians				
HDP	1.88	0.49	0.71	0.54
LightlySup	2.11	0.59	0.53	0.51
Students				
HDP	1.38	0.66	0.70	0.64
LightlySup	1.89	0.61	0.50	0.51
All				
HDP	1.63	0.58	0.70	0.59
LightlySup	2.00	0.60	0.51	0.51

sures we report (recall and F), which are to have an automated method analogous to subject indexing. Librarians had much higher precision on the LightlySup task

(0.59 vs. 0.49). This difference between conditions did not show up for students, who had slightly higher precision on HDP (0.66 versus 0.61). A key difference between librarians and students was the much higher rate at which librarians assigned topics, suggesting a preference for finer-grained models, as has been reported elsewhere for subject matter experts [7]. Our design penalizes LightlySup in that by the time subjects did HDP they had had more practice on the task, and LightlySup assigned more topics per site (including two sites with 4 topics), so that probability of error was higher. Nevertheless, the overall results were roughly equivalent.

6   Conclusions

Supervised HDP yields topics that are more fine-grained, and more distinct, than topics produced by HDP. We tested the use of a controlled vocabulary in worldwide use by librarians for a subdomain pertaining to human rights, but the method can use any *a priori* semantic dimension. It is necessary to produce weighted (word,label) pairs. We tried various methods to do so, including mutual information between the subject terms used to catalog a website and words in the website text. This had noisy results due to a much larger number of (word,label) associations. In future work, we will compare different external resources, and continue to explore degrees of supervision.

Acknowledgements. We thank Terence H. Catapano, Melanie Wacker and Stuart Marquis for introducing us to the human rights archive, and we thank them and the additional librarians and students who participated in the study.

References

1. AlSumait, L., Barbará, D., Gentle, J., Domeniconi, C.: Topic Significance Ranking of LDA Generative Models. In: Buntine, W., Grobelnik, M., Mladenić, D., Shawe-Taylor, J. (eds.) ECML PKDD 2009, Part I. LNCS, vol. 5781, pp. 67–82. Springer, Heidelberg (2009)
2. Andrzejewski, D., Zhu, X.: Latent dirichlet allocation with topic-in-set knowledge. In: Proceedings of the NAACL HLT 2009 Workshop on Semi-Supervised Learning for Natural Language Processing, pp. 43–48 (2009)
3. Blei, D.M., McAuliffe, J.D.: Supervised topic models. In: Advances in Neural Information Processing Systems, NIPS (2007)
4. Blei, D.M., Ng, A., Jordan, M.: Latent dirichlet allocation. JMLR 3, 993–1022 (2003)
5. Escobar, M.D., West, M.: Bayesian density estimation and inference using mixtures. Journal of the American Statistical Association 90, 577–588 (1995)
6. Jaccard, P.: Nouvelles recherches sur la distribution florale. Bulletin de la Société Vaudoise des Sciences Naturelles 44, 223–270 (1908)
7. Mimno, D., Wallach, H., Talley, E., Leenders, M., McCallum, A.: Optimizing semantic coherence in topic models. In: Proceedings of the 2011 Conference on Empirical Methods in Natural Language Processing, Edinburgh, Scotland, pp. 262–272 (July 2011)
8. Perotte, A., Bartlett, N., Elhadad, N., Wood, F.: Hierarchically supervised latent dirichlet allocation. In: Advances in Neural Information Processing Systems, NIPS (2011)
9. Ramage, D., Hall, D., Nallapati, R., Manning, C.D.: Labeled lda: a supervised topic model for credit attribution in multi-labeled corpora. In: Proceedings of the 2009 Conference on Empirical Methods in Natural Language Processing, pp. 248–256 (2009)
10. Teh, Y.W., Jordan, M.I., Beal, M.J., Blei, D.M.: Hierarchical dirichlet processes. Journal of the American Statistical Association 101(476), 1566–1581 (2006)

User-Driven Automatic Resource Retrieval Based on Natural Language Request

Edgar Camilo Pedraza, Julián Andrés Zúñiga,
Luis Javier Suarez-Meza, and Juan Carlos Corrales

Grupo de Ingeniería Telemática (GIT), Universidad del Cauca,
Calle 5 No 4-70, Popayán, Colombia
{epedraza,gzja,ljsuarez,jcorral}@unicauca.edu.co

Abstract. In this paper we propose an innovative approach for User-driven retrieval of Telecom and IT resources over converged environments, which brings together traditional NLP techniques with search and selection process based on lightweight semantic technologies, aimed at ordinary end users. The preliminary experiments show promising results in contrast to traditional approaches.

Keywords: Telecom and IT resources, NLP Techniques, User-driven retrieval.

1 Introduction

Currently, the Web has evolved significantly in relation to telecommunications, due to the emergence of different tools and APIs for the development of services, which enable developers to more easily create services from existing ones and distributed data on the Web [1,2]. In addition, as Web has become more user-centric, the concept of UGS (User Generated Services) has emerged. It gives the possibility to end-users, generally with no technical or programming skills, to create their own services in a fully personalized fashion.

In this context, the ability to retrieve services that accomplish demanding requirements of users, it has led to the development of several approaches focused on service discovery [3]. These approaches are not aimed at ordinary users, since they must specify their requests by complex expressions and it becomes even more complex in dynamic and heterogeneous contexts, as converged environments (Web + Telco) [4,5,6]. Most of the current retrieval approaches are targeted only to Web services, leaving out several other kinds of resources, such as: data, Telecom capabilities [4], Widgets, APIs and so on, which possibly can provide better compliance of user's requirements.

In order to address the above drawbacks, we have proposed an approach that support the automatic retrieval of resources in converged environments [5], with the aim of allowing that end-users (i.e., without technical or programming skills) specify their requests in natural language (NL). So that the search time of resources can be reduced, leading to a better experience and user satisfaction.

The rest of this paper is organized as follows: related works are outlined in Section 2. Our proposal for modeling system entities is introduced in Section 3. A description of the NLP module algorithms proposed is introduced in Section 4. We provide details about our implementation and evaluation in Section 5. Finally in Section 6 we conclude the paper.

G. Bouma et al. (Eds.): NLDB 2012, LNCS 7337, pp. 203–209, 2012.

2   Related Work

There are approaches that consider the resource retrieval aimed at end-users [3,7], whose purpose is to replace formal expression with simple requests made in natural language through different processes of linguistic analysis: text segmentation, irrelevant word removal, stemmming and grammatical correction [3], information retrieval [7] or matching the structure of the request with predefined patterns using keywords. The projects attempts to calculate the similarity between the obtained terms from the user request with the available resources in repositories. In [7], the service repository has a small scale of web and telecommunication services, on the other hand, in [3], the repository is limited only to web services. The different approaches present disadvantages, some of them use common ontologies that may not reflect the vocabulary used by end-users, also restrict the use of NL to simple requests and do not allow the extraction of a services flow from the petition. In contrast, our approach allows to make flexible requests in English and the creation of a generic control flow from keywords of the request made by the user, finally we propose the use of folksonomies to increase the dynamicity in adaption to changes that emerge in user's vocabulary, for which, we implement several NLP tasks that allow advanced analysis of the requests made in NL [5].

3   Model of System Entities

In [5] it is possible to see our proposed architecture for automatic services retrieval in converged environments, which receives as input the user's request made in NL from a mobile device, and gives as output a resource ranking which is related by a generic control flow, in order to meet the user's request. In the following we describe in more detail the models defined.

3.1   Semantic Description Model

In this paper we define three main entities: the user $U = \{u, ..., u_n\}$, the resources $R = \{r_1, ..., r_m\}$, and descriptions of those resources and their functional parameters, the tags $T = \{t, ..., t_k\}$. It is worth noting that in our system we took advantage of the user's request, to implicitly annotate the resources with relevant tags, since users usually do not perform it. Thus, users "annotate" the resources with tags, creating a triple association between the user, the resource and the tag. The triple association define the semantic description of our proposal, and can be defined by a set of annotations $A \subseteq U \times R \times T$[8].

3.2   User Model

From the study of work-related is possible to observe that there is no meta-model that describes the user's requests in NL, Therefore, in the Figure 1 we present our proposal for this concern. From user's request we identify three different words classifications: (i) *Control words*, which will serve to establish the final generic control flow of desired services; (ii) *Functional words*, those words are also classified into Behavior and

Input/Output categories, which represent operational parameters used to retrieval process; and (iii) *Non-Functional words*, which represent non-functional properties of resources. According to the meta-model of the user's request is possible to define the following:

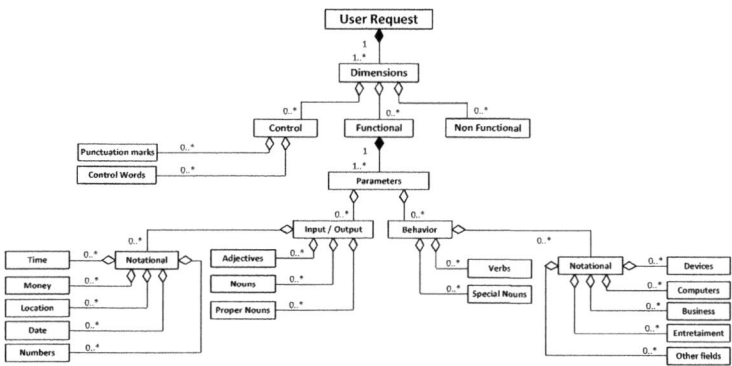

Fig. 1. User's request Meta-model

User's Request (Q_0 and Q). Let Q_0 the *informal user request* (which represents the desired results from users) expressed in natural language before its analysis. A *formal user request Q* is defined as a tuple $Q = (userID, F, NF, C, \lambda)$ where *userID* is the identifier of the user that perform the request, F is a non-empty finite set of elements that represent *Functional Words*, such that $F = \{f_1, ..., f_n\}$. NF is a finite set of elements that represent *Non-Functional Words*, such that $NF = \{nf_1, ..., nf_m\}$. F and NF are distinct, $F \cap NF = \emptyset$. C is a pair $C = (W, P)$ where W is a non-empty finite set of words that denote *Control Words* (e.g., If, which, and, or, etc.), such that $W = \{c_1, ..., c_m\}$, and P is a finite set of elements that represent *Punctuation Marks* (, . ; :). $\lambda = (F \times C)$ is the function that records the sentence meaning/structure, which is helpful to ensure that the results obtained by our system meet the user's request.

Elements both F and C can represent *Resources $R = \{r\}$*. A *Resource* is a basic unit of retrieval phase. Generally, r represents one or more abstract resources from a subset of existing real services. NF is considered a *Parameter $P = \{p\}$* used to refine the ranking of retrieved resources (This represents the non-functional properties of services, e.g. QoS[5]).

In this sense, according to above definitions, both *Resources r* and *Parameters p* can be described by a specific set of tags $d(r) \subseteq T$ and $d(p) \subseteq T$ respectively.

Tag Query (q). In general, a tag query $q \subseteq T$ selects a subset $\Re$ of an resource set $R = \{r\}$ such that each resource in $\Re$ is described by all tags in q, taking into account sub-tag relationships between tags, i.e, if a tag $t_1 \in T$ is a sub-tag of $t_2 \in T$, denoted $t_1 \sim t_2$ [9]. Therefore, according to this, formally we have:

$$\Re_f(r) = \{r \in R \mid \forall t \in q_f \; \exists t' \in d(r) : t' \sim t \land \forall f_j \in F, \exists q_f : q_f \subseteq T\} \quad (1)$$

Where: $\Re_f(r)$ is an resource subset of all resource set $R = \{r\}$ such that each resource in this subset is described by all tags in q_f (set of *functional word* tags). Thus, for each $f \in F$ there is a $q_f \subseteq T$. The above property is also used to $\Re_c(r)$ and $\Re_{nf}(p)$. Where $\Re_c(r)$ is an resource subset of all resource set $R = \{r\}$ such that each resource in this subset is described by all tags in q_c (set of *control word* tags). Thus, for each $c \in W$ there is a $q_c \subseteq T$; and $\Re_{nf}(p)$ is an parameter subset of all parameter set $P = \{p\}$ such that each parameter in this subset is described by all tags in q_{nf} (set of *non-functional word* tags). Thus, for each $n_f \in NF$ there is a $q_{nf} \subseteq T$.

3.3   Resource Model

At a logical level, resources R_l are defined as a set $R_l = \{R_{li} | R_{li} = (name_i, T_i)\}$ with $name_i$ being the unique name of the resource r_{li} and $T_i \in \tau$ represents a description based on tags of the r_{li} (from the some user). However, at an executable level, $R_e = \{R_{ei} | R_{ei} = (In_i, Out_i, Op_i)\}$ is a non-empty set of resources, where: $In_i = \{in_{i0}, ..., in_{ij}\}$, $Out_i = \{out_{i0}, ..., out_{ik}\}$ and $Op_i = \{op_{i0}, ..., op_{il}\}$ are respectively the sets of *input*, *output*, and *operations* of an resource r_{ei}. Thus, the set of *Resources R* is defined as: $R = R_l \cup R_p$. We distinguish two kinds of resources in this work: *Source*, which fetch data from the web or the local machine. They don't have inputs, i.e., $In_i = \emptyset$; and *Typical*, which consume data in the input and produce processed data in the output. Therefore, In_i, $Out_i \neq \emptyset$.

4   NLP Module Algorithms

Based on the models presented in the previous section, below we describe the algorithms that we define to support each of the tasks considered as relevant in the *NLP-module* implemented in our system of resource retrieval [5]. The system has as input the user's request made in natural language, and the purpose of the module is the identification and selection of terms represented by the User's request meta-model. These terms are classified into several categories and used for obtaining resources. The resources are representation of a subset of existing real services (web and telecommunication services) according to the Resource model.

4.1   Named Entity Recognition Algorithm

Name Entity Recognition (NER) task recognizes proper names (entities) in a corpus (user request) and associates them with common types (location, person name, organization, etc). The NER algorithm uses a Gazetter and rules defined in JAPE (Java Annotation Patterns Engine) [10]. The Gazetteer makes lists lookup according to matches found in the corpus with lists of defined entities. With JAPE, we define rules for the recognition of regular expressions. The adaptation of the algorithm consists in identifying those entities that define a service interface: the Input or Output terms of a service or its Behavior (Operations), also Control terms that allow the creation of a generic flow of resources (see Figure 1).

4.2   Service Interface Entities Enrichment with WSD Algorithm

Word Sense Disambiguation (WSD) task consists in selecting the correct sense of a polysemous word in a context. The system requires the execution of the Part Of Speech Tagger (POST) in the user's request, the POST tags words according to its grammatical category (e.g.,she "pronoun", loves "verb"), and uses a default rulset and a lexicon to identify the correct grammatical category for an input word. This task is necessary because the algorithm gets the sense of the word depending on whether the word to disambiguate is a verb or a noun. The WSD task in our proposal uses an unsupervised Lesk algorithm proposed by Lesk [11], which obtains a list of synonymous words of the right sense and its definition. As extension, we compare the different list of synonymous words with the retrieved keywords from the user's request; in case of coincidence, the algorithm adds the first noun of each definition of the right sense as an entity of type input-output or behavior, according to the common type given by the NER algorithm to the retrieved keyword, to enrich the obtained terms from the Ner algorithm.

5   Experimental Study

In this section, we present a brief description of the implementation about presented algorithms; all tests were done in Java language on an Intel Core 2 Duo with 2.30GHz processor, 4.00 GB of RAM under the OS Linux Ubuntu.

The evaluation for the *Word Sense Disambiguation Task* was focused on the accuracy to disambiguate a given word. To perform the test, we used a corpus (utilized in Senseval 3[1]), which consists of 6 words from the 57 available with descriptions of web and telecommunications resource interfaces. The sense inventory is obtained from Wordnet 1.7.1 and for each word, fifty instances of its context were used to disambiguate. Thus, the test compares the correct sense given by Senseval and the chosen one by the system. The evaluation reports results below 50% in precision and recall, this is mainly because the context of the word to disambiguate had common words for the different possible senses of that word, making the process difficult, but the assessment also showed results that exceed the participating systems of Senseval-2, whose best representative had 51.2% of precision and recall measures.

To evaluate the *Named Entity Recognition Task*, we rely on the measures defined in [12] and use a corpus obtained from different descriptions of the Android Market Services[2]. Entities considered in this corpus are: *Behavior* (message, call, maps, email, music, send, video, messenger and sms), *Input/Output* (location, person, number and time) and *Control* (before, however, if, or, and while). In addition, for the evaluation, experts performed manual annotations and annotated the entities that describe the service interface (Input/Output, Behavior) according to the types of entities or words specified. Results of this evaluation obtained precision and recall measures above 50%, however the measurement of categories such as location, is considered as bad results, since the number of entity made by the system and correct entity by the system were low compared to the number of entity given by the user. The differences in these values was due to the lack of location entities in the gazetteer lists that were consistent with the corpus,

---

[1] Available: http://www.senseval.org/senseval3/data.html
[2] Available at: https://market.android.com/

and also due to the rule that made the location annotation in a word when the word was after the words "in" or "at", this rule was too generic to distinguish a real entity and not a false positive.

6   Conclusion and Future Work

In this paper we introduce an approach that brings together traditional NLP techniques with the resource (Telecom, IT and content) retrieval based on lightweight semantic. Our proposal involves: modeling the request of users made in natural language, in order to facilitate the specification of their requirements and allow its subsequent treatment; definition of several NLP techniques for each of the tasks involved in the NLP; and evaluating of these NLP techniques, in order to measure the properties of the proposed algorithms and determining whether the system accomplishes the functionality for which it was designed.

Thus, the assessment shows good performance in general terms, considering the complexity of each module separately. However, some rules and filters established to detect specific feature or categories of input requests can be enhanced taking into account more cases with which may further raise the level of accuracy.

The paper presents a detailed description and evaluation about functionality of the implemented NLP module into our prototype, leaving out other ones (such as: Matching, Inference and Recommender modules) of the entire system, which will be performed in future work.

Acknowledgments. The authors would like to thank Universidad del Cauca, COL-CIENCIAS and TelComp2.0 Project for supporting the Research of the M.Sc. Student Luis Javier Suarez.

References

1. Sari, S.W.: Diversity of the mashup ecosystem. In: WEBIST, Noordewijkerhout, The Netherlands (May 2011)
2. Tapia, B., Torres, R., Astudillo, H.: Simplifying mashup component selection with a combined similarity- and social-based technique. In: Mashups 2011, pp. 8:1–8:8. ACM, New York (2011)
3. Pop, F.-C., Cremene, M., Vaida, M.-F., Riveill, M.: On-demand service composition based on natural language requests. In: WONS 2009, pp. 41–44. IEEE Press, Piscataway (2009)
4. Carlin, J.M.E., Trinugroho, Y.B.D.: A flexible platform for provisioning telco services in web 2.0 environments. In: NGMAST 2010, pp. 61–66. IEEE Computer Society, Washington, DC (2010)
5. Pedraza, E.C., Zuniga, J.A., Suarez Meza, L.J., Corrales, J.C.: Automatic service retrieval in converged environments based on natural language request. In: SERVICE COMPUTATION 2011, Rome, Italy, September 25, pp. 52–56 (2011) 978-1-61208-152-6
6. Bond, G., Cheung, E., Fikouras, I., Levenshteyn, R.: Unified telecom and web services composition: problem definition and future directions. In: IPTComm 2009, pp. 13:1–13:12. ACM, New York (2009)

7. Kirati, S.: A demonstration on service compositions based on natural language request and user contexts. Master's thesis, Norwegian University of Science and Technology (NTNU), Department of Telematics (June 2008)
8. Bouillet, E., Feblowitz, M., Feng, H., Liu, Z., Ranganathan, A., Riabov, A.: A folksonomy-based model of web services for discovery and automatic composition. In: Proceedings of the 2008 IEEE International Conference on Services Computing, vol. 1, pp. 389–396. IEEE Computer Society, Washington, DC (2008)
9. Helic, D., Strohmaier, M., Trattner, C., Muhr, M., Lerman, K.: Pragmatic evaluation of folksonomies. In: WWW 2011, pp. 417–426. ACM, New York (2011)
10. Cunningham, H.: Jape: a java annotation patterns engine. Research Memorandum CS – 99 – 06, Department of Computer Science, University of Sheffield (May 1999)
11. Lesk, M.: Automatic sense disambiguation using machine readable dictionaries: how to tell a pine cone from an ice cream cone. In: SIGDOC 1986, pp. 24–26. ACM, New York (1986)
12. Sun, B.: Named entity recognition: Evaluation of existing systems (2010)

Integrating Lexical-Semantic Knowledge to Build a Public Lexical Ontology for Portuguese

Hugo Gonçalo Oliveira[1], Leticia Antón Pérez[1,2], and Paulo Gomes[1]

[1] CISUC, University of Coimbra, Portugal
[2] Higher School of Computer Engineering, University of Vigo, Spain
{hroliv,pgomes}@dei.uc.pt, leticiaap86@gmail.com

Abstract. Onto.PT is a new public wordnet-like lexical ontology for Portuguese. To make its creation faster and to deal with information sparsity, Onto.PT is created automatically and integrates information extracted from existing lexical resources. After introducing Onto.PT's construction approach, this paper gives a closer look on this resource and provides information on its availability.

Keywords: wordnet, lexical ontology, lexical knowledge base, thesaurus, information extraction, synsets, semantic relations.

1 Introduction

The need to perform natural language processing (NLP) tasks where, besides recognising words and their interactions, it is crucial to understand the meaning of text, lead to the creation of broad-coverage lexical-semantic resources. Despite some terminological issues, these resources may be seen as lexical ontologies, as they have properties of a lexicon as well as properties of an ontology [9,10].

Today, lexical ontologies are key resources for NLP, which is evidenced for English, where Princeton WordNet [3] is widely used in the achievement of a wide range of tasks, including word sense disambiguation [1], text summarisation [15], or question answering [13]. Given the positive impact of a wordnet in the development of NLP tools, the wordnet model spread to other languages[1].

For Portuguese, however, wordnet-based projects are either, or both, proprietary or handcrafted, and are thus being developed slowly (see [18] for an overview on Portuguese lexical knowledge bases). So, as others did for enriching Princeton WordNet [8,12], or for creating new lexical knowledge bases [16,21,7], we moved our efforts to the automatic construction of Onto.PT, a new lexical ontology for Portuguese.

In opposition to most wordnets, Onto.PT is populated automatically, with lexical-semantic information extracted from textual sources. Even though less reliable than manual approaches, its creation is faster, its maintenance easier,

---

[1] A list with the wordnets of the World can be found in
http://www.globalwordnet.org/gwa/wordnet_table.html

G. Bouma et al. (Eds.): NLDB 2012, LNCS 7337, pp. 210–215, 2012.
© Springer-Verlag Berlin Heidelberg 2012

and its growth potential higher. On the other hand, as WordNet, Onto.PT is freely available, and is structured on synsets and semantic relations.

The construction of Onto.PT is based on three independent steps, briefly presented in section 2, adaptable, for instance, to the automatic construction of wordnets in other languages. In section 3 we overview the current version of Onto.PT and its availability. We conclude with cues for further work.

2   The Construction of Onto.PT

Onto.PT is built in a modular approach, divided in three automatic steps, exemplified in figure 1 for a given textual definition:

1. **Extraction of semantic relations** [4]: Definitions are processed according to handcrafted grammars, designed to exploit their regularities and extract semantic relations. Relations are represented as term-based triples – t_1 RELATED-TO t_2 – where t_1 and t_2 are lexical items and RELATED-TO is the name of a relation between one sense of t_1 and one sense of t_2.
2. **Thesaurus enrichment** [5]: When possible, synonymy triples are integrated into suitable synsets of an existing thesaurus, according to the average similarity between the items in the triples and in the synsets. Then, the graph established by the remaining synonymy triples is exploited in the discovery of new synsets, later added to the thesaurus.
3. **Ontologisation of semantic relations** [6]: Non-synonymy triples are ontologised [14], which sets a move from relations based on words to relations based on meanings. Each item argument of a triple is assigned to the most suitable synset in the thesaurus. This results in a synset-based triple. Suitable synsets are selected from the most similar pairs of synsets, such that one contains the first item of the triple and the other contains the second.

gado	s.m.	conjunto de animais criados para diversos fins; rebanho
(cattle	noun	set of animals raised for various purposes; flock)
	tb_triple_1	= rebanho SINONIMO_DE gado (flock SYNONYM_OF cattle)
	tb_triple_2	= animal MEMBRO_DE gado (animal MEMBER_OF cattle)
	$synset_1$	= (manada, rebanho, mancheia, boiada)
$synset_1 + tb\_triple_1$		= (manada, rebanho, mancheia, boiada, gado)
	$synset_2$	= (bicho, animal, alimal, béstia, minante)
	sb_triple_1	= $synset_2$ MEMBRO_DE $synset_1$

Fig. 1. Example of the three construction steps

In steps 2 and 3, similarity is based on the adjacencies of the lexical items in the graph established by the relations extracted in step 1. More details about the construction steps, including their evaluation, can be found elsewhere [4,5,6]. We believe that this is an interesting way of coping with information sparsity

– it enables the acquisition of knowledge from different heterogeneous sources, and harmoniously integrates all the information in a common lexical ontology, as it is structured on synsets and semantic relations between them.

In the current version of Onto.PT, five public domain lexical resources were integrated. We were inspired by the creation of the lexical resource PAPEL [7], and integrated PAPEL 3.0 with lexical-semantic relations extracted from Dicionário Aberto [20] and Wiktionary.PT[2], both dictionaries. Furthermore, we used two handcrafted electronic thesauri, TeP 2.0 [11], for Brazilian Portuguese, and OpenThesaurus.PT[3], for European Portuguese. As it is the largest, TeP is currently used as starting point in step 2.

3   A Public Lexical Ontology for Portuguese

This section describes the contents of Onto.PT and how they are shared with the community.

3.1   Contents

The current version of Onto.PT has about 150,000 lexical items, organised in about 110,000 synsets ($\approx$65k nouns, $\approx$27k verbs, $\approx$18k adjectives, $\approx$2k adverbs) and about 178,000 relational triples between synsets (including $\approx$88k hypernymy, $\approx$8k part-of, $\approx$8k member-of, $\approx$12k causation, $\approx$17k purpose-of and $\approx$37k property-of). Table 1 presents the types of relations in Onto.PT, which are the same as in PAPEL. As in the latter resource, each semantic relation has a different predicate according to the part-of-speech (POS) of its arguments.

Table 1. Semantic relations of Onto.PT

Relations	Predicates
Hypernym	n hiperonimoDe n
Part	n parteDe n n parteDeAlgoComProp adj adj propDeAlgoParteDe n
Member	n membroDe n n membroDeAlgoComProp adj adj propDeAlgoMembroDe n
Contains	n contidoEm n n contidoEmAlgoComProp adj
Material	n materialDe n
Causation	n causadorDe n n causadorDeAlgoComProp adj adj propDeAlgoQueCausa n n causadorDaAccao v v accaoQueCausa n
Place	n localOrigemDe n
Antonym	adj antonimoAdjDe adj

Relations	Predicates
Producer	n produtorDe n n produtorDeAlgoComProp adj adj propDeAlgoProdutorDe n
Purpose	n fazSeCom n n fazSeComAlgoComProp adj v finalidadeDe n v finalidadeDeAlgoComProp adj
Quality	n temQualidade n n devidoAQualidade adj
State	n temEstado n n devidoAEstado adj
Manner	adv maneiraPorMeioDe n adv maneiraComProp adj
Manner without	adv maneiraSem n adv maneiraSemAccao v
Property	adj dizSeSobre n adj dizSeDoQue v

---

[2] http://pt.wiktionary.org/

[3] http://openthesaurus.caixamagica.pt/

3.2   Availability

Onto.PT is available as a RDF model, as this is a W3C standard for describing information as triples. The structure of the ontology is based on the W3C RDF representation of WordNet [2]. There are four classes for synsets of the four POS (NomeSynset, VerboSynset, AdjectivoSynset, AdverbioSynset) and we have defined all the types of extracted semantic relations. The ontology was populated with the synsets and the relational triples. Synsets were ordered according to the average frequency of their lexical items in several Portuguese corpora, provided by the AC/DC project [19]. Also, inside each synset, lexical items were ordered according to their frequency in the same resource.

OntoBusca is a web interface for querying a triple store with the RDF ontology and thus explore Onto.PT easily. This interface is currently very similar to WordNet Search[4]. It is thus possible to search for a word to obtain all the synsets containing it, which may then be expanded for accessing semantic relations.

Figure 2 is an example of querying OntoBusca for the word *envolvente* and expanding some returned synsets. The POS of the synsets is provided and it is possible to navigate through direct and indirect relations. This lexical item has one noun and two adjective synsets – one meaning attractive/captivating, and another meaning something that surrounds. The first adjective synset is the property

OntoBusca
Onto.PT

Introduza a sua pesquisa:

[envolvente]   (Busca)

- S: (adj) **envolvente**, glamouroso, enfeitiçante, sedutor, apaixonante, charmoso, magnetizante, sirénico, mago, fascinador, fádico, atrativo, glamoroso, atractivo, encantador, mágico, feiticeiro, fascinante, cativante, atraente, tentador, atraidor
  - propriedadeDoQue
    - S: (v) ter_qualidade_de_atrair
    - S: (v) exercer_atracção
    - S: (v) seduzir, mesmerizar, encantar, magnetizar, encandear, fascinar, hipnotizar
    - S: (v) ter_charme
    - S: (v) ter_glamour
    - S: (v) persuadir, convencer, atrair, incutir, aliciar, instigar, seduzir
    - descricaoDaPropriedade
    - accaoFinalidadeDe
      - S: (s) galanteio, galanice, galanteria, cortejo, galantaria
        - meioParaAccao
        - resultadoDaAccao
        - hiperonimoDe
          - S: (s) namorisco, flerte, flirt, namorilho, namorico
        - hiponimoDe
      - S: (s) encarva
      - S: (s) coquete
      - accaoQueCausa
      - S: (v) relevar, convir, dar, importar, concernir, afetar, interessar
  - propriedadeDeAlgoQueTemParte
- S: (adj) **envolvente**, ambiente, envolvedor, capcioso, invaginante
- S: (s) **envolvente**

Fig. 2. Results for the query *envolvente* in OntoBusca, partly expanded

---

[4] http://wordnetweb.princeton.edu/perl/webwn

(propriedadeDoQue) of something that attracts, has glamour, or has charm. The synset *persuadir, ... , seduzir* (to persuade/seduce) is expanded to get that it is the purpose (accaoFinalidadeDe) of courting/wooing, which is an hypernym-of a flirt.

4   Conclusion and Further Work

We have presented Onto.PT, a new lexical ontology for Portuguese, created automatically from textual sources. We believe that, in a near future, Onto.PT can be viewed as a public alternative in the paradigm of Portuguese lexical-knowledge bases.

Onto.PT is in development, and we will keep working on the improvement of its quality and coverage. In the current version, Onto.PT integrates knowledge from five resources. We will continue evaluating Onto.PT, especially in the achievement of NLP tasks. So far, we have implemented several methods for knowledge-based word sense disambiguation [1], where Onto.PT is used both as a sense inventory and as a knowledge base for disambiguation. Identifying the synset corresponding to the meaning of a word in context is the first step for tasks such as writing suggestion, machine translation, or query expansion. On the latter, Onto.PT has recently been used in the information retrieval task Págico [17], for providing synonyms of topic categories and thus improve recall.

Other cues for future work involve computing the confidence of extracted relations; or taking advantage of inference features, provided by ontology description languages. Resources generated in the scope of this project, including the whole ontology, are freely available from http://ontopt.dei.uc.pt.

Acknowledgements. Hugo Gonçalo Oliveira is supported by the FCT grant SFRH/BD/44955/2008, co-funded by FSE.

References

1. Agirre, E., Lacalle, O.L.D., Soroa, A.: Knowledge-based WSD on specific domains: performing better than generic supervised wsd. In: Proc. of 21st International Joint Conference on Artifical Intelligence, IJCAI 2009, pp. 1501–1506. Morgan Kaufmann Publishers Inc., San Francisco (2009)
2. van Assem, M., Gangemi, A., Schreiber, G.: RDF/OWL representation of Word-Net. W3c working draft, World Wide Web Consortium (June 2006), http://www.w3.org/TR/2006/WD-wordnet-rdf-20060619/
3. Fellbaum, C. (ed.): WordNet: An Electronic Lexical Database (Language, Speech, and Communication). The MIT Press (1998)
4. Gonçalo Oliveira, H., Antón Pérez, L., Costa, H., Gomes, P.: Uma rede léxico-semântica de grandes dimensões para o português, extraída a partir de dicionários electrónicos. Linguamática 3(2), 23–38 (2011)
5. Gonçalo Oliveira, H., Gomes, P.: Automatically Enriching a Thesaurus with Information from Dictionaries. In: Antunes, L., Pinto, H.S. (eds.) EPIA 2011. LNCS, vol. 7026, pp. 462–475. Springer, Heidelberg (2011)
6. Gonçalo Oliveira, H., Gomes, P.: Ontologising relational triples into a portuguese thesaurus. In: Proc. 15th Portuguese Conf. on Artificial Intelligence, pp. 803–817. EPIA 2011. APPIA, Lisbon, Portugal (2011)

7. Gonçalo Oliveira, H., Santos, D., Gomes, P.: Extracção de relações semânticas entre palavras a partir de um dicionário: o PAPEL e sua avaliação. Linguamática 2(1), 77–93 (2010)
8. Harabagiu, S.M., Moldovan, D.I.: Enriching the WordNet taxonomy with contextual knowledge acquired from text. In: Natural Language Processing and Knowledge Representation: Language for Knowledge and Knowledge for Language, pp. 301–333. MIT Press, Cambridge (2000)
9. Hirst, G.: Ontology and the lexicon. In: Staab, S., Studer, R. (eds.) Handbook on Ontologies. International Handbooks on Information Systems, pp. 209–230. Springer (2004)
10. Huang, C.R., Prévot, L., Calzolari, N., Gangemi, A., Lenci, A., Oltramari, A.: Ontology and the lexicon: a multi-disciplinary perspective (introduction). In: Ontology and the Lexicon: A Natural Language Processing Perspective. Studies in Natural Language Processing, ch.1, pp. 3–24. Cambridge University Press (April 2010)
11. Maziero, E.G., Pardo, T.A.S., Di Felippo, A., Dias-da-Silva, B.C.: A Base de Dados Lexical e a Interface Web do TeP 2.0 - Thesaurus Eletrônico para o Português do Brasil. In: VI Workshop em Tecnologia da Informação e da Linguagem Humana (TIL), pp. 390–392 (2008)
12. Navigli, R., Velardi, P., Cucchiarelli, A., Neri, F.: Extending and enriching WordNet with OntoLearn. In: Proc. 2nd Global WordNet Conf., GWC 2004, pp. 279–284. Masaryk University, Brno (2004)
13. Pasca, M., Harabagiu, S.M.: The informative role of WordNet in open-domain question answering. In: Proc. NAACL 2001 Workshop on WordNet and Other Lexical Resources: Applications, Extensions and Customizations, Pittsburgh, USA, pp. 138–143 (2001)
14. Pennacchiotti, M., Pantel, P.: Ontologizing semantic relations. In: Proc. 21st Intl. Conf. on Computational Linguistics and 44th Annual Meeting of the ACL, pp. 793–800. ACL (2006)
15. Plaza, L., Díaz, A., Gervás, P.: Automatic summarization of news using wordnet concept graphs. International Journal on Computer Science and Information System (IADIS) V, 45–57 (2010)
16. Richardson, S.D., Dolan, W.B., Vanderwende, L.: MindNet: Acquiring and structuring semantic information from text. In: Proc. of 17th Intl. Conf. on Computational Linguistics, COLING 1998, pp. 1098–1102 (1998)
17. Rodrigues, R., Gonçalo Oliveira, H., Gomes, P.: Uma abordagem ao Página baseada no processamento e análise de sintagmas dos tópicos. Linguamática 4(1), 31–39 (2012)
18. Santos, D., Barreiro, A., Freitas, C., Gonçalo Oliveira, H., Medeiros, J.C., Costa, L., Gomes, P., Silva, R.: Relações semânticas em português: comparando o TeP, o MWN.PT, o Port4NooJ e o PAPEL. In: Textos Seleccionados. XXV Encontro Nacional da Associação Portuguesa de Linguística, pp. 681–700. APL (2010)
19. Santos, D., Bick, E.: Providing Internet access to Portuguese corpora: the AC/DC project. In: Proc. 2nd Intl. Conf. on Language Resources and Evaluation, LREC 2000, pp. 205–210 (2000)
20. Simões, A., Farinha, R.: Dicionário Aberto: Um novo recurso para PLN. Vice-Versa, pp. 159–171 (December 2011)
21. Wandmacher, T., Ovchinnikova, E., Krumnack, U., Dittmann, H.: Extraction, evaluation and integration of lexical-semantic relations for the automated construction of a lexical ontology. In: Proc. of 3rd Australasian Ontology Workshop (AOW). CRPIT, vol. 85, pp. 61–69. ACS, Gold Coast (2007)

From Ontology to NL: Generation of Multilingual User-Oriented Environmental Reports

Nadjet Bouayad-Agha[1], Gerard Casamayor[1], Simon Mille[1], Marco Rospocher[2],
Horacio Saggion[1], Luciano Serafini[2], and Leo Wanner[1,3]

[1] DTIC, Universitat Pompeu Fabra
[2] Fondazione Bruno Kessler
[3] Institució Catalana de Recerca i Estudis Avançats

Abstract. Natural Language Generation (NLG) from knowledge bases (KBs) has repeatedly been subject of research. However, most proposals tend to have in common that they start from KBs of limited size that either already contain linguistically-oriented knowledge structures or to whose structures different ways of realization are explicitly assigned. To avoid these limitations, we propose a three layer OWL-based ontology framework in which domain, domain communication and linguistic knowledge structures are clearly separated and show how a large scale instantiation of this framework in the environmental domain serves multilingual NLG.

1 Introduction

In this paper, we tackle the problem of generation of user-oriented multilingual environmental information from ontologies in the context of a personalized environmental decision support service.

Text generation from ontologies (or more generally, knowledge bases, KBs) is a common research topic in Natural Language Generation (NLG). However, until recently, the proposals tended to have in common that they start from KBs of limited size which either already contain linguistically-oriented knowledge structures—as, e.g., the Upper Model [1] or the MIAKT ontology [2], or explicitly assign to the conceptual knowledge structures their possible linguistic realizations; e.g., [3,4]. In other words, they either intermingle application-neutral concept configurations with application-oriented configurations, which is prohibitive when the same KB serves as resource for several different applications, or they require manual intervention to adjust the linguistic realization when the KB is extended by new configurations, which is prohibitive if the generator is supposed to scale up. In the last few years, large application-neutral ontologies have been considered as source, e.g., [5,6,7]. In [7], we proposed to distinguish between domain knowledge and domain communication knowledge in a two-layer ontology which was used for the generation of short football summaries. In what follows, we present a multiple layer ontology framework which improves our previous proposal and illustrate its application to the generation of multilingual user-oriented environmental reports. The framework consists of three ontology layers that reflect [8]'s distinction of the three types of knowledge needed by NLG: the *domain* ontology, the *domain communication* ontology, and the *communication* ontology. The implementation and

G. Bouma et al. (Eds.): NLDB 2012, LNCS 7337, pp. 216–221, 2012.

instantiation of this multi-layered ontology framework is presented in details in the next section. In Section 3 we discuss how these ontologies guide text planning and linguistic generation. In Section 4 we present our evaluation before concluding in Section 5.

2   Multi-layered Ontology as Starting Point for Generation

The three layers of our knowledge representation model are implemented in OWL, the state of the art Semantic Web ontology language.

The Domain Ontology.   The domain ontology models domain-specific knowledge relevant to the considered application settings of a personalized environmental decision support service that collects its data from the web. It captures thus the concepts, relations, and individuals related in particular to (i) *environmental data* (e.g., temperature, wind speed, birch pollen, CO_2); (ii) *environmental measurement providers*: properties of the web-based providers of environmental data measurements; (iii) *geographical information*; (iv) *environment-related user requests*: the user requests supported by the system; (v) *environment-affected user profile aspects*: for instance, user age, user sensitiveness to some environmental conditions (e.g., birch pollen), user diseases related to environmental conditions (e.g., asthma). For the construction of the ontology, techniques for automatic ontology extension are used [9]. Each time a decision support request is submitted to the system, the base ontology is instantiated, in successive steps, with content adequate to fulfil the user request. First, a complete description of the user request is instantiated in the base ontology. This description includes the type of request, the profile of the user involved in the request, and the time period and geographical location to which the request applies. From the request, the system determines, using Description Logics (DL) reasoning, the raw environmental data that are to be retrieved from web-based providers and distilled into the domain ontology.

The Domain Communication Ontology.   The domain communication ontology contains additional personalized content, spanning from data aggregation (e.g., minimum, maximum, and mean value aggregation of data, computed over the time-period considered in the request), qualitative scaling of numerical data, and user tailored recommendations and warnings triggered by the environmental data inserted in the ontology (e.g., the triggering of a recommendation to a pollen sensitive user in case of abundant pollen levels). The computation of this inferred content is performed by the *decision support* module by combining some complementary reasoning strategies, including DL reasoning and rule-based reasoning. A two layer reasoning infrastructure is currently in place. The first layer exploits the HermiT reasoner for the OWL DL reasoning services. The second layer is stacked on top of the previous layer. It uses the Jena RETE rule engine, which performs the rule-based reasoning computation.

The ontology models logico-semantic relations (LSRs) between domain entities such as `Implication`, `Cause` or `Violation of Expectation`. LSRs are formalized as concepts in the ontology, with their arguments defined as object properties, such that for each type of LSR an OWL-class is defined. Computationally, their instantiation co-occurs with the instantiation of the inferred assertions and is realized partly by ad-hoc inference rules or ontology-based computations and partly by extending rules for inferring personalized content (which is due to their dependence on domain knowledge).

The Communication Ontology. The communication ontology models the concepts and relations needed by the two text planning modules in generation: content selection (CS) and discourse structuring (DS). Therefore, the communication ontology defines a class Schema with an n-ary schema component object property whose range can be any individuals of the domain and domain communication KBs. Similar to [7], we assume the output of the DS module to be a well-formed text plan which consists of (i) elementary discourse units (EDUs) that group together individuals of the domain and domain communication ontologies, (ii) discourse relations between EDUs and/or individuals of the domain and domain communication ontologies, (iii) sentence units that group together EDUs to be realized in the same sentence, and (iv) precedence relations between EDUs. This structure translates in the communication ontology into three top classes: Sentence with an n-ary sentence component property, EDU with an n-ary EDU component relation and a linear precedence property, and Discourse Relation with nucleus and satellite relation. Figure 1 shows an instantiation of the DS concepts and relations in the environmental domain for pollen rating and associated recommendation message. In essence, this is an output text plan.[1]

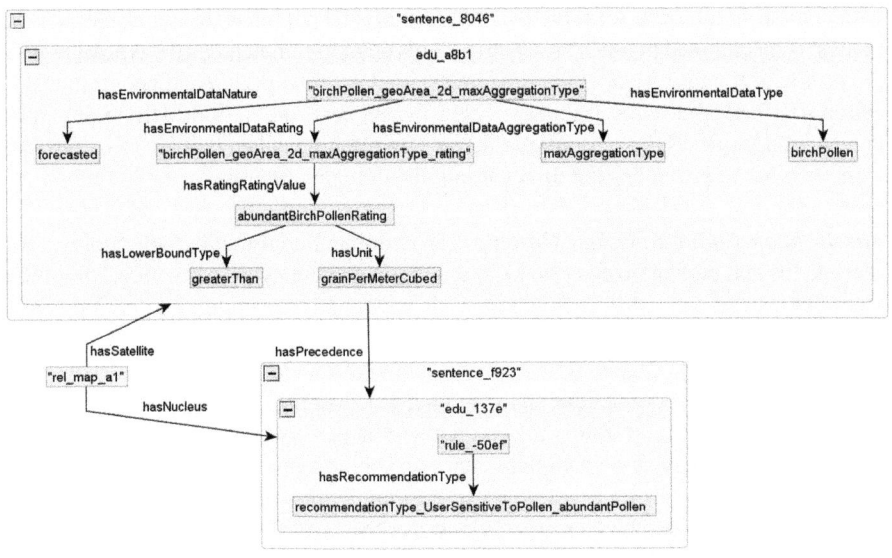

Fig. 1. Sample Text Plan

---

[1] For sake of simplicity, EDU and Sentence components are represented using a "bubble" rather than a component object property from each EDU/Sentence to each of its components.

3   Ontology-Based Natural Language Report Generation

As is common in report generation, our architecture is a standard pipeline architecture 'text planning ⇒ linguistic generation', with text planning being internally subdivided into content selection (CS) and discourse structuring (DS). Both CS and DS tasks are schema- (or template-) based: the concepts and relations of the communication ontology that define them are instantiated using SPARQL query rules.

Content Selection. CS operates on the output of the decision support module, which populates the domain and domain communication ontologies with the knowledge relevant to the submitted user query and profile. It selects the content to be included in the report and groups it by topic, instantiating a number of schemas for each topic, for example a schema for Air Quality (AQ) related information, that is AQ index minimum and maximum values and ratings, responsible pollutant(s), conclusions (i.e., warnings and recommendations) for the AQ, and any LSRs between the schema's components. The inclusion of a given individual in a schema can be subject to some restrictions defined in the queries; for example, if the minimum and maximum AQ index ratings are identical, then only the maximum AQ index rating is selected.

Discourse Structuring. DS is carried out by a pipeline of four rule-based submodules: (i) Elementary Discourse Unit Determination, (ii) Mapping LSRs to discourse relations, (iii) Sentence Unit Determination, and (iv) EDU Ordering. The heuristics used are determined by domain communication experts. One such heuristic for ordering is that conclusions are presented last following the default partial order AQ-pollutants > exceedances > pollen. This default order is changed depending on the presence of conclusions for pollutants/pollen the user is sensitive to, with warnings placed before recommendations.

Linguistic Generation. Our linguistic generation module is based on a multilevel linguistic Meaning-text Theory Model (MTM), such that the generation consists of a series of mappings between structures of adjacent strata (from the conceptual stratum to the linguistic surface stratum). Starting from the conceptual stratum, for each pair of adjacent strata (e.g., mapping conceptual to abstract semantic structure) and each target language, a transition grammar is defined. For details, see [10].

The generator receives as input the language-independent Conceptual Structure (ConStr) [11], derived from the text plan. In a sense, ConStr can thus be considered a projection of selected fragments of the ontologies onto a linguistically motivated structure. Figure 2 shows a sample ConStr derived from the text plan in Figure 1.[2] Thus, the relation hasEnvironmentalDataType and the class of its target node trigger the establishment of the relation between measurement and birch pollen in the ConStr. In the ontology, the measurement values are represented as attribute/numeric value pairs of the corresponding nodes (and thus the measurement showing in the mapped conceptual structure in Figure 2 is not visible in Figure 1). The use of numeric values for different measurements is language and context independent, which ensures great flexibility and efficiency with respect to lexicalization [10].

---

[2] Note that 'Sentence2' is a legally fixed and therefore canned message.

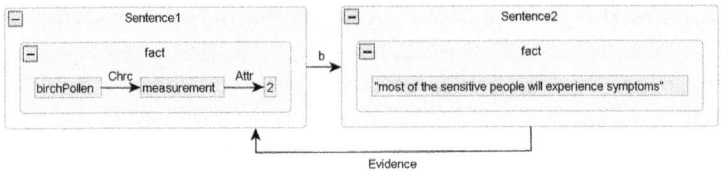

Fig. 2. A sample conceptual structure for the sentences *The birch pollen count will be abundant. As a result, most of the sensitive people will experience symptoms.*

Consider an example English bulletin generated from an instantiated ontology:

> Situation in the selected area between 28/04/2011 (20h00) and 29/04/2011 (00h00). The ozone information threshold value ($180\mu g/m^3$) was exceeded between 23h00 and 00h00 ($193\mu g/m^3$) and the ozone target value ($120\mu g/m^3$) between 15h00 and 23h00 ($123\mu g/m^3$). The minimum temperature was -1°C and the maximum temperature 9°C. The wind was weak (N), the humidity high and the rain light. There is no data available for carbon monoxide and thoracic particles.
>
> **Ozone warning:** Sensitive groups, like children, asthmatics of all ages and elderly persons suffering from coronary heart disease or chronic obstructive pulmonary disease, may experience symptoms. [..]

4   Evaluation

To evaluate the appropriateness of CS, we compared six gold content plans with their generated and baseline (majority select/not-select class) counterparts. The baseline showed an average F-score (that is, the average between the F-score and the inverse F-score, the latter reflecting how good the generated or the baseline approach is at not selecting the same contents discarded by the experts in the gold plans) of 0.74, and the generated content an average F-score of 0.88. The quality of English and Finnish texts was evaluated using a two-panel blind approach [12] in which one panel of two experts wrote the gold standard texts for six generated content plans and another panel of five experts performed a blind evaluation of the generated, gold and baseline texts according to four criteria rated on a 5-point scale (with 5 being the best score). The criteria were intelligibility, information packaging, ordering and accuracy. The baseline was obtained by filling in template sentences for both English and Finnish. For Finnish, on average across all the criteria, the gold texts were rated markedly better than the generated texts, which in their turn were rated markedly better than the baseline (4.5 vs 4.2 vs 3.7 for gold, generated and baseline respectively). This trend is not followed by the English texts, where the gold set (4.4) is rated better than the generated and baseline sets, which are given the same average rating (4.2). This difference between Finnish and English may have to do with the fact that experts are native Finnish speakers, although their level of English, at least in their domain, is considered high. Another more likely explanation are the intricacies of the Finnish language (in particular its rich morphology), which means fill-in templates may not work as well as for English.

5   Conclusion

We presented briefly a three-layer ontology framework and have shown how this framework benefits NLG. The ontologies, their user query-based population and the mentioned NLG modules are implemented in a service-based architecture described in [13]. The proposed approach has the advantages that it supports dynamic population of application-neutral ontologies, it allows for a clean separation of domain, domain communication and communication knowledge and it allows the codification of text planning-relevant aspects as part of the (communication) ontology. With the conceptual structure being modelled in the next version of our generator using the communication ontology as well, all content-oriented knowledge will be codified in the standard ontology language OWL. This will further increase the portability and scalability of our generator.

References

1. Bateman, J.: Enabling technology for multilingual natural language generation: the KPML development environment. Natural Language Engineering 3, 15–55 (1997)
2. Bontcheva, K., Wilks, Y.: Automatic Report Generation from Ontologies: The MIAKT Approach. In: Meziane, F., Métais, E. (eds.) NLDB 2004. LNCS, vol. 3136, pp. 324–335. Springer, Heidelberg (2004)
3. Hovy, E.: Generating Natural Language under Pragmatic Constraints. Lawrence Erlbaum, Hillsdale (1988)
4. Paris, C.: User Modelling in Text Generation. Frances Pinter Publishers, London (1993)
5. Mellish, C., Pan, J.: Natural language directed inference from ontologies. Artificial Intelligence 172, 1285–1315 (2010)
6. Bouttaz, T., Pignotti, E., Mellish, C., Edwards, P.: A policy-based approach to context dependent natural language generation. In: Proceedings of the 13th European Workshop on Natural Language Generation, Nancy, France, pp. 151–157 (2011)
7. Bouayad-Agha, N., Casamayor, G., Wanner, L., Díez, F., López Hernández, S.: FootbOWL: Using a Generic Ontology of Football Competition for Planning Match Summaries. In: Antoniou, G., Grobelnik, M., Simperl, E., Parsia, B., Plexousakis, D., De Leenheer, P., Pan, J. (eds.) ESWC 2011, Part I. LNCS, vol. 6643, pp. 230–244. Springer, Heidelberg (2011)
8. Kittredge, R., Korelsky, T., Rambow, O.: On the need for domain communication knowledge. Computational Intelligence 7, 305–314 (1991)
9. Tonelli, S., Rospocher, M., Pianta, E., Serafini, L.: Boosting collaborative ontology building with key-concept extraction. In: Proceedings of 5th IEEE International Conference on Semantic Computing, IEEE ICSC 2011 (2011)
10. Lareau, F., Wanner, L.: Towards a generic multilingual dependency grammar for text generation. In: King, T., Bender, E. (eds.) Proceedings of the GEAF 2007 Workshop, pp. 203–223. CSLI, Stanford (2007)
11. Sowa, J.: Knowledge Representation. Brooks Cole, Pacific Grove (2000)
12. Lester, J.C., Porter, B.: Developing and empirically evaluating robust explanation generators: The knight experiments. Computational Linguistics Journal 23, 65–101 (1997)
13. Wanner, L., Vrochidis, S., Tonelli, S., Moßgraber, J., Bosch, H., Karppinen, A., Myllynen, M., Rospocher, M., Bouayad-Agha, N., Bügel, U., Casamayor, G., Ertl, T., Kompatsiaris, I., Koskentalo, T., Mille, S., Moumtzidou, A., Pianta, E., Saggion, H., Serafini, L., Tarvainen, V.: Building an Environmental Information System for Personalized Content Delivery. In: Hřebíček, J., Schimak, G., Denzer, R. (eds.) ISESS 2011. IFIP AICT, vol. 359, pp. 169–176. Springer, Heidelberg (2011)

Web-Based Relation Extraction
for the Food Domain

Michael Wiegand, Benjamin Roth, and Dietrich Klakow

Spoken Language Systems, Saarland University, D-66123 Germany
{michael.wiegand,benjamin.roth,dietrich.klakow}@lsv.uni-saarland.de

Abstract. In this paper, we examine methods to extract different domain-specific relations from the food domain. We employ different extraction methods ranging from surface patterns to co-occurrence measures applied on different parts of a document. We show that the effectiveness of a particular method depends very much on the relation type considered and that there is no single method that works equally well for every relation type. As we need to process a large amount of unlabeled data our methods only require a low level of linguistic processing. This has also the advantage that these methods can provide responses in real time.

1 Introduction

There has been only little research on natural language processing in the food domain even though there is a high commercial potential in automatically extracting relations involving food items. For example, such knowledge could be beneficial for virtual customer advice in a supermarket. The advisor might suggest products available in the shop that would potentially complement the items a customer has already in their shopping cart. Additionally, food items required for preparing a specific dish or typically consumed at a social occasion could be recommended. The advisor could also suggest an appropriate substitute for a product a customer would like to purchase if that product is out of stock.

In this paper, we explore different methods, such as simple manually designed surface patterns or statistical co-occurrence measures applied on different parts of a document. Since these methods only require a low level of linguistic processing, they have the advantage that they can provide responses in real time. We show that these individual methods have varying strength depending on which particular food relation is considered.

Our system has to solve the following task: It is given a *partially instantiated relation*, such as *Ingredient-of(FOOD-ITEM=?, pancake)*. The system has to produce a ranked list of possible values that are valid arguments of the unspecified argument position. In the current example, this would correspond to listing ingredients that are necessary in order to prepare *pancakes*, such as *eggs*, *flour*, *sugar* and *milk*. The entities that are to be retrieved are always food items. Moreover, we only consider binary relations. The relation types we examine (such as *Ingredient-of*) are domain specific and to the best of our knowledge have not been addressed in any previous work.

G. Bouma et al. (Eds.): NLDB 2012, LNCS 7337, pp. 222–227, 2012.
© Springer-Verlag Berlin Heidelberg 2012

2   Data and Resources

For our experiments we use a crawl of *chefkoch.de*[1] as a domain-specific dataset. *chefkoch.de* is the largest web portal for food-related issues in the German language. We obtained the crawl by using *Heritrix* [1]. The plain text from the crawled set of web pages is extracted by using *Boilerpipe* [2]. The final domain-specific corpus consists of 418,558 webpages (3GB plain text). In order to have an efficient data access we index the corpus with *Lucene*.[2]

3   The Different Relations

In this section, we will briefly describe the four relation types we address in this paper. Due to the limited space of this paper, we just provide English translations of our German data in order to ensure general accessibility.

- **Suits-to(FOOD-ITEM, EVENT)** describes a relation about food items that are typically consumed at some particular cultural or social event. Examples are <*roast goose, Christmas*> or <*popcorn, cinema visit*>.
- **Served-with(FOOD-ITEM, FOOD-ITEM)** describes food items that are typically consumed together. Examples are <*fish fingers, mashed potatoes*>, <*baguette, ratatouille*> or <*wine, cheese*>.
- **Substituted-by(FOOD-ITEM, FOOD-ITEM)** lists pairs of food items that are almost identical to each other in that they are commonly consumed or served in the same situations. Examples are <*butter, margarine*>, <*anchovies, sardines*> or <*Sauvignon Blanc, Chardonnay*>.
- **Ingredient-of(FOOD-ITEM, DISH)** denotes some ingredient of a particular dish. Examples are <*chickpea, falafel*> or <*rice, paella*>.

4   Method

4.1   Surface Patterns (PATT)

For this work, we only considered manually compiled patterns. Our objective was to have some very few generally applicable and, if possible, precise patterns. As a help for building such patterns, we looked at mentions of typical relation instances in our corpus, e.g. <*butter, margarine*> for *Substituted-by* or <*mince meat, meat balls*> for *Ingredient-of*.

The formulation of such patterns is difficult due to the variety of contexts in which a relation can be expressed. This was further confirmed by computing lexical cues automatically with the help of statistical co-occurrence measures, such as the *point-wise mutual information*, which were run on automatically extracted sentences containing mentions of our typical relation instances. The output of that process did not reveal any additional significant patterns.

---

[1] www.chefkoch.de
[2] lucene.apache.org/core

Our final patterns exclusively use lexical items immediately before, between or after the argument slots of the relations. Table 1 illustrates some of these patterns. The level of representation used for our patterns (i.e. word level) is very shallow. However, these patterns are precise and can be easily used as a query for a search engine. Other levels of representation, e.g. syntactic information, would be much more difficult to incorporate. Moreover, in our initial exploratory experiments, we could not find many frequently occurring patterns using these representations to help us find relation instances that could not be extracted by our simple patterns. Additionally, since our domain-specific data comprise informal user generated natural language, the linguistic processing tools, such as syntactic parsers, i.e. tools that are primarily built with the help of formal newswire text corpora, are severely affected by a domain mismatch.

The extraction method PATT comprises the following steps: Recall from the task description in Section 1 that we always look for a list of values for an unspecified argument in a partially instantiated relation (PIR) and that the unspecified argument is always a food item. Given a PIR, such as *Substituted-by(butter, FOOD-ITEM=?)*, we partially instantiate each of the pertaining patterns (Table 1) with the given argument (e.g. *FOOD-ITEM instead of FOOD-ITEM* becomes *FOOD-ITEM instead of butter*) and then check for any possible food item (e.g. *margarine*) whether there exists a match in our corpus (e.g. *margarine instead of butter*). The output of this extraction process is a ranked list of those food items for which a match could be found with any of those patterns. We rank by the frequency of matches. Food items are obtained using GermaNet [3]. We collected all those lexical items that are contained within the synsets that are hyponyms of *Nahrung* (English: *food*).

Table 1. Illustration of the manually designed surface patterns

Relation Type	#Patterns	Examples
Suits-to	6	FOOD-ITEM at EVENT; FOOD-ITEM on the occasion of EVENT; FOOD-ITEM for EVENT
Served-with	8	FOOD-ITEM and FOOD-ITEM; FOOD-ITEM served with FOOD-ITEM; FOOD-ITEM for FOOD-ITEM
Substituted-by	8	FOOD-ITEM or FOOD-ITEM; FOOD-ITEM (FOOD-ITEM); FOOD-ITEM instead of FOOD-ITEM
Ingredient-of	8	DISH made of FOOD-ITEM; DISH containing FOOD-ITEM

4.2   Statistical Co-Occurrence (CO-OC)

The downside of the manual surface patterns is that they are rather sparse as they only fire if the exact lexical sequence is found in our corpus. As a less constrained method, we therefore also consider statistical co-occurrence. The rationale behind this approach is that if a pair of two specific arguments co-occurs significantly often (at a certain distance), such as *roast goose* and *Christmas*, then there is a likely relationship between these two linguistic entities.

As a co-occurrence measure, we consider the *normalized Google distance (NGD)* [4] which is a popular measure for such tasks. The extraction procedure of CO-OC is similar to PATT with the difference that we do not rank food items by the frequency of matches in a set of patterns but the correlation score with the given entity. For instance, given the PIR *Suits-to(FOOD-ITEM=?, Christmas)*, we compute the scores for each food item from our (food) vocabulary and *Christmas* and sort all these food items according to the correlation scores.

We believe that this approach will be beneficial for relations where the formulation of surface patterns is difficult – this is typically the case when entities involved in such a relation are realized within a larger distance to each other.

4.3   Relation between Title and Body of a Webpage (TITLE)

Rather than computing statistical co-occurrence at a certain distance, we also consider the co-occurrence of entities between title and body of a webpage. We argue that entities mentioned in the title represent a predominant topic and that a co-occurrence with an entity appearing in the body of a webpage may imply that the entity has a special relevance to that topic and denote some relation. The co-occurrence of two entities in the body is more likely to be co-incidental. None of those entities needs to be a predominant topic. If our experiments prove that the co-occurrence of entities occurring in the title and body of a webpage is indicative of a special relation type, we would have found an extraction method for a relation type that (similar to CO-OC) bypasses difficult/ambiguous surface realizations that present a significant obstacle for detection methods that explicitly model those surface realizations, such as PATT (Section 4.1).

The extraction procedure of this method selects those documents that contain the given argument of a PIR (e.g. *lasagna* in *Ingredient-of(FOOD-ITEM=?, lasagna)*) in the title and ranks food items that co-occur in the document body of those documents according to their frequency. We do not apply any co-occurrence measure since the number of co-occurrences that we observed with this method is considerably smaller than we observed with CO-OC. This makes the usages of those measures less effective.

5   Experiments

We already stated in Section 1 that the unspecified argument value of a partially instantiated relation (PIR) is always of type FOOD-ITEM. This is because these PIRs simulate a typical situation for a virtual customer advisor, e.g. such an advisor is more likely to be asked what food items are suitable for a given event, i.e. *Suits-to(FOOD-ITEM=?, EVENT)*, rather than the opposite PIR, i.e. *Suits-to(FOOD-ITEM, EVENT=?)*. The PIRs we use are presented in Table 2.[3] For each relation, we manually annotated a certain number of PIRs as our gold standard (see also Table 2). The gold standard including its annotation guidelines are presented in detail in [5]. Since our automatically generated output are ranked lists of food items, we use *precision at 10 (P@10)* and *mean reciprocal rank (MRR)* as evaluation measures.

---

[3] Since the two relation types *Served-with* and *Substituted-by* are reflexive, the argument positions of the PIRs do not matter.

Table 2. Statistics of partially instantiated relations in gold standard

Partially Instantiated Relations (PIRs)	#PIRs
Suits-to(FOOD-ITEM=?, EVENT)	40
Served-with(FOOD-ITEM, FOOD-ITEM=?)	58
Substituted-by(FOOD-ITEM, FOOD-ITEM=?)	67
Ingredient-of(FOOD-ITEM=?, DISH)	49

Table 3 compares the different individual methods on all of our four relation types. (Note that for CO-OC, we consider the best window size for each respective relation type.) It shows that the performance of a particular method varies greatly with respect to the relation type on which it has been applied. For *Suits-to*, the methods producing some reasonable output are CO-OC and TITLE. For *Served-with*, PATT and CO-OC are effective. For *Substituted-by*, the clear winner is PATT. For *Ingredient-of*, TITLE performs best. This relation type is difficult to model with PATT. It is also interesting to see that TITLE is much better than CO-OC. From our manual inspection of relation instances extracted with CO-OC we found that this method returns instances of any relation type that exclusively involves entities of type FOOD-ITEM (i.e. *Served-with*, *Substituted-by* and *Ingredient-of*).[4] TITLE, on the other hand, produces a much more unambiguous output. It very reliably encodes the relation type *Ingredient-of*. This is also supported by the fact that TITLE performs poorly on *Served-with* and *Substituted-by*.

Table 3. Comparison of the different individual methods

Method	Suits-to		Served-with		Substituted-by		Ingredient-of	
	P@10	MRR	P@10	MRR	P@10	MRR	P@10	MRR
PATT	0.023	0.133	**0.343**	**0.617**	**0.303**	**0.764**	0.076	0.331
CO-OC	**0.340**	**0.656**	0.310	0.584	0.172	0.553	0.335	0.581
TITLE	0.300	0.645	0.171	0.233	0.049	0.184	**0.776**	**0.733**

Table 4. The 10 most highly ranked food items for some automatically extracted relations; *: denotes match with the gold standard

Suits-to(?, picnic)	Served-with(mince meat, ?)	Substituted-by(beef roulades, ?)	Ingredient-of(?, falafel)
sandwiches*	onions	goulash*	chickpea*
fingerfood	leek	marinated beef*	cooking oil*
noodle salad*	zucchini*	roast*	water
meat balls*	bell pepper	roast beef*	coriander*
potato salad*	noodle casserole	braised meat*	onions*
melons*	feta cheese	cutlet*	flour*
fruit salad*	spinach	rabbit*	salt*
small sausages	rice*	rolling roast*	garlic*
sparkling wine	tomatoes	rolled pork	peas*
baguette*	sweet corn	game	shortening*

---

[4] Note that the entity type DISH in *Ingredient-of* is a subset of FOOD-ITEM.

Table 4 illustrates some automatically generated output using the best configuration for each relation type. Even though not all retrieved entries match with our gold standard, most of them are (at least) plausible candidates. Note that for our gold standard we aimed for high precision rather than completeness.

6 Conclusion

In this paper, we examined methods for relation extraction in the food domain. We have shown that different relation types require different extraction methods. Since our methods only require a low level of linguistic processing, they may serve for applications that have to provide responses in real time.

Acknowledgements. The work presented in this paper was performed in the context of the Software-Cluster project EMERGENT and was funded by the German Federal Ministry of Education and Research (BMBF) under grant no. "01IC10S01".

References

1. Mohr, G., Stack, M., Ranitovic, I., Avery, D., Kimpton, M.: An Introduction to Heritrix, an open source archival quality web crawler. In: Proc. of IWAW (2004)
2. Kohlschütter, C., Fankhauser, P., Nejdl, W.: Boilerplate Detection using Shallow Text Features. In: Proc. of WSDM (2010)
3. Hamp, B., Feldweg, H.: GermaNet - a Lexical-Semantic Net for German. In: Proc. of ACL workshop Automatic Information Extraction and Building of Lexical Semantic Resources for NLP Applications (1997)
4. Cilibrasi, R., Vitanyi, P.: The Google Similarity Distance. IEEE Transactions on Knowledge and Data Engineering 19(3), 370–383 (2007)
5. Wiegand, M., Roth, B., Lasarcyk, E., Köser, S., Klakow, D.: A Gold Standard for Relation Extraction in the Food Domain. In: Proc. of the LREC (2012)

Blog Distillation
via Sentiment-Sensitive Link Analysis

Giacomo Berardi, Andrea Esuli, Fabrizio Sebastiani, and Fabrizio Silvestri

Istituto di Scienza e Tecnologie dell'Informazione,
Consiglio Nazionale delle Ricerche, 56124 Pisa, Italy
firstname.lastname@isti.cnr.it

Abstract. In this paper we approach blog distillation by adding a link analysis phase to the standard retrieval-by-topicality phase, where we also we check whether a given hyperlink is a citation with a positive or a negative nature. This allows us to test the hypothesis that distinguishing approval from disapproval brings about benefits in blog distillation.

1   Introduction

Blog distillation is a subtask of blog search. It is defined as the task of ranking in decreasing order of relevance the set of blogs in which the topic expressed by the query q is a recurring and principal topic of interest. Blog distillation has been intensively investigated within the TREC Blog Track [4], where participants have experimented with various combinations of (i) methods for retrieval by topicality and (ii) sentiment analysis methods. Retrieval by topicality is needed since topicality is a key aspect in blog distillation, while sentiment analysis is needed since blogs tend to contain strongly opinionated content.

We test a method for blog distillation in which, on top of a standard system for retrieval by topicality, we add a link analysis phase meant to account for the reputation, or popularity, of the blog. However, due to the highly opinionated nature of blog contents, many hyperlinks express a *rebuttal*, and not an approval, of the hyperlinked post on the part of the hyperlinking post. In this work we test the hypothesis that distinguishing hyperlinks expressing approval from ones expressing rebuttal may benefit blog distillation. We thus define a *sentiment-sensitive link analysis* method, i.e., a random-walk method on which the two types of hyperlinks have a different impact. We detect the sentimental polarity of a given hyperlink (i.e., detect if it conveys a positive or a negative endorsement) by performing sentiment analysis on a text window around the hyperlink.

2   Sentiment-Sensitive Link Analysis for Blog Ranking

In the following we discuss the model we have adopted to compute the sentiment-sensitive, link-analysis-based ranking of blogs and blog posts. We rely on a graph-based model in which nodes represent either blogs or blog posts, and the weights attached to nodes represent their importance.

G. Bouma et al. (Eds.): NLDB 2012, LNCS 7337, pp. 228–233, 2012.

A Graph-Based Model of the Blogosphere. Let $\mathcal{P} = \{p_1, p_2, \ldots, p_n\}$ be a set of blog posts, partitioned into a set of blogs $\mathcal{B} = \{b_1, b_2, \ldots, b_m\}$. Let $G_P = (\mathcal{P}, E_P)$ be a graph, where the set of nodes $\mathcal{P}$ is as above and E_P is a set of edges corresponding to hyperlinks between posts, i.e., edge e_{xy} from post $p_x \in \mathcal{P}$ to post $p_y \in \mathcal{P}$ denotes the presence of at least one hyperlink from p_x to p_y. Similarly, let $G_B = (\mathcal{B}, E_B)$ be a graph, where the set of nodes $\mathcal{B}$ is as above and E_B is a set of edges corresponding to hyperlinks between blogs, i.e., edge e_{ij} from blog $b_i \in \mathcal{B}$ to blog $b_j \in \mathcal{B}$ denotes the presence of at least one hyperlink from a post $p_x \in b_i$ to the homepage of blog b_j. Let $w_P : E_P \rightarrow \mathbb{R}$ and $w_B : E_B \rightarrow \mathbb{R}$ be two weighting functions (described in detail in the next section), where $\mathbb{R}$ is the set of the reals. Informally, the weight assigned to an edge models the importance the corresponding hyperlink confers onto the hyperlinked post (for E_P) or blog (for E_B).

Let q be a query. Our blog distillation method comprises a first step consisting in running a standard (i.e., text-based) retrieval engine on $\mathcal{P}$, yielding a ranked list of the k top-scoring posts for q. Let $L = l_1, l_2, \ldots, l_k$ be this ranked list (with l_1 the top-scoring element), and let $s_1, s_2, \ldots, s_k$ be the scores returned for the posts in L by the retrieval engine.

Weighting the Hyperlinks. We define the weighting functions w_P and w_B on the basis of a sentiment-based analysis, whose aim is to determine if an edge e_{xy} denotes a *positive* or a *negative* attitude of post p_x towards post p_y (for w_P) or of post $p_x \in b_i$ towards blog b_j (for w_B). A positive (resp., negative) value of $w_P(e_{xy})$ will indicate a positive (resp., negative) attitude of the linking document p_x toward the linked document p_y, and the absolute value of $w_P(e_{xy})$ will indicate the intensity of that attitude. The same goes for w_B.

For determining $w_P(e_{xy})$, all the hyperlinks from p_x to p_y are taken into account; similarly, for determining $w_B(e_{ij})$, all the hyperlinks from any post $p_x \in b_i$ to the homepage of blog b_j are taken into account. For determining the impact of a hyperlink on the weighting function, the anchor text and the sentence in which the anchor text is embedded are analysed. This analysis begins with POS-tagging the sentence in order to identify *candidate sentiment chunks,* i.e., all the sequences of words matching the (RB|JJ)+ and NN+ patterns.

We assign a sentiment score to each term in the chunk by using SentiWordNet (SWN) [2] as the source of sentiment scores. From SentiWordNet we have created a word-level dictionary (that we call SentiWordNet$_w$) in which each POS-tagged word w (rather than each word sense s, as in full-fledged SentiWordNet) is associated to a score $\sigma(w)$ that indicates its sentimental valence, averaged across its word senses. We have heuristically obtained this score by computing a weighted average $\sigma(w) = \sum_i \frac{1}{i}(Pos(s_i(w)) - Neg(s_i(w)))$ of the differences between the positivity and negativity scores assigned to the various senses $s_1(w), s_2(w), \ldots$ of w. In this weighted average the weight is the inverse of the sense number, thus lending more prominence to the most frequent senses.

All candidate sentiment chunks only composed by terms that have been assigned a sentiment score equal to 0 (i.e., sentiment-neutral words) are discarded from consideration. Each of the remaining *sentiment chunks,* together with a portion of text preceding it, is then checked for the presence of *sentiment modifiers*, i.e., negators (e.g., "no", "not") or intensifiers / downtoners (e.g., "very", "strongly", "barely", "hardly"). As the resource for sentiment modifiers we have used the appraisal lexicon defined in [1]. When a modifier is found, the score of the word that follows it is modified accordingly (e.g., "very good" is assigned a doubly positive score than "good"). We then assign a sentiment score to a chunk by simply summing the sentiment scores of all the words in a chunk, taking into account the modifiers as described above.

In order to determine a sentiment score for the hyperlink, we compute a weighted sum of the sentiment scores of all the chunks that appear in the sentence containing the anchor text. We use weights that are a decreasing function of the distance of the chunk from the anchor text, according to the assumption that that the closer a chunk is to the hyperlink, the more it is related to it. This distance is itself computed as a weighted sum, where each token between the anchor text and the chunk has its own weight depending on its type; for instance, mood-changing particles such as "instead" are assigned a higher weight.

Finally, we compute $w_P(e_{xy})$ as the mean of the sentiment scores assigned to the links from p_x to p_y; if both positive and negative links are present, they thus compensate each other. Similarly, we compute $w_B(e_{ij})$ as the mean of the values associated to hyperlinks from posts in b_i to the homepage of blog b_j.

Using the w_P function we then split G_P into two graphs $G_P^+ = (\mathcal{P}, E_P^+)$ and $G_P^- = (\mathcal{P}, E_P^-)$, where E_P^+ and E_P^- are the sets of edges e_{xy} such that $w_P(e_{xy}) \geq 0$ and $w_P(e_{xy}) < 0$, respectively. Analogously, we split G_B into $G_B^+ = (\mathcal{B}, E_B^+)$ and $G_B^- = (\mathcal{B}, E_B^-)$, where E_B^+ and E_B^- are the sets of edges e_{ij} such that $w_B(e_{ij}) \geq 0$ and $w_B(e_{ij}) < 0$, respectively.

Ranking the Nodes. We use the graphs G_P^+, G_P^-, G_B^+, G_B^- in order to compute the ranking of posts and blogs based on sentiment-sensitive link analysis. We use an algorithm known as *random walk with restart* (RWR), also known as *personalized* (or *topic-sensitive*) *random walk* [3], which differs from more standard random walk algorithms such as PageRank for the fact that the $\mathbf{v_P}$ and $\mathbf{v_B}$ vectors (see below) are not uniform. The values in the latter vectors are sometimes referred to as the *restart probabilities*. Two RWR computations are run on G_P^+ and G_B^+, respectively, yielding $\mathbf{r_P}$ and $\mathbf{r_B}$ (i.e., the vectors of scores for posts and blogs) as the principal eigenvectors of the matrices $\mathbf{P} = (1-a) \cdot \mathbf{A_P^+} + \frac{a}{k} \cdot \mathbf{v_P}$ and $\mathbf{B} = (1-a) \cdot \mathbf{A_B^+} + \frac{a}{k'} \cdot \mathbf{v_B}$, where $\mathbf{A_P^+}$ (resp., $\mathbf{A_B^+}$) is the adjacency matrix associated with graph G_P^+ (resp., G_B^+), and a is the damping factor (i.e., the factor that determines how much backlinks influence random walks). Vector $\mathbf{v_P}$ is the preference vector for blog posts, i.e., a vector whose entries corresponding to the k pages in L are set to 1 and whose other entries are set to 0. Vector $\mathbf{v_B}$ is obtained in a slightly different way. We first group together posts belonging to the same blog and build a vector $\mathbf{\bar{v}_B}$ whose entries count the number of retrieved posts belonging to the blog corresponding to the entry. Vector $\mathbf{\bar{v}_B}$ is then normalized into vector $\mathbf{v_B}$ and, in this case, k' is the number of entries in $\mathbf{v_B}$ greater than zero. In order to find the principal eigenvectors of $\mathbf{P}$ and

B we solve the eigenproblems $\mathbf{r}_P^+ = \mathbf{P} \cdot \mathbf{r}_P^+ = (1-a) \cdot \mathbf{A}_\mathbf{P}^+ \cdot \mathbf{r}_P + \frac{a}{k} \cdot \mathbf{v}_\mathbf{P}$ and $\mathbf{r}_B^+ = \mathbf{B} \cdot \mathbf{r}_B^+ = (1-a) \cdot \mathbf{A}_\mathbf{B}^+ \cdot \mathbf{r}_B + \frac{a}{k'} \cdot \mathbf{v}_\mathbf{B}$. We calculate the vectors of negative scores $\mathbf{r}_P^- = \mathbf{A}_\mathbf{P}^- \cdot \mathbf{r}_P^+$ and $\mathbf{r}_B^- = \mathbf{A}_\mathbf{B}^- \cdot \mathbf{r}_B^+$, and we normalize them so that the sum of their negative components is -1. $\mathbf{A}_\mathbf{P}^-$ and $\mathbf{A}_\mathbf{B}^-$ contain the negative values associated with the edge weights of G_P^- and G_B^-. Finally, the scoring vectors $\mathbf{r}_P$ and $\mathbf{r}_B$ that result from taking into account both positive and negative links are given by $\mathbf{r}_P = (1-\theta) \cdot \mathbf{r}_P^+ + \theta \cdot \mathbf{r}_P^-$ and $\mathbf{r}_B = (1-\theta) \cdot \mathbf{r}_B^+ + \theta \cdot \mathbf{r}_B^-$, where θ is a coefficient for tuning the relative impact of positive and negative links on the overall score of a post (or blog). In our tests we have set $\theta = \frac{1}{2}$.

Scoring Blogs against a Query. We are now in a position to describe our blog distillation method. Let q be a query whose aim is to rank the blogs in descending order of relevance to the query. Our method consists of three steps.

As described in Section 2, in the 1st step a standard (i.e., text-based) retrieval engine is run on $\mathcal{P}$, yielding a ranked list of the k top-scoring posts for q. Let $L = l_1, l_2, \ldots, l_k$ be this ranked list (with l_1 the top-scoring element), and let $s_1, s_2, \ldots, s_k$ be the corresponding scores returned by the retrieval engine.

The 2nd step consists of combining the retrieval scores s_x with the link-based scores. We use the combination $w_x = (1-\alpha) \cdot s_x + \alpha \cdot \mathbf{r}_{P_x}$ for all $x \in [1, k]$, where α is a coefficient used to balance the weights of link-based and text-based scores and $\mathbf{r}_{P_x}$ is the x-th component of the vector $\mathbf{r}_P$ as computed in Section 2.

The 3rd step consists of merging the scores computed for each post according to the blog the post itself comes from. Obviously the choice of the merging function is an important issue. We simply weight each blog b_i using the average of the scores w_x for each post $p_x \in b_i$ plus the static score $\mathbf{r}_{Bi}$ of blog b_i smoothed using the actual number of posts retrieved for blog b_i. More formally, we use the same α coefficient as above to balance the weight of average post score and the link-based score of blog b_i; to score blog b_i we use the equation $\omega_i = \frac{(1-\alpha)}{|b_i|} \cdot \sum_{x:p_x \in b_i} w_x + \alpha \cdot \mathbf{r}_{Bi} \cdot \frac{|L \cap b_i|}{|b_i|}$ where by $|L \cap b_i|$ we denote the number of posts of blog b_i retrieved as top-k posts in the list L. Eventually, our blog retrieval system returns blogs sorted by the ω_i scores.

3 Experiments

We have tested our method on the Blogs08 collection used in the 2009 and 2010 editions of the TREC Blog Track [4]. Blogs08 consists of a crawl of 1,303,520 blogs, for a total of 28,488,766 blog posts, each identified by a unique URL. We have followed the protocol of the 2009 Blog Track, using the 50 queries of 2009.

We have processed all the hyperlinks via our hyperlink polarity scoring method (see Section 2). The processing of hyperlinks relative to the graph of blog posts resulted in the identification of 8,947,325 neutral links (i.e., polarity weight equal to zero), 1,906,182 positive links, and 1,780,281 negative links.

As the system for performing the first step of our method we have used Terrier [5]. With Terrier we have built a baseline for the comparison of our results, based on ranking the documents with respect to each query via the well-known

Table 1. Ranking effectiveness on the Blogs08 collection. "Base" indicates the baseline; "$\alpha = r$" indicates our method, where r is the value of α which has returned the best result; "Mean" indicates the average performance of our method across the four tested values of α. **Boldface** indicates the best result.

k	1,000			2,000			3,000			4,000		
	Base	$\alpha = 0.85$	Mean	Base	$\alpha = 0.75$	Mean	Base	$\alpha = 0.80$	Mean	Base	$\alpha = 0.65$	Mean
MAP	0.1775	0.1802	0.1801	0.1914	0.1943	0.1942	0.1958	**0.1986**	0.1983	0.1949	0.1977	0.1976
P@10	0.2878	**0.2898**	0.2880	0.2796	0.2837	0.2819	0.2653	0.2735	0.2719	0.2592	0.2673	0.2650
bPref	0.2039	0.2056	0.2050	0.2203	0.2222	0.2220	0.2226	**0.2247**	0.2242	0.2210	0.2230	0.2224

BM25 weighting method. These rankings have been used also as the input to the random-walk-based reranking phase of our method. For each query we retrieve the first hundred posts, which are set as the nodes with non-null restart probabilities given as input to the RWR method. As the damping factor a (see Section 2) we have used the value 0.85, a fairly typical value in these methods [3]. The value used for stopping the iterative process is set to $\epsilon = 10^{-9}$.

We have evaluated the results of our experiments via the well-known *mean average precision* (MAP), *binary preference* (bPref), and *precision at 10* (P@10).

We have tested various values for parameters α (see Section 2) and k (the number of posts retrieved in the 1st step of our method). For α we have tested the values 0.65, 0.75, 0.80, 0.85; preliminary experiments with values outside this range had shown clear deteriorations in effectiveness. For k we have tested the values 1000, 2000, 3000, 4000; here too, preliminary experiments with values outside this range had shown clear deteriorations.

The different values for the α parameter do not yield substantial differences in performance. This can be gleaned from Table 1 by looking at the differences among best and mean values obtained by our method (2nd and 3rd column of each block), which are always very small.

The results do show an improvement over the retrieval-by-topicality baseline, but this improvement is very small. A closer inspection of the results reveals that this is due to the sparsity of the graphs resulting from the collection of posts retrieved in the first step: these posts are often isolated nodes in the posts graph, which means that the random-walk step only affects a small subset of the results. Increasing the value of the k parameter determines an increase of the improvement over the baseline, since more posts are affected by the random walk scores, but the relative improvement is still small anyway. We have seen that increasing k beyond the value of 4000, instead, introduces a higher number of irrelevant posts in the results, which decreases the magnitude of the improvement.

We have also tried different values for θ (see Section 2), but this has not brought about any substantial improvement. The main reason is that, due to the above mentioned sparsity of the posts graph, the impact of the link analysis phase on the final ranking is low anyway, regardless of how this link analysis balances the contribution of the positive and negative links.

In order to obtain a further confirmation of the fact that using link analysis does not impact substantively on the final ranking, we have run an experiment in which the RWR random-walk method has been replaced by standard PageRank.

Table 2. MAP evaluations for our random walk with restart (RWR) and PageRank (PR), with $k = 3,000$. **Boldface** indicates best results for each weight.

α	0.65	0.75	0.80	0.85
RWR	**0.1981**	**0.1983**	**0.1986**	0.1983
PR	0.1978	**0.1983**	0.1985	**0.1988**

Here, $\mathbf{r_P}$ and $\mathbf{r_B}$ are computed similarly to our method but for the fact that the entire G_P and G_B graphs are used, i.e., without differentiating positive and negative links. Edges are weighted with uniform probability of transition equal to $1/outdegree(node)$. The differences in the final results are negligible (see Table 2). The very slight improvement brought about by PageRank over our method is probably due to the fact that the two graphs used by PageRank have a larger proportion of non-isolated nodes than each of the graphs on which our method operates. Our algorithm, however, is faster than PageRank, because it works on the reduced transition matrices given by G_B^+ and G_P^+.

4   Conclusions

We can conclude that the hypothesis according to which it makes sense to distinguish positive from negative endorsements in blog analysis has neither been confirmed nor disconfirmed. To see this, note that the literature on blog search has unequivocally shown that the best results are obtained when sentiment analysis is performed not on the entire blogosphere, but on the subset of blogs which have been top-ranked by a standard retrieval-by-topicality engine [4]. The essential conclusion that can be drawn from our results is that the blogs (and their posts) retrieved in the retrieval-by-topicality phase contain too few hyperlinks / endorsements, no matter whether positive or negative, for a link analysis phase to have a substantial impact on retrieval.

References

1. Argamon, S., Bloom, K., Esuli, A., Sebastiani, F.: Automatically Determining Attitude Type and Force for Sentiment Analysis. In: Vetulani, Z., Uszkoreit, H. (eds.) LTC 2007. LNCS, vol. 5603, pp. 218–231. Springer, Heidelberg (2009)
2. Baccianella, S., Esuli, A., Sebastiani, F.: SentiWordNet 3.0: An enhanced lexical resource for sentiment analysis and opinion mining. In: Proceedings of LREC 2010 (2010)
3. Haveliwala, T.H.: Topic-sensitive PageRank: A context-sensitive ranking algorithm for Web search. IEEE TKDE 15(4), 784–796 (2003)
4. Macdonald, C., Santos, R.L., Ounis, I., Soboroff, I.: Blog Track research at TREC. SIGIR Forum 44(1), 58–75 (2010)
5. Ounis, I., Amati, G., Plachouras, V., He, B., Macdonald, C., Lioma, C.: Terrier: A high performance and scalable information retrieval platform. In: Proceedings of the SIGIR 2006 Workshop on Open Source Information Retrieval, Seattle, US (2006)

Performing Groupization in Data Warehouses: Which Discriminating Criterion to Select?

Eya Ben Ahmed, Ahlem Nabli, and Faïez Gargouri

Sfax University
eya.benahmed@gmail.com,
{ahlem.nabli,faiez.gargouri}@fsegs.rnu.tn

Abstract. In this paper, we aim to optimally identify the analyst'groups in data warehouse. For that reason, we study the similarity between the selected queries in the analytical history. Four axis for group identification are distinguished: (*i*) the function exerted, (*ii*) the granted responsibilities to accomplish goals, (*iii*) the source of groups identification, (*iv*) the dynamicity of discovered groups. A semi-supervised hierarchical algorithm is used to discover the most discriminating criterion. Carried out experiments on real data warehouse demonstrate that groupization improves upon personalization for several group types, mainly for function-based groupization.

Keywords: personalization, groupization, semi-supervised hierarchical clustering, OLAP log files, data warehouse.

1 Introduction

In order to support the analyst in his decision making process, several user-centric data warehouses approaches are introduced. Two main streams of them are distinguished, namely [3]: (*i*) Approaches privileging implicit intervention through the customizability of the response to OLAP query according to the analyst preferences. Such strategy is called *personalization* [2,5,13,14]; (*ii*) Approaches favoring explicit intervention through applying *recommendation* to better support the analyst on his decision making process even he does not exactly discern the data warehouse schema [6,7].

Unlike the contributions of [1], [4] and [6] where a learning of analyst preferences from the OLAP log files is performed, the analyst preferences are explicitly identified in the literature. In our previous work [4], we introduced an innovative approach for building multiview analyst profile from multidimensional query logs. However, this learning of analyst preferences is closely dependent on the number of analysts. Indeed, the growing number of data warehouse analysts may prevent this automatic discovery of preferences. To overcome this limitation, we expand the personalization process through combining similar individual's data analysts to improve the process of preferences learning. This extension of this mechanism through considering data related to all group members instead of data associated to single user is called "*groupization*". An emergent challenge

G. Bouma et al. (Eds.): NLDB 2012, LNCS 7337, pp. 234–240, 2012.

of groupization is the efficient identification of groups. It is well-recognized that several criteria are more useful than others for group detection based on their past activities. Few works in the literature focused on the group identification within communities [10], particularly in data warehouse area.

For example, given a stock exchange data warehouse, performing OLAP may help diverse economic agents in their decisions, namely, the *portfolio managers*, the *investors*, the *private managers* and the *private investors*. In fact, many listed companies are integrated in this data warehouse therefore huge number of analysts may conduct their OLAP analysis. We propose to group analysts having similar preferences together, while this solution will reduce the gap between the result and the expected response to launched query using group information of analysts. However, the identification of analysts groups remains a complex problem. The collected data disclosed that whether people grouped by several properties were similar in the queries they selected. Indeed, analysts can be gathered according to different criteria, namely: the *function exerted*, the *granted responsibilities* to accomplish defined goals, the *source of group identification*, the *dynamic identification* of groups.

In this paper, we focus on identifying relevant criteria for analysts group detection from data warehouse. Its distinguishing contribution is building groups in the data warehouse context. Its crucial originality is the automatic discovery of the most discriminating criteria for data warehouse analyst groupization.

The paper is organized as follows. Section 2 presents our contribution for group identification in data warehouse. Section 3 reports experimental results showing the best discriminating group identification criterion in the stock exchange data warehouse. Finally, Section 4 concludes and points out avenues of future work.

2 $\mathcal{GID}$ Approach for Analysts $\mathcal{G}$roup $\mathcal{I}$dentification in $\mathcal{D}$ata Warehouses

The enhancing scrutiny of analysts behaviors leads to identify relevant criteria for group segmentation based on analysts' past activities. Four analysis axes may be distinguished:

1. *Function*: we assume that analysts working at the same position have similar preferences, two alternatives are plausible:(i) the restriction to the current function; (ii) taking the experience and expertise into account;
2. *Responsibilities* to accomplish the goals: the analysts of the warehouse have communally goals that may be operational for short-term, tactical for medium-term or strategic for long-term. The example of two portfolio managers who do not assume the same responsibilities, accordingly, they will have distinctive goals throughout time illustrates this idea;
3. *Source* of group identification: (i) is explicit when an analyst chooses to belong to given group or (ii) this action is performed implicitly;
4. *Dynamicity* of the identified groups: in other words the, identified groups will be updated dynamically or remained static.

To discover the most discriminating factor among identified axes, we detail our approach $\mathcal{GID}$ for analysts group identification in data warehouses. We analyze the similitude between decision makers through scrutinizing their past analytical histories. Therefore, a hierarchical semi-supervised clustering algorithm is applied on log files because the supervision of the learning process will guide the possible mergers between objects according to the constraints dictated by each groupization criterion.

To measure the similarity between OLAP log files, we use the Jaccard distance because it significantly suits the large documents. Thus, we extend it to fit the OLAP log files in the multidimensional context. Our inter-OLAP log files Jaccard distance is backboned on the MDX query structure, basically on similarity between used facts, measures, dimension attributes, as well as slicer specification members. The later is used in the Where clause and restricts the result data. Any dimension that does not appear on an axis in the SELECT clause can be named on the slicer. The similarity metric between two OLAP log files is given by the number of common queries having common facts, measures, dimensions and members of the common specification in both historical files, divided by the total number of queries in both files, it is computed using the following formula:

$$Sim(LF_i, LF_j) = \frac{C_{Queries(LF_i,LF_j)}}{\sum Queries(LF_i) + \sum Queries(LF_j) - C_{Queries(LF_i,LF_j)}}. \tag{1}$$

With $C_{Queries(LF_i,LF_j)}$: Number of common queries in two log files LF_i of analyst i and LF_j of analyst j, $\sum Queries(LF_i)$: Sum of all existing queries in the log file LF_i.

The process of $\mathcal{GID}$ is summarized as follows:

1. *Initialization*: Consider each file as a cluster;
2. *Treatment*: Compute the similarity matrix using the Jaccard measure adapted to the context of the multidimensional OLAP log files;
3. *Constraints checking*: Two types of constraints are considered: (*i*) *Must-link constraints*: indicating that the two clusters must be in the same group, thus the distance between them is replaced by the greatest distance, (*ii*) *Cannot-link constraints*: indicating that the two clusters should not be merged, thus the distance between them is adjusted to zero;
4. *Assignment*: Merge the two clusters with the maximum value of adapted Jaccard distance;
5. *Update*: Similarity matrix update;
6. *Iteration*: Repeat steps 3, 4 and 5 until the found number of clusters reaches the number of clusters as input parameter.

We apply this process four times in respect to each identified criterion. Indeed, the semi-supervised hierarchical clustering algorithm based on complete link schema introduced by Klein et *al.* is used to cluster the log files of each specified criterion, namely, the function, the responsibilities, the source and the dynamicity. In fact, this algorithm merges at each step the two closest clusters with the highest similarity. This similarity is computed using distance measure between log files. This measure is adjusted according to the types of constraints including: (i) *Must-link constraints*: the distance between the pairs is changed to the greater distance, (ii) *Cannot-link constraints*: the distance between the pairs that should not be merged is replaced by zero.

To better illustrate our contribution $\mathcal{GID}$, we present in the following an illustration of its process through the following example. We consider the function-based criterion of groupization so the number of clusters is set to four. The log files are illustrated by the table 1. First, the initialization is performed and five clusters emerged. Then, the computation of distances between the five clusters provides the similarity matrix. Let us consider the number of common queries between C_1 and C_2 is 876, our adapted Jaccard distance is computed as follows:

$$J(C_1, C_2) = \frac{C_{Queries(C_1,C_2)}}{\sum Queries(C_1) + \sum Queries(C_2) - C_{Queries(C_1,C_2)}} \qquad (2)$$

$$= \frac{876}{2742 + 2518 - 876} = 0.2.$$

Table 1. Number of stored queries in log files for each analyst

	Portfolio Manager C_1	Portfolio Manager C_2	Private Manager C_3	Investor C_4	Private Investor C_5
Number of queries in log file	2742	2518	1313	847	563

We assume that the similarity matrix obtained in our case is given by the table 2. The integration of constraints is accomplished. In fact, *Must-link* constraints are related to analysts working at the same position and *Cannot-link* constraints concern different functions. During constraints checking, the distance between C_1 and C_2 will be adjusted to 0.8 according to *Must-link* constraint and the distance between the rest of clusters will be replaced by zero in respect to *Cannot-link* constraint. The assignment will merge the two clusters C_1 and C_2 with the highest similarity measure. Up to four clusters, the algorithm stops.

Table 2. Similarity matrix provided by the first iteration

Distance	C_1	C_2	C_3	C_4	C_5
C_1	0	0.2	0.5	0.1	0.8
C_2	0	0	0.4	0.2	0.6
C_3	0	0	0	0.3	0.2
C_4	0	0	0	0	0.5
C_5	0	0	0	0	0

3   Experimental Study

In this section, we report the experiments carried out on stock market data warehouse in order to identify the best discriminating criterion of groupization. In conducted experiments, we used the OLAP log files. Each log file contains approximately 5000 stored queries.

We choose to compare our algorithm to ID3 [9] and NAIVE BAYES [12] classification methods using Weka platform 3.6.5 edition[1]. Based on generated results of groupization, the expert denotes the OLAP log files. The class of each OLAP file is assigned according to the output cluster generated using our approach. A CSV file (the input file of Weka) is generated for each OLAP log file. It is used as an input for chosen classification algorithms.

In our context, three key metrics may be employed to assess the best criterion of groups detection: (i) The *True Positive rate* (TP) measures the proportion of examples classified as class X, among all examples which truly belong to class X. It is equivalent to the *recall*; (ii) The *False Positive rate* (FP) measures the proportion of examples classified as class X then they belong to another class, whereas TF is related to the rate of *accuracy*; (iii)The *receiver operating characteristic* (ROC) is the relationship between the rate of TP and FP [11].

Through these experiments, we scrutinize the effectiveness of each criterion-based groupization by comparing the derived results.To show which the most discriminating criteria for groupization, we present a comparison of ROC metric for each discovered criterion shown by the table 3. By comparing the studied criteria of groupization, we notice that the function criterion is the most discriminating. Indeed, the ROC of function-based clustering is equal to 96.3% for ID3 and 96.7% for NAIVE BAYES algorithm. The second ranking criterion is the source of group identification. In fact, the ROC of the source of identification is 93.2% for ID3 and 93.4% for NAIVE BAYES. The third ranking criteria is the dynamicity and finally the responsibilities.

In summary, we highlight the importance of the function exerted by the analyst facing the source of identification. The latter may be considered as a fairly important criterion because any analyst knows his colleagues and is the ablest to choose to which group he may belong. As for the dynamic criterion, it is well-recognized that the identification may take changes into consideration such as probable variations in the socket market. Finally, the granted responsibilities aiming to accomplish given goals over time is the least important criterion because such goals are constantly changing, hence the granted responsibilities are constantly varying.

Table 3. Comparison of ROC for each criterion

Training set	ROC of ID3	ROC of Naive Bayes
Function-based clusters	0.963	0.967
Responsibilities-based clusters	0.841	0.842
Source of identification-based clusters	0.932	0.934
Dynamicity-based clusters	0.939	0.94

---

[1] http://www.cs.waikato.ac.nz/~ml/weka/

4   Conclusion

In this paper, we explored the problem of groupization in data warehouses. We analyzed the similarity of query choices from the historic activities of data warehouses analysts. A hierarchical semi-supervised clustering algorithm is used to identify the best discriminating criteria from the four founded axes, namely the function, the planed goals, the source of group identification and the dynamicity of the group. Our findings demonstrate that the function is the most discriminating criterion for analysts groups identification.

Future issues for the present work mainly concern: (i) the coupling of several criteria of identification for enhancing the quality of detected groups of analysts; (ii) the proposal of our personalized hierarchical semi-supervised clustering algorithm to better fit our identification context; (iii) exploration of alternative methods to employ supervision in guiding the unsupervised clustering, e.g., supervised feature clustering.

References

1. Aligon, J., Golfarelli, M., Marcel, P., Rizzi, S., Turricchia, E.: Mining Preferences from OLAP Query Logs for Proactive Personalization. In: Eder, J., Bielikova, M., Tjoa, A.M. (eds.) ADBIS 2011. LNCS, vol. 6909, pp. 84–97. Springer, Heidelberg (2011)
2. Bellatreche, L., Giacometti, A., Marcel, P., Mouloudi, H., Laurent, D.: A personalization framework for OLAP queries. In: International Workshop on Data Warehousing and OLAP (DOLAP 2005), pp. 9–18 (2005)
3. Ben Ahmed, E., Nabli, A., Gargouri, F.: A Survey of User-Centric Data Warehouses: From Personalization to Recommendation. The International Journal of Database Management Systems (IJDMS) 3(2), 59–71 (2011)
4. Ben Ahmed, E., Nabli, A., Gargouri, F.: Building MultiView Analyst Profile From Multidimensional Query Logs: From Consensual to Conflicting Preferences. The International Journal of Computer Science Issues (IJCSI) 9(1) (to appear, January 2012)
5. Favre, C., Bentayed, F., Boussaid, O.: Evolution et personnalisation des analyses dans les entrepts de donnes: une approche oriente utilisateur. In: XXVme Congrs Informatique des Organisations et Systmes d'information et de Dcision, INFORSID 2007, Perros-Guirec, pp. 308–323 (2007)
6. Giacometti, A., Marcel, P., Negre, E.: A Framework for Recommending OLAP Queries. In: ACM Eleventh International Workshop on Data Warehousing and OLAP (DOLAP 2008), California, US, pp. 307–314 (2008)
7. Jerbi, H., Ravat, F., Teste, O., Zurfluh, G.: Applying Recommendation Technology in OLAP Systems. In: Filipe, J., Cordeiro, J. (eds.) ICEIS 2009. LNBIP, vol. 24, pp. 220–233. Springer, Heidelberg (2009)
8. Klein, D., Kamvar, S.D., Manning, C.D.: From instance-level constraints to space-level constraints: making the most of prior knowledge in data clustering. In: Proceedings of the 19th International Conference on Machine Learning (ICML 2002), CA, USA, pp. 307–314 (2002)
9. MacQueen, J.: Some Methods for Classification and Analysis of Multivariate Observations. In: 5th Berkeley Symp. Math. Statist. Prob. (1967)

10. Morris, M.R., Teevan, J.: Understanding Groups' Properties as a Means of Improving Collaborative Search Systems. In: 8th Workshop on Collaborative Information Retrieval, JCDL 2008, Pittsburgh, USA (2008)
11. Provost, F., Fawcett, T., Kohavi, R.: The case against accuracy estimation for comparing induction algorithms. In: Proceeding of the 50th International Conference on Machine Learning, Madison, Wisconsin USA, pp. 445–453 (1998)
12. Quinlan, J.R.: Induction of decision trees. Machine Learning, 81–106 (1986)
13. Ravat, F., Teste, O.: Personalization and OLAP Databases. Annals of Information Systems. New Trends in Data Warehousing and Data Analysis 3, 71–92 (2008)
14. Rizzi, S.: New Frontiers in Business Intelligence: Distribution and Personalization. In: Catania, B., Ivanović, M., Thalheim, B. (eds.) ADBIS 2010. LNCS, vol. 6295, pp. 23–30. Springer, Heidelberg (2010)
15. Teevan, J., Morris, M.R., Bush, S.: Discovering and Using Groups to Improve Personalized Search. In: Proceedings of Web Search and Data Mining, WSDM (2009)

The Study of Informality as a Framework for Evaluating the Normalisation of Web 2.0 Texts

Alejandro Mosquera and Paloma Moreda

University of Alicante, DLSI,
Alicante, Spain
{amosquera,moreda}@dlsi.ua.es

Abstract. The language used in Web 2.0 applications such as blogging platforms, realtime chats, social networks or collaborative encyclopaedias shows remarkable differences in comparison with traditional texts. The presence of informal features such as emoticons, spelling errors or Internet-specific slang can lower the performance of Natural Language Processing applications. In order to overcome this problem, text normalisation approaches can provide a clean word or sentence by transforming all non-standard lexical or syntactic variations into their canonical forms. Nevertheless, because the characteristics of each normalisation approach there exist different performance metrics and evaluation procedures. We hypothesize that the analysis of informality levels can be used to evaluate text normalization techniques. Thus, in this study we are going to propose a text normalisation evaluation framework using informality levels and its application to Web 2.0 texts.

Keywords: Informality, Normalisation, Web 2.0.

1 Introduction

Social media has become one of the most popular online platforms for information exchange between users. The language used in these Web 2.0 applications such as blogging platforms, realtime chats, social networks or collaborative encyclopaedias shows remarkable differences in comparison with traditional texts. It is common to find emoticons, spelling errors, Internet-specific slang or non-standard lexical variants [3] in Web 2.0 texts. There are also shared features with texting or chat language, such as the abuse of contractions or letter dropping, mainly because message size limits imposed by Twitter or SMS applications. As a consequence of these informal features, the automatic processing of small and noisy texts has been proven challenging, thus lowering the performance of Natural Language Processing (NLP) tools [12].

This problem has drawn the attention of the NLP community recently and different solutions have been proposed and developed. They have in common the purpose of handling these informal features and their approaches range from

G. Bouma et al. (Eds.): NLDB 2012, LNCS 7337, pp. 241–246, 2012.

statistic translation [1], spell checking [2] and a combination of lexical and phonetic edit distances [5] to normalise noisy texts. The process of text normalisation basically cleans an input word or sentence by transforming all non-standard lexical or syntactic variations into their canonical forms.

Nevertheless, because the characteristics of each normalisation approach there exist different performance metrics and evaluation procedures. These measures focus on the accuracy and correctness of the cleaned output rather than reflecting a qualitative information about the texts after and before the normalisation process. For example, the normalisation of words such as *nrm4lisati0n*, that involves the detection of phonetic substitutions, transliterations and grapheme omission, would be more challenging than the normalisation of *teh* that only needs a simple character transposition. Moreover, these metrics usually require the annotation of relatively large corpora, that is a costly and time-consuming process. This creates a need for new methods and resources that can be used to perform a more qualitative evaluation of text normalisation tools.

We hypothesize that making use of the concept of informality, text normalisation techniques can be evaluated by analysing the informality level change after and before the cleaning process. Specifically, we propose the analysis of informality variation as a framework for the evaluation of text normalization tools.

We aim to make two contributions in this study. First, we have implemented TENOR, a normalisation system for noisy English texts, with focus on the substitution of Web 2.0 lexical variants. Second, a normalisation evaluation framework based on the analysis of informality levels have been developed and its obtained results and validity have been discussed.

This article is organized as follows: In Section 2 we review the state of the art. Section 3 describes our informality evaluation framework. In Section 4, the obtained results are analysed. Finally, our main conclusions and future work are drawn in Section 5.

2   Related Work

The most simple evaluation method for text normalisation is the human judgement of the non-standard input and its corresponding canonic output. This manual supervision besides being a time-consuming and inherently subjective task, usually requires multiple annotators to calculate the agreement of the resulting score.

There are formula-based measures that help with the evaluation task and correlate moderately with the human judgement. These scores represent the quality of the output providing a meaningful score but they can be still subject to interpretation. The optimal choice of the evaluation metric usually depends on the normalisation strategy.

One of the most commonly used scores is the word error rate (WER) [6]. It derives from the Levenshtein edit distance taking into account the number of substitutions, deletions and inserts at word level between a reference text and its automatic transcription. The score is obtained by normalising the edit distance by the length of the reference text.

The BLEU algorithm (Bilingual Evaluation Understudy) [9] is a measure frequently used for evaluating the quality of texts which have been normalised using the paradigm of machine-translation. This metric treats one sentence as the reference transcription and the other as the candidate transcription, calculating the geometric mean of n-gram counts precision. The NIST score [8] extends the BLEU score by taking into account different n-gram weighting. Both scores are multiplied by a brevity penalty factor to avoid bias in short texts.

Another popular measures in Information Retrieval (IR) or Text Classification applications such as Precision, Recall and F1 can also be used in text normalisation tasks [13]. Assuming the out-of-vocabulary words (OOV) as a target and the in-vocabulary-words (IV) as non-target for normalisation the problem can be reduced to a binary class evaluation.

3   Our Evaluation Framework for Text Normalisation

This section describes an evaluation framework for text normalisation approaches. Our proposed framework, based on informality analysis [7], calculates the informality variation before and after the normalisation process. In order to do this, we make use of unsupervised machine learning techniques.

3.1   Informality Analysis

Using a multidimensional informality analysis approach based on clustering text features [7] we have evaluated four informality dimensions: Complexity, Emotiveness, Expressiveness and Incorrectness. In order to do this, a subset of the CAW2.0 dataset[1] has been used to process 7656 texts from different Web 2.0 sources. In addition we have included texts from news comments of TheGuardian, an on-line newspaper, and wall updates from Facebook, a social network.

3.2   Text Normalisation Process

We have implemented TENOR, a normalisation system for short English texts using a combination of lexical and phonetic edit distances. In order to do this, OOV words are detected with a dictionary lookup. This custom-made lexicon was built over the expanded Aspell dictionary and then augmented with domain-specific knowledge from the Spell Checking Oriented Word Lists (SCOWL)[2] package.

The OOV words are matched against a phone lattice using the double metaphone algorithm [10] to obtain a list of substitution candidates. With the Gestalt pattern matching algorithm [11] a string similarity score is calculated between the OOV word and its candidate list. Candidates with similarity values lower than 60 have been discarded empirically.

Nevertheless, there are acronyms and abbreviated forms that can not be detected properly with phonetic indexing techniques *(lol - laugh out loud)*.

---

[1] http://caw2.barcelonamedia.org
[2] http://wordlist.sourceforge.net/

For this reason, a custom-made list of 196 entries was compiled into an exception dictionary with common Internet abbreviations and slang collected from online sources[3].

Moreover, a number transliteration lookup table and several heuristics such as word-lengthening compression, emoticon translation and simple case restoration were applied to improve the normalisation results. Finally, a trigram language model was used in order to enhance the clean candidate selection.

4   Analysis

Applying the informality analysis process detailed at 3.1, three clusters have been discovered corresponding with three informality levels (C1, C2 and C3). After the normalisation process using our TENOR tool, the clustering step have been repeated on the cleaned data, with the same cluster distribution but obtaining different results.

4.1   TENOR Qualitative Evaluation

We have found that the C1 level is the less informal cluster followed by C2, a transition partition to the C3 level, where the most informal texts are grouped (see Figure 1).

Performing an analysis between raw and normalised cluster pairs, we have noticed that there is a little variation on the C1 level (see Figure 1-a). Thus the application of normalisation techniques to low informal texts had almost no impact on this informality level. Regarding the C2 level, the emotiveness dimension was decreased significantly (see Figure 1-b). This can be explained as a side-effect after the normalisation step, there were many emotion-related words that were poorly written and previously missed, now they can be detected and clustered into a higher informality level. At the C3 level can be appreciated an increase of slang and offensive words because the same reason (see Figure 1-c).

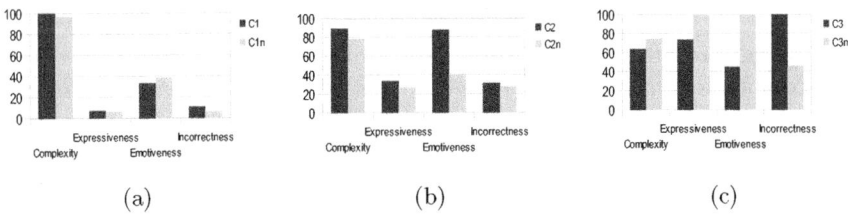

(a)   (b)   (c)

Fig. 1. Dimension variation between raw (Cn and normalised (Cxn) informality levels

4.2   TENOR Quantitative Evaluation

The informality level variation of our dataset follows a higher-to-lower informality tendency. The increased average value of informality dimensions at C3 level

---

[3] http://en.wiktionary.org/wiki/Appendix:English_internet_slang

seen in the results is not caused by an increase of its size but a level reordering instead. With a total 3% gain and a 29% loss of instances at the levels C1 and C3 respectively. Moreover, the intermediate level C2 remains almost with the same number of texts with less than a 1% increase (see Figure 2-a).

Taking into account the OOV distribution there are significant differences between the Web 2.0 applications used in this study. Texts from very informal sources such as The Guardian comments and Kongregate chats have more than 30% of OOV words before normalisation, otherwise a less informal source such as Ciao has texts with less than a 10% OOV rate (see Figure 2-b).

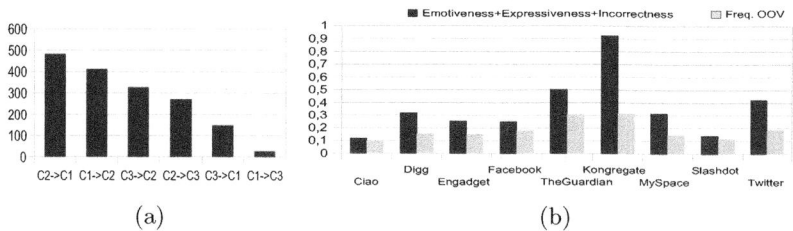

(a)   (b)

Fig. 2. a) Number of informality level changes after the normalisation process. b) Average frequency of OOV words by application.

5   Conclusions and Future Work

In this research we have described a proposal of evaluation for the normalisation of Web 2.0 texts. This automatic evaluation using informality levels avoids the need of annotated corpora and is less prone to subjective interpretations. In addition we have developed TENOR, an state-of-the-art text normalisation approach

The information gathered in this study demonstrates that our approach can provide two different types of evaluation, both at qualitative and quantitative levels. The broad informal features that can be found in Web 2.0 texts are not limited to misspellings or typos that can be targeted by a lexical or syntactical normalisation. These results provide information not available with score-based or manual evaluations, with aim to augment their evaluation results rather than replacing them.

Problems detected in this study such as the lack of a metric to measure the informality variation are left to a future work. In addition, we plan to enhance TENOR and adapt it to other languages.

Acknowledgements. This paper has been partially supported by Ministerio de Ciencia e Innovación - Spanish Government (grant no. TIN2009-13391-C04-01), and Conselleria d'Educación - Generalitat Valenciana (grant no. PROMETEO/2009/119, ACOMP/2010/286 and ACOMP/2011/001)

References

1. Aw, A., Zhang, M., Xiao, J., Su, J.: A phrase-based statistical model for sms text normalization. In: Proceedings of the COLING/ACL, pp. 33–40 (2006)
2. Choudhury, M., Saraf, R., Jain, V., Sarkar, S., Basu, A.: Investigation and modeling of the structure of texting language. In: Proceedings of the IJCAI-Workshop on Analytics for Noisy Unstructured Text Data, pp. 63–70 (2007)
3. Crystal, D.: Language and the Internet. Cambridge Univ. Press (2001)
4. Dempster, A.P., Laird, N.M., Rubin, D.B.: Maximum likelihood from incomplete data via the EM algorithm. Journal of the Royal Statistical Society: Series B (Statistical Methodology) 39, 1–22 (1977)
5. Han, B., Baldwin, T.: Lexical normalisation of short text messages: Makn sens a #twitter. In: Proceedings of the 49th Annual Meeting of the Association for Computational Linguistics: Human Language Technologies, pp. 368–378. Association for Computational Linguistics, Portland (2011)
6. Hunt, M.J.: Figures of Merit for Assessing Connected Word Recognisers. Speech Communication 9, 239–336 (1990)
7. Mosquera, A., Moreda, P.: Informality levels in Web 2.0 texts, the Facebook case study. In: Proceedings of LREC, Workshop @NLP can u tag #user_generated_content?!, Istambul, TR (2012)
8. NIST. Automatic Evaluation of Machine Translation Quality Using N-gram Co-Occurence Statistics (2002),
   http://www.nist.gov/speech/tests/mt/mt2001/resource/
9. Papineni, K., Roukos, S., Ward, T., Zhu, W.-J.: BLEU: a method for automatic evaluation of machine translation. In: Proceedings of 40th Annual Meeting of the Association for Computational Linguistics, pp. 311–318 (2002)
10. Philips, L.: The double metaphone search algorithm. C/C++ Users Journal 18, 38–43 (2000)
11. Ratcliff, J., Metzener, D.: Pattern matching: The gestalt approach. Dr. Dobb's Journal 13(7), 46–72 (1988)
12. Ritter, A., Cherry, C., Dolan, B.: Unsupervised modeling of Twitter conversations. In: HLT 2010: Human Language Technologies: The 2010 Annual Conference of the North American Chapter of the Association for Computational Linguistics, Los Angeles, USA, pp. 172–180 (2010)
13. Tang, J., Li, H., Cao, Y., Tang, Z.: Email data cleaning. In: KDD 2005: Proceeding of the Eleventh ACM SIGKDD International Conference on Knowledge Discovery in Data Mining, pp. 489–498. ACM, New York (2005)

Modeling Math Word Problems
with Augmented Semantic Networks

Christian Liguda[1] and Thies Pfeiffer[2]

[1] German Research Center for Artificial Intelligence (DFKI GmbH),
Cyber-Physical Systems, Bremen
Christian.Liguda@dfki.de
[2] Artificial Intelligence Group, Faculty of Technology, Bielefeld University
tpfeiffe@techfak.uni-bielefeld.de

Abstract. Modern computer-algebra programs are able to solve a wide
range of mathematical calculations. However, they are not able to understand and solve math text problems in which the equation is described
in terms of natural language instead of mathematical formulas. Interestingly, there are only few known approaches to solve math word problems
algorithmically and most of employ models based on frames. To overcome
problems with existing models, we propose a model based on augmented
semantic networks to represent the mathematical structure behind word
problems. This model is implemented in our Solver for Mathematical
Text Problems (SoMaTePs) [1], where the math problem is extracted
via natural language processing, transformed in mathematical equations
and solved by a state-of-the-art computer-algebra program. SoMaTePs
is able to understand and solve mathematical text problems from German primary school books and could be extended to other languages
by exchanging the language model in the natural language processing
module.

1 Introduction

Example 1. A primary school has won 30 table tennis rackets at a competition.
To each racket belong 3 balls. Two more rackets are donated by a teacher.
Everything should be shared among 8 classes. How many table tennis rackets
and balls do each class receive? (translated from [2], sentence two and three were
interchanged, question added)

Word problems like this describe situations in which mathematical relationships
are embedded within a short story [3]. For this reason word problems are different from pure arithmetics, because the calculation must first be extracted from
the given text. Therefore it is necessary to build a formal model which represents the mathematical structure. This is especially the case when information
is distributed over several sentences, possibly not in temporal order, as with the
relation between rackets and balls in Example 1 (sentence one, two and three).

G. Bouma et al. (Eds.): NLDB 2012, LNCS 7337, pp. 247–252, 2012.
© Springer-Verlag Berlin Heidelberg 2012

Math word problems are designed to teach humans how to extract and formalize mathematical structures from an information source. This makes math word problems an interesting subject to study information extraction from natural language texts.

In this paper, which is a follow-up of a workshop paper [4], we give a review of former systems built to solve word problems in Section 2. In the following section we present our model for representing the structure of word problems which is based on augmented semantic networks (ASN). Afterwards we show how to solve a word problem when the corresponding ASN is given and how our system can handle some errors in the model. We conclude with a discussion of our system in Section 4 and highlight questions to be addressed in future work.

2   Related Work

The first system able to solve math text problems was the program STUDENT [5]. Its approach to natural language processing was based on basic sentence templates of the form *?x* plus *?y* or the product of *?x* and *?y*. Together with some preprocessing and splitting of complex sentences, a word problem like *"If the number of customers Tom gets is twice the square of 20% of the number of advertisements he runs, and the number of advertisements he runs is 45, what is the number of customers Tom gets?"* was transformed to prefix notation and solved by techniques for simultaneous equations. STUDENT was able to solve some word problems from high school algebra books, but was limited to those using specific sentence structures and keywords. Because STUDENT took whole parts of a sentence as name for a variable it was not confused by irrelevant information, because they were hold in variables that were not referenced later. On the other hand STUDENT had obviously no strategies to deal with different concepts like time, money or distance.

The system WORDPRO [7] was based on a psychological model developed by Kintsch and Greeno [6]. The main idea was to organize relevant information about sets in frames with the slots Object, Quantity, Specification and Role (see Table 1). In the slot Specification information about owner, time or location of the set could be stored. The slot Role could contain roles like superset

Table 1. Frame representation of a set [6]

Slot	Value
Object	<cnoun>
Quantity	<number>, SOME, HOWMANY
Specification	<owner>, <location>, <time>
Role	start, transfer[In/Out], result, superset...

or transIn. A rule based system was used for problem solving. Each rule (also named as schema) checked whether the slots of the frames met some preconditions, and if so, a calculation was done and the result was added to a slot. This way, WORDPRO solved some problems of addition and subtraction, but it was based on propositions only, with no natural language interface. By including information about location, time and owner, WORDPRO could deal with some irrelevant information.

The combination of frames for representation and schemata for deriving the solution is found in nearly every later system. The latest is MSWPAS [8], which is able to solve multi-step word problems containing addition and subtraction for sets. However, their multi-step approach is very restricted. If, for example, a set is related to more than one other set, it could have different roles for each relation, but such multi-step problems cannot be represented in MSWPAS. Furthermore, their schema does not consider the relation to other schemata, but is restricted to a local view on the frames. Hence, problems of the following kind: *"Bob gives 7 marbles to Tim. Afterwards Bob has half the marbles he had at the beginning. How much marbles had Bob at the beginning?"* will activate no schema, as the number of marbles Bob has at the beginning as well as at the end are unknown.

Mukherjee and Garain give a more detailed review about the mentioned systems (except MSWPAS) and others not mentioned here [9].

3 Formal Representation of a Problem

We propose an augmented semantic network for representing the structure of a math word problem in which the concepts are represented by nodes and their interrelations by edges. With the representation of word problems using an ASN we aim at making relations to other concepts explicit [10]. This will also allow us to solve the problem of the established approaches based on schema concepts with multiple relations (see Section 2).

In this section we provide a brief overview of the representation, model-to-equations transformation and equations solving.

3.1 Representation

Node: Nodes represent the concepts, which are measurable quantities. Different kinds of nodes with distinct attributes are used. At the moment SoMaTePs is able to represent sets, by nodes of the type `set` with the attributes `Object`, `Quant`, `Location` and `Owner`, and an amount of money, by nodes of the type `money` with the attributes `Object`, `Value` and `Owner`.

Situation History: An important extension to standard semantic networks is the situation history. In our representation, every attribute for a measurable quantity is represented as a list of tuples. Each tuple holds a situation counter as index and a value. This way, a node can represent dynamic changes of an attribute in the course of the story told in the math text problem. The system uses one situation counter per node which is incremented whenever a relation changes an attribute. This augmentation is crucial for handling intermediate values.

Edge: In contrast to normal semantic networks where edges connect nodes, an edge in the ASN is between nodes, their attributes and specific situations. Therefore the edges contain information to what attributes and situations they belong. Furthermore, some edges represent relations between concepts in which additional information are important. The relation `factor` is an example for this type of relation. Given the text *"To each racket belong 3 balls."*, the cardinality (here 3) of the relation needs to be represented. At the moment SoMaTePs has different types of edges for addition, subtraction, difference between values, transfer between nodes, part-whole and factor relations.

Example: In Fig. 1 the ASN of the word problem of Example 1 is given. Unknown values for the attribute `Quant` in a situation or unknown multiplicative factors are labeled as `SOME`.

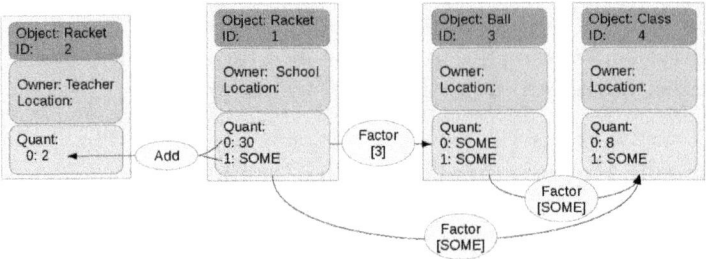

Fig. 1. Augmented semantic network of the word problem of Example 1. The edges of the network span between situation specific entries of attributes and can be augmented by additional parameters.

3.2   Model-to-Equations Transformation

As we have shown in Section 2 (for a more elaborated discussion see [4]), solving a model with a rule based system leads to several problems. Hence our system uses a different approach in which every relation is transformed into a set of equations. More than one equation is generated if an edge is connected to more than one attribute in one situation, as in Fig. 1, where the attribute `Quant` of the set *Racket* with the ID 1 in the model contains two situations. Therefore the `Factor` relation is transformed into the following two equations:

$$Y_0 = 3 * 30, \quad Y_1 = 3 * X_1$$

The variables Y_0 and Y_1 represent the two values for the attribute `Quant` of the set *Ball*. X_1 is a variable for the unknown value of the attribute `Quant` of the set *Racket* after the addition. A central variable management unit assures globally unique names for all variables.

The generated set of equations can be solved with standard techniques of computer algebra. For this we use the external open source computer algebra program Sage [11]. By doing so, our system is able to to solve multi-step word problems and is able to solve complex equation systems in which dependencies between equations have to be taken into account. By making the relations explicit, differentiating between attributes and situation history, giving each unknown value an unique variable name and transforming relations into equations, our model is furthermore very robust to irrelevant informations.

3.3   Iterative Refinement

In the process of constructing the ASN, ambiguities may rise because of various reasons, most commonly underspecification. This makes the model unsolvable, as it then contains too little or inconsistent information. Here is one example: The model given in Fig. 1 is underspecified, because the value for the attribute Quant in situation 1 of the set *Class* is unknown. Originally only the value in situation 0 is known and situation 1 was added, because the set *Racket* changed over time and therefore every other set which is in a relation to the set *Racket* has to notice this change. Depending on the given problem, the added situations in the related sets have either an unknown value or the same value as before. Both possibilities can be found in Fig. 1. The value for the attribute Quant of the set *Ball* changes over time, while the value for the attribute Quant of the set *Class* remains unchanged. In such situations the created model needs to be refined using heuristics. In a first step, the model is analyzed, possible underspecifications are detected and a list of heuristics correcting these weaknesses is created. Afterwards the model is refined by iteratively applying the heuristics, evaluating the model and checking the results. If the model has been solved, the process terminates, if not, the last heuristic is revoked and the refinement process continues. SoMaTePs is able to handle five different types of heuristics and can also correct models which contain several of them.

4   Discussion

The presented system for solving math word problems goes beyond its predecessors. An essential difference is the transition from a frame-based approach to augmented semantic networks. This way, the relationships between individual concepts can be represented more precisely, which allows a more robust detection of irrelevant information. The augmented semantic network is extended to support the representation of the temporal order of actions by implementing a state system for attributes and values of nodes. The scalability of this approach to different types of math problems is ensured by a defined mapping from the semantic network to a system of equations which is then solved by an external computer-algebra program. This allows math problems of higher complexity to be solved than what is representable in schemes. In fact SoMaTePs is able to solve all classes of problems that could be solved by the schema-based systems mentioned in Section 2 (based on the examples that are mentioned in the corresponding papers). In addition to that SoMaTePs could solve around 20 other classes of word problems from a school book [2] which were in most cases not solvable by other systems. Even if a wide range of math word problems from primary school books could already be modeled, SoMaTePs is currently limited to problems of the domains set and money. Therefore the model needs to be extended to the concepts weight, time, distance and volume in future work and should then be evaluated against a corpus of primary school math text problems.

SoMaTePs could be used to get insights on human cognitive capabilities which might be gained by constructing a system for understanding math word problems. Furthermore SoMaTePs can enable tutoring system to solve and explain math text problems previously unknown to the system.

SoMaTePs is released as an open source project hosted at SourceForge.net [1].

References

[1] Liguda, C., Pfeiffer, T.: Somateps (2012),
    `http://sourceforge.net/projects/somateps/` (visited January 31, 2012)
[2] Bergmann, H., Hauer, E., Heuchert, D., Kaufmann, A., Kuehne-Zuern, D., Schneider, A., Schumacher, K., Teifke, R., Usemann, K., Woerwag, K.: 200 Textaufgaben wie in der Schule. Klett, Stuttgart (2010)
[3] Franke, M., Ruwisch, S.: Didaktik des Sachrechnens in der Grundschule. Spektrum Akad. Verl., Heidelberg (2010)
[4] Liguda, C., Pfeiffer, T.: A question answer system for math word problems. In: Proceedings of the First International Workshop on Algorithmic Intelligence (2011), `http://pub.uni-bielefeld.de/publication/2423329`
[5] Bobrow, D.G.: A question-answering system for high school algebra word problems. In: Proc. of the Fall Joint Computer Conference, Part I. AFIPS 1964, October 27-29, pp. 591–614. ACM, New York (1964)
[6] Kintsch, W., Greeno, J.G.: Understanding and solving word arithmetic problems. Psychological Review 92(1), 109–129 (1985)
[7] Fletcher, C.: Understanding and solving arithmetic word problems: A computer simulation. Behavior Research Methods 17, 565–571 (1985), doi:10.3758/BF03207654
[8] Yuhui, M., Ying, Z., Guangzuo, C., Yun, R., Ronghuai, H.: Frame-based calculus of solving arithmetic multi-step addition and subtraction word problems. In: International Workshop on Education Technology and Computer Science, vol. 2, pp. 476–479 (2010)
[9] Mukherjee, A., Garain, U.: A review of methods for automatic understanding of natural language mathematical problems. Artif. Intell. Rev. 29, 93–122 (2008)
[10] Simmons, R.F.: Semantic networks: Their computation and use for understanding English sentences. Freeman, San Francisco (1973)
[11] Stein, W., Joyner, D.: Sage: System for algebra and geometry experimentation. Communications in Computer Algebra, SIGSAM Bulletin (2005)

Improving Document-Level Sentiment Classification Using Contextual Valence Shifters

Sara A. Morsy and Ahmed Rafea

The American University in Cairo
{sara_am,rafea}@aucegypt.edu

Abstract. Traditional sentiment feature extraction methods in document-level sentiment classification either count the frequencies of sentiment words as features, or the frequencies of modified and unmodified instances of each of these words. However, these methods do not represent the sentiment words' linguistic context efficiently. We propose a novel method and feature set to handle the contextual polarity of sentiment words efficiently. Our experiments on both movie and product reviews show a significant improvement in the classifier's performance (an overall accuracy increase of 2%), in addition to statistical significance of our feature set over the traditional feature set. Also, compared with other widely-used feature sets, most of our features are among the key features for sentiment classification.

1 Introduction

Sentiment Classification has been emerging as a new application in Natural Language Processing (NLP), machine learning and text mining, due to the huge opinionated web content people generate, such as reviews, blogs, and forums. Some studies have treated this problem like topic-based text-classification problems by applying machine learning algorithms. But since the overall sentiment of a document can also be determined by the polarities of the different sentiment words used in the document, other studies have focused on developing sentiment dictionaries (lexicons). Some recent studies have proposed hybrid approaches by taking the frequencies of sentiment words found in the document [6] as well as the frequencies of their unmodified and modified instances [5] as features into the classifier. However, these methods were inefficient and created huge feature spaces.

Therefore, we propose a new method and feature set to refine the traditional sentiment feature extraction method provided by [5,6] to efficiently handle the effect of the linguistic context of sentiment words, i.e. contextual valence shifters, in order to improve the classifier's performance in document-level sentiment classification.

2 Related Work

Early research in document-level sentiment classification has used both machine learning [1,8] and semantic orientation [10] approaches separately. More recent

G. Bouma et al. (Eds.): NLDB 2012, LNCS 7337, pp. 253–258, 2012.

studies have combined both approaches into different hybrid approaches and showed better results than a non-hybrid approach [5,6].

Dang et al. built an SVM classifier with n-grams and stylistic features and used the frequency of sentiment words as sentiment features for classification of online reviews [6]. Their experiments showed a slight improvement when adding sentiment features. Since the polarity of sentiment words may be modified by negators, intensifiers, diminishers and other polarity shifters (known as *Contextual Valence Shifters* [7]), Kennedy and Inkpen attempted to add the effect of these valence shifters by creating 4 different features for each sentiment word: unmodified, negated, intensified, and diminished frequencies of each sentiment word [5]. For example, the word *good* is represented with 4 features: *good*, *not_good*, *intensified_good*, and *diminished_good*. However, this method does not efficiently handle the linguistic context of sentiment words, since the classifier cannot learn that these 4 features are in any way related. Moreover, creating 4 features for each sentiment word increases the feature space in an unnecessary way.

Therefore, we present an alternative method to extract sentiment features and to handle the contextual polarity of words and we propose a new smaller feature set for contextual valence shifters so that the classifier would efficiently learn how to take the contextual polarity of sentiment words into account.

3   Feature Set and Method Description

3.1   Contextual Valence Shifters Feature Set

Our proposed feature set consists of 16 features. The first 12 features count the frequencies of intensified (int), negated (neg), and polarity-shifted (shifted) sentiment words grouped together according to the polarity and subjectivity type (intensity) of the sentiment words modified by them (polarity can be either *positive* or *negative* and subjectivity type can be either *strong* or *weak*). For instance, one of these features is *int_pos_strongsubj*, which counts the frequency of words with positive prior polarity and strong subjectivity type preceded by an intensifier. The remaining 4 features count the frequencies of sentiment words grouped together with their final (modified) polarity and type, e.g. *pos_strongsubj* counts the frequency of all words with final positive polarity and strong subjectivity [2].

3.2   Feature Extraction

To extract sentiment words, we lemmatize the words before looking them up in the sentiment lexicon. After extracting each sentiment word, we search for the valence shifters. We use three boolean variables; namely *intensified*, *negated*, and *shifted* to calculate the final polarity of sentiment words.

To extract intensifiers, which could be adjectives or adverbs such as: '*very good*' or '*huge* problem', we check if the sentiment word is preceded by an intensifier. We also check if there is an intensifier that is an object of a sentence that has the sentiment word as its subject, e.g. *This problem is huge*. Then, only the types of sentiment words with weak subjectivity type are intensified and we set the variable *intensified* to TRUE.

To extract negated terms, such as *not* and *never*, we follow a method similar to [4]. We search for the negation term that appears in a "neg" dependency or in a negation term list, then we get the First Common Ancestor between the negation term and the word immediately after it. We then assume the descendant leaf nodes that appear after the negation term and within a window of size 6 only to be affected by the negation term. Also, we omit the negation terms that appear in expressions that do not represent negation, e.g. *not only*. We use the *shift negation* method [10] by negating the polarities of sentiment words as follows: *positive strong* and *negative weak* are negated to *positive weak*; *positive weak* and *negative strong* are negated to *negative weak*. Then, we set the variable *negated* to TRUE.

To extract polarity shifters, we search before the sentiment word to check if it is preceded by a positive polarity shifter, e.g. *abate* the damage, or a negative polarity shifter, e.g. *lack of* interest. If so, the variable *shifted* is set to TRUE.

After extracting the valence shifters, we modify the polarity and type of sentiment words preceded by any of these shifters, according to the 3 variables we use (*intensified, negated* and *shifted*). Since each of these variables has a possible value of TRUE or FALSE, we have 8 different combinations of their values, and so there are 8 different possibilities for the final polarity of the sentiment word. For instance, if we have *negated* = TRUE, *intensified* = TRUE and *shifted* = FALSE, e.g. *he is not very good*, the negation term *not* negates the whole intensified expression *very good*. Hence, the polarity of the sentiment word will be first intensified (from *positive weak* to *positive strong*) and then negated (to *positive weak*) (in this case, the final polarity is the same as the prior polarity, but this will not be the case if the sentiment word's type is strong) [2]. Finally, we increment the word's sentiment feature by 1 (or decrement it by 1 if it is negated) and we increment the frequency of sentiment words with the same polarity and type as the final polarity and type of the sentiment word by 1.

4   Experimental Setup

We compare the performance of our proposed features against the traditional sentiment features that count the frequencies of all sentiment words (baseline). We refer to the baseline as **F1** and our features as **ref_F1+F2**, where F1 represents the traditional sentiment feature set, ref_F1 represents the refined (modified) sentiment feature set and F2 represents the proposed feature set for contextual valence shifters. We then combine these features with other proposed feature sets (stylistic features and n-grams, referred to as **F3**) used in [6] and perform feature selection to evaluate the usefulness of our feature set against the other sets.

We use two of the widely-used datasets in sentiment classification. The first dataset is the Polarity Dataset version 2.0[1], which contains 1,000 positive and 1,000 negative movie reviews. The second dataset is the multi-domain dataset, which consists of 1,000 positive and 1,000 negative reviews in each of the: books, DVDs, electronics and kitchen appliances domains [3]. We use the Subjectivity dictionary, and lists of intensifiers and other valence shifters provided by [9]. The Subjectivity

---

[1] http://www.cs.cornell.edu/people/pabo/movie-review-data/

dictionary provides each sentiment word with its prior polarity and subjectivity type as well as other useful information, such as its Part-Of-Speech. We utilize the Stanford tagger[2] and parser[3]. We use the Weka Suite software[4] for classification and Support Vector Machines (SVM) as the classification algorithm with default kernel settings. For lemmatization, we use the MontyLingua lemmatizer[5].

We perform two sets of experiments. In experimental set 1 (**E1**), we run the classifier using: 1) **F1**; and 2) **ref_F1+F2**. We also measure the statistical significance of ref_F1+F2 against F1 by performing the two-tailed *t*-test. In experimental set 2 (**E2**), we add F3 to our proposed features and the baseline. Then, we perform feature selection on these features using the Information Gain (IG) heuristic with a threshold of 0.0025 (as proposed by [6]) on the baseline (F1+F3) and all features (ref_F1+F2+F3 or all) to extract the key features for sentiment classification from all the features used. For the two experiment sets, we measure the accuracy of the classifier. We run the classifier using 10-fold cross-validation. We perform the two-tailed *t*-test by using a sequence of 40 randomly chosen train-test partitions of each dataset, with 1900 reviews for training and 100 reiews for testing, as proposed in previous work [1].

5   Results and Analysis

5.1   Results of E1

Table 1 (under E1) shows the accuracy of the classifier using the baseline and ref_F1+F2. There is a clear improvement in the accuracy, as there is an increase in the accuracy from 0.8-5% with an overall increase by more than 2% among all the datasets. In addition, the results of the two-tailed *t*-test of ref_F1+F2 against F1 are shown in table 2. The *t*-ratio is large in each of the books, electronics and kitchen appliances datasets with a very small *p* value, which shows the statistical significance of our proposed feature sets against the baseline in these 3 datasets. Also, the mean accuracies in both the Polarity and multi-domain datasets using ref_F1+F2 are higher than the mean accuracies using F1. These results clearly show that our proposed sentiment feature extraction method is better than the traditional method in sentiment classification.

5.2   Results of E2

The results of this experiment are shown in table 1 (under E2). The results show that our proposed feature set outperform the baseline when combined with other features, except for the DVDs dataset which remained the same, with an overall increase of about 1%. It is important to note that most of the 16 features in our proposed feature set F2 were selected among the key features for the classification,

---

[2] http://nlp.stanford.edu/software/tagger.shtml
[3] http://nlp.stanford.edu/software/lex-parser.shtml
[4] http://www.cs.waikato.ac.nz/ml/weka/
[5] http://web.media.mit.edu/~hugo/montylingua/

Table 1. Accuracy of different feature sets in E1 and E2

Feature Set	Polarity Dataset	Multi-domain Dataset				Average
		Books	DVDs	Electronics	Kitchen	
E1						
F1(baseline)	79%	72.95%	77.10%	76.05%	75.55%	76.13%
ref_F1+F2	**79.85%**	**77.90%**	**77.35%**	**78.70%**	**79.55%**	**78.67%**
E2						
F1+F3	87.05%	84.20%	84.60%	83.55%	84.85%	84.85%
all	**87.15%**	**85.50%**	84.60%	**84.85%**	**86.55%**	**85.73%**

Table 2. The two-tailed t-test results for ref_F1+F2 against F1

Measure	Polarity Dataset	Multi-domain Dataset			
		Books	DVDs	Electronics	Kitchen
t-ratio	0.78	7.01	0.54	2.22	1.94
P value	0.44	2.1E-8	0.59	0.03	0.06
Mean Acc. Diff.	0.8	5.45	0.6	2.3	1.5

Table 3. Top IG-Ranked Features for the books dataset (with our proposed features in bold)

No.	Feature	IG
1	**neg_strongsubj**	0.0693
2	**intensifier_neg_strongsubj**	0.0513
3	**neg_weaksubj**	0.0362
4	**pos_strongsubj**	0.0356
5	no	0.0311
6	**negation_pos_strongsubj**	0.0291

with some of them having the highest IG. For example, table 3 shows the top IG-ranked features for the books dataset. This indicates the importance and relevance of our proposed feature set compared with the other previously-proposed features in classifying the documents.

6   Conclusion and Future Work

Our experiments show the improvement of the classifier's performance and the statistical significance of our proposed feature set over the baseline. It is important to note that our model outperformed the baseline with better results in product reviews than in movie reviews, due to the nature of movie reviews, which can contain a lot of objective sentences about the characters or the movie that contain sentiment words. These sentiment words should not be counted when determining the overall sentiment of the author [5].

One direction for future work could be to extend the proposed feature set for contextual valence shifters by adding diminishers and creating 4 other features to represent diminished words grouped together according to the polarity and type of the sentiment words modified by them. Another direction is to perform topic extraction to extract topic sentences so that we can classify the sentiment of each topic being talked about in the document or only the topic of interest.

Acknowledgment. This work was supported by ITIDA.

References

1. Abbasi, A., Chen, H., Salem, A.: Sentiment analysis in multiple languages: Feature selection for opinion classification in Web forums. ACM Trans. Inf. Syst. 26, 1–34 (2008)
2. Morsy, S.: Recognizing contextual valence shifters in document-level sentiment classification. Masters Thesis, The American University in Cairo (2011)
3. Blitzer, J., Dredze, M., Pereira, F.: Biographies, bollywood, boom-boxes and blenders: Domain adaptation for sentiment classification. In: Proc. Assoc. Computational Linguistics, pp. 440–447 (2007)
4. Carrillo de Albornoz, J., Plaza, L., Gervás, P.: A hybrid approach to emotional sentence polarity and intensity classification. In: Proceedings of the Fourteenth Conference on Computational Natural Language Learning (CoNLL 2010), pp. 153–161 (2010)
5. Kennedy, A., Inkpen, D.: Sentiment classification of movie reviews using contextual valence shifters. Computational Intelligence 22, 110–125 (2006)
6. Dang, Y., Zhang, Y., Chen, H.: A lexicon-enhanced method for sentiment classification: an experiment on online product reviews. IEEE Intelligent Systems 25, 46–53 (2010)
7. Polanyi, L., Zaenen, A.: Contextual valence shifters. Computing Attitude and Affect in Text: Theory and Applications 20, 1–10 (2006)
8. Whitelaw, C., Garg, N., Argamon, S.: Using appraisal groups for sentiment analysis. In: Proceedings of the 14th ACM International Conference on Information and Knowledge Management (CIKM 2005), pp. 625–631 (2005)
9. Wilson, T., Wiebe, J., Hoffmann, P.: Recognizing contextual polarity in phrase-level sentiment analysis. In: Proceedings of the Conference on Human Language Technology and Empirical Methods in Natural Language Processing (HLT 2005), vol. 22, pp. 347–354 (2005)
10. Taboada, M., Brooke, J., Tofiloski, M., Voll, K., Stede, M.: Lexicon-based methods for sentiment analysis. Computational Linguistics 37, 267–307 (2011)

Comparing Different Methods
for Opinion Mining in Newspaper Articles

Thomas Scholz[1], Stefan Conrad[1], and Isabel Wolters[2]

[1] Heinrich-Heine-University, Institute of Computer Science, Düsseldorf, Germany
{scholz,conrad}@cs.uni-duesseldorf.de
[2] pressrelations, Development Department, Düsseldorf, Germany
isabel.wolters@pressrelations.de

Abstract. Adapting opinion mining for news articles is a challenging field and at the same time it is very interesting for many analyses, applications and systems in the field of media monitoring. In this paper, we illustrate specifics in this area in comparison with sentiment analysis of product reviews. Likewise, we introduce new methods for the determination of the sentiment polarity in statements, which are extracted from news articles. Our evaluation on a real world data set of a German Media Response Analysis (MRA) shows that these methods perform better than existing approaches and resources.

Keywords: Opinion Mining, Sentiment Score, Media Response Analysis.

1 Motivation

The amount of potentially useful online articles increased substantially in the last years and this trend is continuing. While research in opinion mining is concentrated on analysing customer and film reviews, we believe that opining mining in news articles is more interesting than in reviews because review systems such as webstores, movie review sites, online booking services already provide a quick overview of the tonality of the products, films and services by overall scores or selected reviews. The offer for such an overview system, which can perform and illustrate a Media Response Analysis (MRA) [11], represents a seperate business segment for analysis services and is very interesting for companies, organisations like political parties, associations or distinguished public figures.

Today, a MRA is carried through with a huge amount of human effort in media monitoring companies. After the news texts which contains a certain search string have been collected by Web crawlers, domain specialists (the so called media analysts) read these texts and mark statements which are relevant to the customers and set the polarity of the opinion. In this case, we speak of tonality. In contrast to customer's reviews or some contributions in social media, news items are not as subjective as these [1]. Also, there can be different

G. Bouma et al. (Eds.): NLDB 2012, LNCS 7337, pp. 259–264, 2012.

polarities of opinion in different parts of one article. Thus, it is our focus to classify statements as positive or negative.

Problem Definition: In this context, opinion mining has the task to find a function t which should determine the tonality y for a given statement s, consisting of s_n words:

$$t : s = (w_1, w_2, ..., w_{s_n}) \mapsto y \in \{pos, neg\} \tag{1}$$

A statement s represents a relevant part of text and can consist of a single sentence or several consecutive sentences (e.g. up to three or four). So, the focus is on the statement level and not on the document level.

2   Related Work

Research to date has tended to focus on sentiment analysis in customer reviews or social media rather than in newspaper articles [7].

In the product review context many approaches use a collection of annotated data to collect important sentiment keywords. While some of the approaches are also extract words like nouns, verbs, adverbs and adjectives [2], most of the attention is given to the adjectives [4,6]. One solution in this area involves the construction of powerful and partly very complex adjective patterns to handle sentiment polarity and the relationship between products and adjectives [2,4].

Kaji and Kitsuregawa [6] have compiled a lexicon from HTML documents. Their idea of creating adjectival phrases is unfortunately not completely adaptable due to particularities of the Japanese language and differences in context (cf. chapter 3), but their methods of selecting these phrases are treated in this paper. Harb et al. [4] extract opinions from blogs and they use association rules for the extraction of suitable adjectives. Furthermore, the use of a sentiment lexicon is very common. In German one of the first public available lexicons is called SentiWS [9].

Opinion mining in newspaper articles has focused on quotations [1,8], because reported speech is typically more subjective than other parts of an article. However, a major problem with this technique is that it can only obtain opinions which are a part of reported speech. If we examine 3120 statements of our real world data set, only 21,25% of the statements contain a quotation and less than 5% have a proportion of quoted text larger than 50% of the whole of the statement.

3   Pre-evaluation

At first, we validate which word classes are the most important by using different machine learning techniques. Our data set consists of 1596 statements including 796 positive and 800 negative statements. We use a $tf * idf$ matrix to weigh the terms in the four classes: nouns, verbs, adjectives and adverbs, which are the important classes for the tonality [2,9].

Category	SVM	Naive Bayes	Decision Tree	k-NN	k-means	Linear Regression
Adjectives	61.76%	61.44%	54.39%	52.82%	50.60%	49.76%
Verbs	73.67%	72.10%	67.01%	52.27%	51.16%	49.76%
Nouns	77.27%	70.06%	69.36%	55.72%	55.55%	77.51%
Adverbs	61.44%	61.60%	59.64%	51.65%	50.91%	59.95%

Fig. 1. Evaluation of the different word classes and learning methods

As figure 1 shows the most important words do not belong to the adjective word group, which is a common assumption in approaches for customer reviews. A typical sentence of a customer review is "The zoom is great.". In newspaper articles the verbs and nouns do play a more important role. Examples of newspaper articles such as "Analysts fear the end of the eurozone." make clear that the tonality is created through the verb "fear" and the noun "end".

4 Determination of Polarity

4.1 Word Based Methods

We compute a tonality score TS_{Method} for each single word in the important categories (adjectives, verbs, nouns, adverbs). Thus, we get four tonality scores for one statement. Each score is the average TS value of all words in one of the four categories. For the TS value, we use methods such as the **chi-square** [6], the **PMI** method [3,6] or **Association Rule Mining** [4]. Furthermore, we propose two new methods: the entropy and the information gain method.

Entropy: By analogy with the chi-square method, the probability of appearance in positive and negative statements is calculated first.

$$P(w|pos) = \frac{f(w, pos)}{f(w, pos) + f(\neg w, pos)} \qquad P(w|neg) = \frac{f(w, neg)}{f(w, neg) + f(\neg w, neg)}$$

$$(2)$$

$f(w, pos)$ is the frequency of the candidate word w in positive statements and $f(\neg w, pos)$ is the same for all candidates words without w. If the probability for word w to appear in positive statements is higher than the score is positive and vice versa.

$$TS_{ENT}(w) = \begin{cases} 1.0 + P(w|pos) * \log_2(P(w|pos)) & \text{if } P(w|neg) \leq P(w|pos) \\ -1.0 + P(w|neg) * \log_2(P(w|neg)) & \text{otherwise} \end{cases}$$

$$(3)$$

Information Gain: Another possibility is the use of the information gain for our tonality value. The information gain is also based on the entropy [10] of a set of statements S which contains positive and negative statements.

$$entropy(S) = -1 * [(P(pos) * \log_2(P(pos)) + P(neg) * \log_2(P(neg))] \qquad (4)$$

If one word is chosen, then the gain of purity can be calculated by considering the two sets $S_{w,pos}$ and $S_{w,neg}$ in which w appears.

$$TS_{IG}(w) = \begin{cases} 1.0 - \sum_{y \in \{pos,neg\}} \frac{|S_{w,y}|}{|S|} * \text{entropy}(S_{w,y}) & \text{if } P(w|neg) \leq P(w|pos) \\ -1.0 + \sum_{y \in \{pos,neg\}} \frac{|S_{w,y}|}{|S|} * \text{entropy}(S_{w,y}) & \text{otherwise} \end{cases}$$

(5)

Lexicon Based: Many approaches [7] use lexicons such as SentiWS [9].

Bigrams: The approach uses the entropy to determine the quality of a certain word (adverbs, adjectives, verbs, nouns, negation particles or entities like persons or organisations). The algorithm searches all bigrams to which the word has got a part-of relationship. If the entropy value is smaller, when the single word is exchanged through all bigrams which contain this word and the tonality scores of these bigrams are used.

4.2   Pattern Based Action Chains

Triplets: A triplet consists of the three basic elements: subject, verb and object. Our triplet model also includes the adjectives, which belong to subjects and objects, and the adverbs, which belong to verbs.

Verb Based Patterns: All full verbs and all associated objects (proper nouns, entities) will be extracted, if the distance of the object and the verb is lower than or equal to x words. Likewise the pattern includes the adjectives belonging to the objects and the adverbs belonging to the verbs.

Similarity Functions for Action Chains: The first one is based on the Jaccard similarity coefficient [5] and the lemmas of the different elements.

$$sim(c_1, c_2) = \frac{n(c_1 \cap c_2)}{n(c_1 \cup c_2)}$$

(6)

$n(c_1 \cap c_2)$ is the number of matching elements and $n(c_1 \cup c_2)$ the number of all elements. The second similarity function compares the tonality values of the lemmas w_1, w_2.

$$\delta(w_1, w_2) = 1 - |TS_{Method}(w_1) - TS_{Method}(w_2)|$$

(7)

For the tonality scoring method TS_{Method} every word based approach above can be used. The best fitting chain of a training set delivers the value for a new one and the score of a statement is the average value of all containing chains.

5   Evaluation

Our test corpus consists of 5500 statements (2750 positive, 2750 negative) from 2097 different news articles. The test corpus was created by up to ten media analysts (professional domain experts in the field of MRA). Their quality assurance ensures an inter-annotator agreement of at least 75% to 80% in a MRA.

We have split the corpus randomly into a part consisting of 1600 statements (800 positive, 800 negative) for extracting dictionaries, triplets, verb based patterns or bigrams. We obtained 4581 words, 4816 triplets, 4809 verb based patterns and 6924 bigrams. We have also constructed smaller collections of 600 and 1100 statements (called $\Delta600$ and $\Delta1100$) from this set for our comparison with SentiWS [9]. The $\Delta1100$ collection should generate dictionaries which has got approximately the same number as lemmas of the SentiWS.

After that we again split the second randomly into two parts. The first part of 780 statements (split ratio 0.2) is used as a training set and the larger second part is our test set. The approach implies a SVM which gets the four different tonality values as attributes for learning. We also construct smaller training sets of 390 (called S-0.1), 195 (S-0.05) and 39 statements (S-0.01). As in the pre-evaluation, SVMs perform better than Naive Bayes et cetera. Our SVM uses the Rapidminer[1] standard implementation with default parameters. But we force the SVM to balance between precision and recall in and between the two classes because media response analyses are usually not so balanced as our test set.

Method	Accuracy	Category	S-0.2	S-0.1	S-0.05	S-0.01
Word Based		Adjective	0.6215	0.6236	0.6224	0.3831
TF-IDF	0.6353	Adverb	0.5987	0.5855	0.5725	0.5804
Chi-square	0.6837	Noun	0.6625	0.6641	0.6659	0.6674
PMI	0.6404	Verb	0.6474	0.6464	0.6499	0.4996
Association Rule Mining	0.5234	Noun&Adj.	0.6718	0.6741	0.6729	0.3825
Entropy	**0.7006**	Noun&Verb	0.6808	0.6804	0.6826	0.6713
Information Gain	0.6955	All	**0.7006**	0.6940	0.6804	0.5872
Bigrams	0.6888	Method		$\Delta1600$	$\Delta1100$	$\Delta600$
Action Chains		Chi-square		0.6837	0.6548	0.6292
Triplet	0.6580	PMI		0.6404	0.6224	0.6196
Triplet(tonality-based)	0.6567	Entropy		**0.7006**	0.6805	**0.6628**
Verb Based Pattern	0.5925	Information Gain		0.6955	**0.6952**	0.6558
Verb Based Pattern(ton.-based)	0.5925	SentiWS (1 size)		-	0.6036	-

Fig. 2. Left: Comparison between the different methods. Up right: The different word classes (entropy method) by different sizes of the training set. Bottom right: Comparison to the SentiWS [9] and different sizes of the created lexicons.

The tables of figure 2 show the results. The best performing methods (figure 2 left) are our entropy and information gain based methods with about 70%. The entropy-based method remains very stable even if the training set is decreased (figure 2 up right). Furthermore, the entropy and information gain methods have over 7 % resp. 9 % higher performance than SentiWS using the $\Delta1100$ set (figure 2 bottom right).

---

[1] Rapid-I: (http://rapid-i.com/)

6   Conclusion and Future Work

In this contribution, we have demonstrated some special characteristics of opinion mining in newspaper articles. The results of the experiments show that our entropy method outperform other word based methods and complex pattern based approaches. Our future work will cover the creation of features which are very specific for opinion mining in news and the automated determination of viewpoints.

Acknowledgments. This work is funded by the German Federal Ministry of Economics and Technology under the ZIM-program (Grant No. KF2846501ED1).

References

1. Balahur, A., Steinberger, R., Kabadjov, M., Zavarella, V., van der Goot, E., Halkia, M., Pouliquen, B., Belyaeva, J.: Sentiment analysis in the news. In: Proc. of the 7th Intl. Conf. on Language Resources and Evaluation, LREC 2010 (2010)
2. Bollegala, D., Weir, D., Carroll, J.: Using multiple sources to construct a sentiment sensitive thesaurus for cross-domain sentiment classification. In: Proc. of the 49th Annu. Mtg. of the ACL: Human Language Technologies, HLT 2011, vol. 1, pp. 132–141 (2011)
3. Church, K.W., Hanks, P.: Word association norms, mutual information, and lexicography. In: Proc. of the 27th Annu. Mtg. on ACL, ACL 1989, pp. 76–83 (1989)
4. Harb, A., Plantié, M., Dray, G., Roche, M., Trousset, F., Poncelet, P.: Web opinion mining: how to extract opinions from blogs? In: Proc. of the 5th Intl. Conf. on Soft Computing as Transdisciplinary Science and Technology, CSTST 2008, pp. 211–217 (2008)
5. Jaccard, P.: Etude comparative de la distribution orale dans une portion des alpes et des jura. Bulletin del la Socit Vaudoise des Sciences Naturelles 37, 547–579 (1901)
6. Kaji, N., Kitsuregawa, M.: Building lexicon for sentiment analysis from massive collection of html documents. In: Proc. of the 2007 Joint Conf. on Empirical Methods in Natural Language Processing and Computational Natural Language Learning, EMNLP-CoNLL 2007 (2007)
7. Pang, B., Lee, L.: Opinion mining and sentiment analysis. Foundations and Trends in Information Retrieval 2(1-2), 1–135 (2008)
8. Park, S., Lee, K., Song, J.: Contrasting opposing views of news articles on contentious issues. In: Proc. of the 49th Annu. Mtg. of the ACL: Human Language Technologies, HLT 2011, vol. 1, pp. 340–349 (2011)
9. Remus, R., Quasthoff, U., Heyer, G.: SentiWS – a publicly available germanlanguage resource for sentiment analysis. In: Proc. of the 7th Intl. Conf. on Language Resources and Evaluation, LREC 2010 (2010)
10. Shannon, C.E.: A mathematical theory of communication. The Bell System Technical Journal 27, 379–423 (1948)
11. Watson, T., Noble, P.: Evaluating public relations: a best practice guide to public relations planning, research & evaluation. PR in practice series. Kogan Page (2007)

Extracting Social Events Based on Timeline and Sentiment Analysis in Twitter Corpus

Bayar Tsolmon, A-Rong Kwon, and Kyung-Soon Lee[*]

Division of Computer Science and Engineering, Chonbuk National University,
567 Baekje-daero, deokjin-gu, Jeonju-si, Jeollabuk-do
561-756 Republic of Korea
{bayar_277,selfsolee}@chonbuk.ac.kr, lifecorrect@naver.com

Abstract. We propose a novel method for extracting social events based on timeline and sentiment analysis from social streams such as Twitter. When a big social issue or event occurs, it tends to dramatically increase in the number of tweets. Users write tweets to express their opinions. Our method uses these timeline and sentiment properties of social media streams to extract social events. On timelines term significance is calculated based on Chi-square measure. Evaluating the method on Korean tweet collection for 30 events, our method achieved 94.3% in average precision in the top 10 extracted events. The result indicates that our method is effective for social event extraction.

Keywords: Event extraction, Timeline, Sentiment, Chi-Square, Twitter.

1 Introduction

Social media has become in recent years an attractive source of up-to-date information and a great medium for exploring the types of developments which most matter to a broad audience [1]. Twitter is a popular real-time micro blogging service that allows its users to share short pieces of information known as "tweets" (limited to 140 characters). Users write tweets to express their opinions about various topics pertaining to their daily lives [2].

There are recent works for event extraction based on sentiment analysis and timeline analysis. Popescu et.al. [1] proposed the 3-regression machine learning models for event extraction. Sayyadi et.al. [3] developed an event detection algorithm by creating a keyword graph and using community detection methods analogous to those used for social network analysis to discover and describe events. Benson et.al. [4] proposed a model based on a set of canonical records, the values of which are consistent with aligned messages. Pak et.al. [5] performed linguistic analysis on the automatically collected corpus for sentiment analysis, explained phenomena and built a sentiment classifier. Lanagan et.al. [6] showed that it is possible to use the publicly available tweets about a sports event to aid in event detection and summary generation for associated media. Zhao et al. [7] have defined the term event as the information flow between a group of social actors on a specific topic over a certain time period.

---

[*] Corresponding author.

G. Bouma et al. (Eds.): NLDB 2012, LNCS 7337, pp. 265–270, 2012.
© Springer-Verlag Berlin Heidelberg 2012

In this paper, we propose a method for extracting social events from tweets documents based on timeline and sentiment analysis. We observed the following characteristics in social media streams such as twitter: (1) The increase in the number of tweets can be the first characteristic of a social event. We use timeline analysis of the twitter data. If an event did not occur, the number of tweets observed could be less than a certain level. But when a social event occurred, the number of tweets has increased dramatically. (2) Expressions using sentiment words can be the second characteristic. Twitter users express their opinions on tweets with sentiment words such as positive or negative words for issues.

2   Extracting Events Based on Timeline and Sentiment Analysis

To extract social events from Twitter corpus, we propose a novel method based on timeline and sentiment features. These features are combined with the basic frequency feature of a word bigram. We used a word bigram which is a pair consisting of a noun or verb to express an event. The window size is set to 3.

2.1   Event Term Extraction Based on Timeline Analysis

Significant term shifts and rapid changes in the term frequency distribution are typical of stories writing their opinions for a new event on time in twitter. On comparing the tweets obtained on that day with other normal days, it can be observed that the rate of increase in tweets is very high. When any new event occurs on a certain day, there might be use of certain terms many times on that day as compared to use of those terms on any other normal day.

In order to give distinctive weight to an event term, term significance is calculated on the basis of timelines. The term significance on that day is measured by the chi-square statistic which measures the lack of independence between a term w and the class $t0$. In other words, the number of occurrences of a particular term on a special day can be correlated to a special event occurring on that day by using this chi-square value. Table 1 gives a 2×2 contingency table. On the contingency table, time $t0$ is each day, which means the total number of tweet documents on that day is equal to the sum of a and b.

Table 1. Contingency table to calculate term significance on timelines

	# of tweets containing term w ($w \in$ D)	# of tweets not containing term w ($w \notin$ D)
# of tweets on time $t0$	a	b
# of tweets before time $t0$	c	d

Chi-square values of a term w computed at time $t0$ can be calculated as follow.

$$ChiSquare(w, t0) = \frac{(a+b+c+d)(ad-bc)^2}{(a+c)(b+d)(a+b)(c+d)} \tag{1}$$

From the above formula we have got the chi-square for the highest frequency of terms used in a particular event. When a term occurs as few times in tweets before time $t0$ and occurs many times in tweets on time $t0$, $ChiSquare(w,t0)$ becomes high.

2.2   Event Term Extraction Based on Sentiment Features

The users write tweets to express their opinions on certain issues or events with positive or negative sentiment. We use a sentiment lexicon to consider context of an event term which contains users' opinions.

The opinion score of a term is measured by combining the frequency of a term and the frequency of opinion words.

$$OpScore(w,t0) = Freq(w,t0) + 2 \cdot OpFreq(w,s,t0) \tag{2}$$

The term frequency on the timeline $t0$ is measured as follows:

$$Freq(w,t0) = \sum_{D \in t0} tf(w,D) \tag{3}$$

where w is a bigram term, $t0$ is a particular date, D is a tweet document written on the particular date ($t0$). When a big social issue or event occurs, it tends to dramatically increase in the number of tweets. The frequency of an event term is directly influenced as the number of tweets increase.

The opinion frequency is measured based on co-occurrence with sentiment words s and a term w as following:

$$OpFreq(w,s,t0) = \sum_{w \in D, s \in D, D \in t0} tf(s,D) \tag{4}$$

where s is any sentiment word in the lexicon. When w occurs in the tweet, the frequency of s is counted. Here the sentiment words s can be context of a term w.

The subjectivity lexicon from Opinion Finder [8] contains 8,221 words with their polarity and strength. We translated these words to Korean by the English to Korean machine translation using Google translator API and revised by human translator. As a result, we selected 1,192 strong positive and strong negative words, which is good for Korean sentiment lexicon.

2.3   Event Term Extraction Based on Timeline and Sentiment Feature

To reflect the term distribution on timelines and user's sentiment on a specific event, event terms are extracted by combining chi-square value and opinion score.

$$ChiOpScore(w,t0) = \lambda \cdot ChiSquare(w,t0) + (1 - \lambda) \cdot OpScore(w,t0) \tag{5}$$

In our experiment, the parameter λ is set to 0.7, which is learned by one (i.e. 'Cheonanham' event[1]) of the event lists (See Table 3).

---

[1] The South Korean Navy ship "Cheonanham" which sank on 26 March 2010.

3   Experiments and Evaluation

We have evaluated the effectiveness of the proposed method on tweet collection. Six issues are chosen and tweet documents for the issues are collected, spanning from November 1, 2010 to March 26, 2011 by Twitter API (all issues and tweets are written in Korean). Table 2 shows the number of tweets related to each issue. Each record in this tweet data set contains the actual tweet body and the time when the tweet was published.

Table 2. Twitter Data Set

Cheonanham	Kim Yu-Na	Park Ji-Sung	Earthquake	Galaxy Tab	iPad
84,195	26,844	131,533	46,795	476,883	553,466

Table 3. Event lists and answers for six issues

Issue	Event	Event List	Date	# of answer
Cheonanham	E1	Cheonanham torpedo attack	2010.11.04	3
	E2	'In-Depth 60 Minute' KBS broadcast	2010.11.17	2
	E3	Bombardment of Yeonpyeong Island	2010.11.23	3
	E4	'In-Depth 60 Minute' Scandal	2011.01.06	5
	E5	A Rumor about prepared for attack	2011.02.03	6
	E6	The 1st anniversary of Cheonanham	2011.03.21	4
Kim Yu-Na	E7	Announcing her new programs	2010.11.30	4
	E8	UNICEF *Goodwill Ambassador*	2010.12.02	3
	E9	Private training center	2010.12.27	6
	E10	Yu-Na's Devil Mask?	2011.01.26	3
	E11	Ultra - mini dress	2011.02.15	3
	E12	"Giselle" domestic training	2011.03.22	3
Park Ji-Sung	E13	Park Ji-Sung Double goal	2010.11.07	5
	E14	Promote World Cup bid	2010.12.02	4
	E15	Season 6th Goal	2010.12.14	4
	E16	Park's facial injuries	2011.01.23	5
	E17	Ji-Sung retirement?	2011.01.31	7
	E18	Cha Beomgeun's confession	2011.02.01	2
Earthquake	E19	Japan earthquake	2010.11.30	3
	E20	Jeju Island earthquake	2011.01.12	2
	E21	Signs of a volcanic eruption in Japan	2011.01.27	4
	E22	New Zealand earthquake	2011.02.22	6
	E23	Japan Tsunami Warning	2011.03.09	3
	E24	Japan Massive Earthquake	2011.03.11	1
Galaxy Tab	E25	Samsung launching Galaxy Tab	2010.11.04	5
	E26	Start Selling Galaxy Tab in Korea	2010.11.13	3
	E27	Galaxy tab Product Placement	2010.11.22	4
iPad	E28	Pre-orders for New iPad postponed	2010.11.09	4
	E29	Sale by subscription	2010.11.17	4
	E30	Start selling iPad in Korea	2011.11.30	4

The comparison methods are as follows:

- **Freq:** Basic event extraction based on term frequency (using Equation 3).
- **ChiSquare:** Event extraction based on timeline analysis (using Equation 1).
- **ChiOpScore:** Event extraction based on timeline analysis and sentiment features (using Equation 5). This is the proposed method.

Table 3 shows that 30 social events for six issues and the number of answers. All the events have taken place at a particular date. The answer set is judged by three human assessors for the top 10 event terms extracted by each method.

We have first categorized tweets based on the date which has the highest number of tweets, and then selected the terms which have the top N chi-square values. Each method ranks those candidate event terms. To evaluate the effectiveness of the methods, the precision in the top 10 extracted events (P@10) is used as an evaluation metric.

Table 4 shows the experimental results for 30 events. For each event, it shows the number of answers for the total answers. Here '1/3' describes that one answer is detected by the proposed method among 3 answers.

Table 4. Performance comparisons

Events	Freq	Chi-Square	Chi-OpScore	Events	Freq	Chi-Square	Chi-OpScore
E1	2/3	2/3	**3/3**	E16	1/5	**5/5**	**5/5**
E2	1/2	**2/2**	**2/2**	E17	5/7	**7/7**	6/7
E3	1/3	**2/3**	**2/3**	E18	**2/2**	**2/2**	**2/2**
E4	3/5	**5/5**	**5/5**	E19	**3/3**	**3/3**	**3/3**
E5	**6/6**	**6/6**	**6/6**	E20	1/2	1/2	**2/2**
E6	3/4	3/4	**4/4**	E21	**4/4**	2/4	**4/4**
E7	2/4	3/4	**4/4**	E22	**6/6**	5/6	5/6
E8	1/3	2/3	**3/3**	E23	2/3	**3/3**	**3/3**
E9	**5/6**	**5/6**	**5/6**	E24	**1/1**	**1/1**	**1/1**
E10	**3/3**	**3/3**	**3/3**	E25	3/5	**4/5**	**4/5**
E11	**2/3**	1/3	**2/3**	E26	**3/3**	**3/3**	**3/3**
E12	2/3	**3/3**	**3/3**	E27	**3/4**	**3/4**	**3/4**
E13	3/5	**5/5**	**5/5**	E28	2/4	**4/4**	**4/4**
E14	**4/4**	**4/4**	**4/4**	E29	3/4	3/4	**4/4**
E15	**4/4**	3/4	**4/4**	E30	1/4	3/4	**4/4**
				Avg. P@10	72.1%	80.8%	**94.3%**

The method *Freq* and *ChiSquare* achieved 72.1% and 80.8%, respectively. The proposed method *ChiOpScore* achieved 94.3% in the average of precision in the top 10 results (avg. P@10). Our extraction method showed improvements in seventeen events such as E1, E2, E4, E8, E16 and E30 compared to *Freq*.

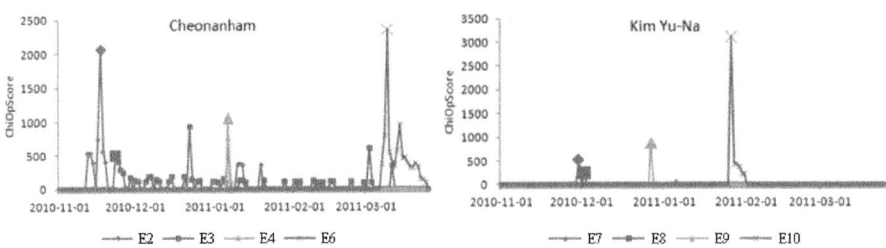

Fig. 1. Extracted events based on *ChiOpScore* method

In Fig. 1, we can see the scores obtained for list of events which have been categorized under different issues. It can be observed that when a special event has occurred on a particular date under an issue, the scores obtained on that date for the special event is very high based on proposed *ChiOpScore* method.

4   Conclusions

We proposed the event extraction method for social issues using chi-square measure on timeline and sentiment features. In order to give distinctive weight to an event term, term significant is measured on timelines. Evaluating on Korean tweet test collection, the results show that chi-square feature on timeline analysis and sentiment features are effective by achieving 94.3% in the average precision in the top 10 results.

Discovering methods for better event expressions and less dependent on the number of tweets are future works.

References

1. Popescu, A.-M., Pennacchiotti, M.: Detecting Controversial Events from Twitter. In: Proceedings of CIKM 2010 (2010)
2. Lai, P.: Extracting Strong Sentiment Trends from Twitter (2010),
   http://nlp.stanford.edu
3. Sayyadi, H., Hurst, M., Maykov, A.: Event Detection and Tracking in Social Streams. In: Proceedings of ICWSM 2009 (2009)
4. Benson, E., Haghighi, A., Barzilay, R.: Event Discovery in Social Media Feeds. In: Proceedings of ACL (2011)
5. Pak, A., Paroubek, P.: Twitter as a Corpus for Sentiment Analysis and Opinion Mining. In: Proceedings of LREC 2010 (2010)
6. Lanagan, J., Smeaton, A.F.: Using Twitter to Detect and Tag Important Events in Live Sports. In: Proceedings of AAAI 2011 (2010)
7. Zhao, Q., Mitra, P., Chen, B.: Temporal and information flow based event detection from social text streams. In: Proceedings of WWW 2007 (2007)
8. Wilson, T., Wiebe, J., Hoffmann, P.: Recognizing contextual polarity in phrase-level sentiment analysis. In: Proceedings of HLT/EMNLP 2005 (2005)

Can Text Summaries Help Predict Ratings?
A Case Study of Movie Reviews

Horacio Saggion[1], Elena Lloret[2], and Manuel Palomar[2]

[1] Department of Information and Communication Technologies,
Universitat Pompeu Fabra,
C/ Tanger 122, 08018 Barcelona, Spain
horacio.saggion@upf.edu
[2] Department of Software and Computing Systems,
University of Alicante, Apdo. de correos, 99, E-03080 Alicante, Spain
{elloret,mpalomar}@dlsi.ua.es

Abstract. This paper presents an analysis of the rating inference task
– the task of correctly predicting the rating associated to a review, in
the context of movie reviews. For achieving this objective, we study the
use of automatic text summaries instead of the full reviews. An extrinsic
evaluation framework is proposed, where full reviews and different types
of summaries (positional, generic and sentiment-based) of several compression rates (from 10% to 50%) are evaluated. We are facing a difficult
task; however, the results obtained are very promising and demonstrate
that summaries are appropriate for the rating inference problem, performing at least equally to the full reviews when summaries are at least
30% compression rate. Moreover, we also find out that the way the review is organised, as well as the style of writing, strongly determines the
performance of the different types of summaries.

Keywords: Content representation and processing, Natural Language
Processing, Text Summarization, Rating Inference, Opinion
Classification.

1 Introduction

One of the main features of the Web 2.0 is the abundance of user-generated
content, which is increasing at a exponential rate. In Pang and Lee [7], it was
reported that 32% of Internet users provide a rating on a product, service, or
person via an on-line rating system[1], and 30% post an on-line comment or review
regarding a product or service. The use of a rating for scoring a product, movie,
restaurant, etc. is a quick way of providing a user's opinion that can help other
users in their decision-making process.

---

[1] In these cases, users give a score, depending on how much they liked a product,
movie, restaurant, hotel, service, etc., normally associated with a scale rating, either
numerical (e.g., 1=worst..5=best) or by means of visual symbols (e.g., stars).

G. Bouma et al. (Eds.): NLDB 2012, LNCS 7337, pp. 271–276, 2012.

Previous research has shown that the rating inference task is especially challenging when dealing with longer reviews [9]. A straightforward manner to obtain short texts is by applying Text Summarization (TS) techniques which besides condensing texts should keep the most relevant information of the document [11].

Therefore, in this paper, we analyze whether the use of summaries are appropriate for the rating inference process, focusing on the movies domain. Although this task is extremely difficult, we show that text summaries are capable of selecting important information for predicting the correct rating associated to a review compared to the full review. Furthermore, we observe that the structure and style of writing strongly influences on the performance prompting for adaptive strategies.

The remaining of the paper is organized as follows. Section 2 summarizes the previous work concerning the rating inference task. Section 3 describes the experimental setup. Further on, Section 4 explains and discuss the results obtained, and finally, some relevant conclusions are drawn and further work is outlined in Section 5.

2   Related Work

The task of rating inference has been addressed in previous research. For instance, Pang and Lee [6] and Goldberg and Zhu [2] used machine learning algorithms for predicting the rating for the movies domain, obtaining an accuracy ranging between 36% and 62%. Qu et al. [8] tackled this task by suggesting a *bag-of-opinions* model, consisting of a root word, its modifiers and negation words, in case they appear in the same sentence. They use this model together with regression techniques to predict the rating associated to either a movie, book or music review.

When trying to predict ratings for product reviews with very short documents (reviews containing at most three sentences), the classification process achieves good results, reporting an accuracy of 80% for binary classification (thumbs up or thumbs down) and around of 75% for a 1 to 5-star rating scale [9]. On the contrary, when dealing with longer texts, accuracy only reaches around 32%.

What all of these approaches have in common is that they use the whole document to predict its rating. To the best of our knowledge, summaries have not been exploited for the rating inference task to a great extent. Only in [5] and [10] preliminary experiments have been carried out to show the effect of summaries on the rating inference task, but in the context of bank reviews.

3   Experimental Setup

3.1   Corpora

As corpus, we selected the movie review corpus described in [6]. This corpus[2] consists of a large collection of movie reviews (more than 5,000) written by

---

[2] This corpus is freely available at
  http://www.cs.cornell.edu/people/pabo/movie-review-data/

four different authors. The reviews are associated with a star-rating scale of three values (1=worst..3=best), representing the author's feeling towards each movie, and each one has 758 words on average. Moreover, most of the reviews (1,914 reviews) were classified as 2-star, corresponding to the 38% of the total documents in the corpus.

3.2 Experiments

For the experiments, single-document summaries of several compression rates (10% to 50%) are generated using COMPENDIUM summarizer [4]. Such summaries can be either generic ($Gen_{TE+TF+CQP}$) or sentiment-based ($Sent_{TE+TF+CQP}$). Generic summaries provide a general overview of what a document is about, whereas sentiment-based summaries capture the author's opinion about a topic. In addition, two positional TS approaches are also defined: *Lead*, which extracts the first sentences of the review; and *Final*, which extracts the last ones. In this manner, we can study whether the structure of the text is also important for our task.

We used the General Architecture for Text Engineering (GATE)[3] [1] for document and summary analysis and rating inference. First of all, the full reviews (or the generated summaries) are analyzed using different GATE components, such as tokenization and part of speech tagging to produce annotated documents. Then, the root of each token is extracted as a feature for document representation. Finally, for the rating inference, we adopt a *Support Vector Machine* (SVM) learning paradigm[4], relying on a bag-of-words model, where the words are the root of the tokens in the input text (either full document or summary). For the experiments, the corpus is split into two groups: 80% of the documents are used for training the SVM with the root of the tokens, whilst 20% are used as testing.

4 Evauation and Discussion

For quantifying the performance of the full reviews and text summaries we compute the *accuracy*, which accounts for the number of reviews that have been classified as correct with respect to the total number of reviews.

Table 1 shows the average accuracy results obtained both for the full review and the different summarization approaches tested, with numbers improving the rating prediction performance of full reviews shown in boldface. As it can be seen, the accuracy value for the full review is above 50%, whereas the accuracy of the summaries ranges from 43% to 53%. Both values are very competitive and they are in line with state-of-the-art research with this standard corpus [6], [2].

In general terms, for compression rates over 30%, the results for the *Lead* summarization approach are very close to the full reviews, whereas for the remaining summarization approaches, they are equal or better. This means that the proposed TS approaches are able to delete noisy information from the full reviews, preserving at the same time relevant information for the rating inference process.

---

[3] http://gate.ac.uk/
[4] We rely on the SVM implementation distributed with the GATE system [3].

Table 1. Accuracy results for the rating inference

Approach	Compression Rate				
	10%	20%	30%	40%	50%
Full review	0.507	0.507	0.507	0.507	0.507
Lead	0.459	0.493	0.489	0.494	0.498
Final	0.502	**0.524**	**0.507**	**0.529**	**0.529**
$\text{Gen}_{TE+TF+CQP}$	0.430	0.465	0.501	0.503	0.501
$\text{Sent}_{TE+TF+CQP}$	0.429	0.488	**0.508**	**0.512**	**0.530**

However, due to small size of the summaries of compression rates of 10% and 20%, they are not enough to be able to predict the overall opinion an author has towards a movie. Moreover, it is worth mentioning that the summaries that were generated using the final part of a document perform equal or better than the full reviews. This observation is very interesting, because it can help to understand the manner of how the reviews are written, where the last sentences contain a strong feeling about the movie. Finally, comparing the summaries generated by COMPENDIUM, we would like to note that the sentiment-based approach obtains slightly better results than the generic one. This is reasonable, since the review may contain opinion vocabulary and expressions, that can help for determining whether the author liked or not the film.

In order to understand in more detail the results obtained, we also carry out an analysis of the structure and writing style of the reviews. To this end, we selected a subset of reviews and we manually analyzed them. We wanted to identify if there was a common structure for all the reviews, and how sentiment was expressed within the documents.

Table 2. Fragment of a movie review

Movie review (2-star): While watching THE WEDDING BANQUET (currently in limited release), I found myself thinking that **it would have a very good short**. This Taiwanese-American comedy about generational and cultural conflict **has fifty minutes of genuine energy stretched into an hour and fifty minute film**. There are **moments of real charm and insight, but ultimately the sluggish pace doesn't hold them together**. [...] *It's difficult not to recommend* THE WEDDING BANQUET, since it's heart seems so firmly in the right place. [...] It has its moments, and the themes of familial conflict may strike a chord with many viewers. However, I just never found myself caring quite enough, and *I found myself looking at my watch too frequently.*

An example of a fragment of a movie review is shown in Table 2. We have remarked different expressions that can be indicative of emotion and personal opinions of the author in boldface. As it can be seen, some opinions are directly expressed while others cannot be seen so straightforward. We identified two common phenomena that appear for all the reviews, regardless of the author, and are very difficult to compute automatically, thus making more challenging the rating inference task:

- **Sarcasm**. Reviews are full of irony and many sarcastic expressions are found. This is very challenging to detect by means of automatic systems, especially if the expression contains terms with positive meaning but they are used to mean the opposite (e.g. *"apparently, the producers of jumanji wanted this film to be a jurassic park for 1995's holiday season"*).

- **Implicit opinions**. Although a sentence does not include any opinion vocabulary, it can express an opinion which is not directly stated. For instance, the sentence *"this is what happens when someone takes what might have been a moderately-entertaining television Christmas special and tries to adapt it for the big screen"* is expressing the author's dissatisfaction but, as it refers to previous content, we have to understand such content in order to capture the real meaning behind it. In other cases, we can find metaphors that can be directly associated to the rating of the film, but are very difficult to detect and understand (e.g., *"the best recommendation I can come up with regarding this movie is to turn on the Nintendo and play a game yourself"*).

In some cases, we can also find some triggers that can indicate the author's feeling (e.g. *"i didn't expect"*, *"i think"*, *"my overall impression"*, etc.). These expressions can be very useful to have an idea of the author's overall satisfaction.

Concerning the structure of the review, it is worth mentioning that each author exhibits a different way of writing and the language and style employed makes the opinions very difficult to detect in some cases. Although not all the authors exhibit the same style of writing, the general structure of the reviews is that at the beginning most of the authors provide a description of what the film is about, whereas at the end of the review they end up with a closing sentence containing their opinion (e.g., *"It's difficult not to recommend"*), and in some cases, a justification of their final decision (e.g., *"I only gave it this rating because it was a too intense for me"*).

5   Conclusion and Future Work

In this paper, an analysis of the rating inference task in the case of movie reviews was presented. Our main contribution concerns the use of automatic summaries, instead of the full reviews, for tackling this task. To this end, different summarization approaches of several sizes were evaluated.

The results obtained showed that text summaries help to filter the noisy information not being directly related to the review, and consequently, they are suitable for the rating inference task. However, we found out that compression rates below 30% did not allow to include enough content for correctly predicting the star of a review. Moreover, concerning the structure of the reviews, it seemed that more strong opinions were provided at the end of the review, and due to this fact, the *Final*, as well as the sentiment-based TS approaches were the best-performing ones, obtaining higher accuracy results than the full reviews. Although, we can concluded that automatic summaries may be beneficial for the rating-inference task, there is still a lot of room for improvement. Therefore, in the future, we would like to analyze in depth new methods and approaches for measuring the correlation between the opinion expressed in the reviews and the real associated rating, and how to face the problem if the rating does not reflect what a user has written.

Acknowledgments. H. Saggion is grateful to a Ramón y Cajal advanced research fellowship from Ministerio de Economía y Competitividad (RYC-2009-04291). Moreover, this research has been partially supported by the projects grant no. TIN2009-13391-C04-01 funded by the Spanish Government, and the projects grant no. PROMETEO/2009/119 and ACOMP/2011/001 funded by the Valencian Government.

References

1. Cunningham, H., Maynard, D., Bontcheva, K., Tablan, V.: GATE: A Framework and Graphical Development Environment for Robust NLP Tools and Applications. In: Proceedings of the 40th Annual Meeting of the Association for Computational Linguistics, Philadelphia, USA (2002)
2. Goldberg, A.B., Zhu, X.: Seeing stars when there aren't many stars: graph-based semi-supervised learning for sentiment categorization. In: Proceedings of the 1st Workshop on Graph Based Methods for NLP, pp. 45–52 (2006)
3. Li, Y., Bontcheva, K., Cunningham, H.: Adapting SVM for Data Sparseness and Imbalance: A Case Study in Information Extraction. Natural Language Engineering 15(2), 241–271 (2009)
4. Lloret, E.: Text Summarisation based on Human Language Technologies and its Applications. Ph.D. thesis, University of Alicante (2011)
5. Lloret, E., Saggion, H., Palomar, M.: Experiments on Summary-based Opinion Classification. In: Proceedings of the Workshop on Computational Approaches to Analysis and Generation of Emotion in Text, pp. 107–115 (2010)
6. Pang, B., Lee, L.: Seeing Stars: Exploiting Class Relationships for Sentiment Categorization with Respect to Rating Scales. In: Proceedings of the Association of Computational Linguistics, pp. 115–124 (2005)
7. Pang, B., Lee, L.: Opinion Mining and Sentiment Analysis. Foundations and Trends in Information Retrieval 2(1-2), 1–135 (2008)
8. Qu, L., Ifrim, G., Weikum, G.: The bag-of-opinions method for review rating prediction from sparse text patterns. In: Proceedings of the 23rd International Conference on Computational Linguistics, pp. 913–921 (2010)
9. Saggion, H., Funk, A.: Extracting Opinions and Facts for Business Intelligence. RNTI E-17, 119–146 (2009)
10. Saggion, H., Lloret, E., Palomar, M.: Using Text Summaries for Predicting Rating Scales. In: Proceedings of the 1st Workshop on Computational Approaches to Subjectivity and Sentiment Analysis (2010)
11. Spärck Jones, K.: Automatic Summarising: The State of the Art. Information Processing & Management 43(6), 1449–1481 (2007)

Towards User Modelling in the Combat
against Cyberbullying

Maral Dadvar[1], Roeland Ordelman[1], Franciska de Jong[1], and Dolf Trieschnigg[2]

[1] Human Media Interaction Group
[2] Database Group, University of Twente,
P.O. Box 217, 7500 AE, Enschede, The Netherlands
{m.dadvar,f.m.g.dejong,r.j.f.ordelman,
r.b.trieschnigg}@utwente.nl

Abstract. Friendships, relationships and social communications have all gone to a new level with new definitions as a result of the invention of online social networks. Meanwhile, alongside this transition there is increasing evidence that online social applications have been used by children and adolescents for bullying. State-of-the-art studies in cyberbullying detection have mainly focused on the content of the conversations while largely ignoring the users involved in cyberbullying. We hypothesis that incorporation of the users' profile, their characteristics, and post-harassing behaviour, for instance, posting a new status in another social network as a reaction to their bullying experience, will improve the accuracy of cyberbullying detection. Cross-system analyses of the users' behaviour - monitoring users' reactions in different online environments - can facilitate this process and could lead to more accurate detection of cyberbullying. This paper outlines the framework for this faceted approach.

1   Introduction

Young people have fully embraced the internet for socializing and communicating. It first started with a simple two-way stream of communication between two people. For example sending and receiving emails. Later on it expanded by having several people communicating at the same time in a particular online environment, such as a chat room, a discussion forum or commenting on the same video. The rise of social networks in the digital domain has led to a new definition of friendships, relationships and social communications. People interact through different social networks, such as Facebook, Twitter, MySpace, and YouTube at the same time. A comment is posted on a friend's video on YouTube, and a reply is received on Twitter. Meanwhile, alongside this vast transition of information, ideas, friendships and comments, an old troubling problem arises with a new appearance in new circumstances: cyberbullying, or online bullying. There is increasing evidence that social media have been used by children and adolescents for bullying [1]. Cyberbullying is defined as an aggressive, intentional act carried out by a group or individual, using electronic forms of contact (e.g. email and chat rooms) repeatedly or over time against a victim who cannot easily defend themself [2]. Cyberbullying can have deeper and longer-lasting effects

G. Bouma et al. (Eds.): NLDB 2012, LNCS 7337, pp. 277–283, 2012.

compared to physical bullying. Online materials spread fast and there is also the persistency and durability of online materials and the power of the written word [1]. The victim and bystanders can read what the bully has said over and over again. Bullying can cause depression, low self-esteem and there have been cases of suicide among teenagers [3].

Cyberbullying is a well-studied problem from the social perspective [1, 4] while few studies have been dedicated to automatic cyberbullying detection [5, 6]. The main focus of these studies is on the content of the text written by the users rather than the users' information and characteristics. State-of-the-art studies have investigated the detection of bullying in a single environment at a single time without considering the further effects and reactions of the user toward this act in other social networks. For instance, if someone gets bullied on Facebook, later on, their Twitter postings can be an indication of their feelings and their state of mind. Focusing on the text itself and finding harassing sentences is not enough to conclude that the act of bullying has taken place. Profanities in a discourse do not necessarily mean that they are being used to bully someone. There are many foul words that are used among teenagers just as a sign of friendship and close relationships. Moreover, being bullied and becoming a victim of cyberbullying is also dependent on the personality of each person. One person may feel bullied, threatened and depressed by sentences that do not cause any bad feelings for someone else. Therefore, even if a sentence is harassing and is used with the intention of bullying someone, it does not necessarily mean that the other party was offended or felt bullied. These subtleties complicate the differentiation of "bullying" detection from "harassment" detection.

Use Case and Applications

The main role of an effective cyberbullying detection system in a social network is to prevent or at least decrease the harassing and bullying incidents in cyberspace. It can be used as a tool to support and facilitate the monitoring task of the online environments. For instance, having a moderator specially in the forums that are mostly used by teenagers is a common thing. But because of the volume of entries in these fora it is impossible for moderators to read everything. A system that gives warnings if something suspicious is detected would greatly help the moderator to focus only on these cases instead of randomly reading the fora. Moreover, cyberbullying detection can be used to provide better support and advice for the victim as well as to monitor and track the bully. Tracking the behaviour of the users involved in an incident, across different social networks in a time frame can help to conclude that whether there is a victim or a bully.

2   State-of-the-Art

For several topics related to cyberbullying detection, research has been carried out based on text mining paradigms, such as identifying online sexual predators [7] and spam detection [8]. However, very little research has been conducted on technical solutions for cyberbullying detection, for which lack of sufficient and appropriate training datasets, privacy issues and ambiguities in definition of cyberbullying can be some of the reasons. The related studies provide some inspiration for cyberbullying detection but their approaches are not directly suitable for this problem. For instance, the main difference between a spam message/email and a harassing one is that the former is usually about a different topic than the topic of discussion. In a recent study on cyberbullying detection Dinakar et al. [6], applied a range of binary and multiclass classifiers on a manually labelled corpus of YouTube comments. Their findings showed that binary classifiers can outperform the detection of textual cyberbullying compared to multiclass classifiers. They have illustrated the application of common sense knowledge in the design of social network software for detecting cyberbullying. The authors treated each comment on its own and did not consider other aspects to the problem as such the pragmatics of dialogue and conversation and the social networking graph. They concluded that, taking such features into account will be more useful on social networking websites and can lead to better modelling of the problem. Yin et al. [5] used a supervised learning approach for detecting harassment. They used content, sentiment, and contextual features of documents to train a support vector machine classifier for a corpus of online posts. In this study only the content of the posts were used to determine either a post was harassing or not, and the characteristics of the author of the posts were not considered. Yin et al. [5] have used the combination of these three features. Their results show improvements over the baselines. In another study with the same dataset the authors tried to identify clusters containing cyberbullying using a rule-based algorithm [9]. A newly emerging field of work, which will be integrated into our study, is the issue of identifying users via interaction over the web. While providing profile information for social networks, browsing the web, users leave large number of traces. This distributed user data can be used as a source of information for systems that provide personalized services for their users or need to find more information about their users [14]. Connecting data from different sources has been used for different purposes, such as standardization of APIs (e.g. OpenSocial[1]) and personalization [10]. In another study authors evaluated cross-system user modelling and its impact on cold-start recommendations on real world datasets from three different social web systems [11]. The social connections of an individual user across different services can be obtained by Google's Social Graph API [2].

3   Proposed Approach

We propose that the incorporation of the users' information, their characteristics, and post-harassing behaviour, alongside the content of their conversations, will improve the accuracy of cyberbullying detection. We will investigate the cyberbullying detection from two perspectives. First, which is the conventional way, the users' behaviour will be considered only in a single environment, for instance, the user's comments on a video on YouTube. We envision an algorithm that would go through the comments' body and would classify them as either bullying or non-bullying. At this phase of the

---

[1] http://code.google.com/apis/opensocial/
[2] http://socialgraph.apis.google.com

experiment we hypothesize that including the users' characteristics – either the bully or the victim - such as age and gender, will improve the detection accuracy. Social studies show that there are differences between males and females in the way that they bully each other. For instance, Argamon et al. [12] found that females use more pronouns (e.g. "you", "she") and males use more noun specifiers (e.g. "the", "that").

Our second perspective would be cross-system analysis of users' behaviour. As we mentioned earlier, a content-based approach is not sufficient to classify a sentence as a bullying one. It also depends on the impact of the content on the person that it has been directed to. One way to understand how a person feels is to study the way that they respond and react to the harassing sentences. This can be the user's next reply to the comment in the same environment or it can be in another form of reaction in another environment, for example, it can be a new tweet on the same user's Twitter profile. By identifying the same users in different social networks we can monitor their behaviour and see how they react after a case of harassment in one initial starting point and whether the harassment has led to a bullying case or not. The above mentioned facts motivated us to concentrate more on the information and behaviour of the users involved in the conversation (bully or victim) rather than only the content of the conversation itself. Our approach is the first attempt to incorporate social networking graphs and user information into both cross-system and single system automatic cyberbullying detection (see Figure 1).

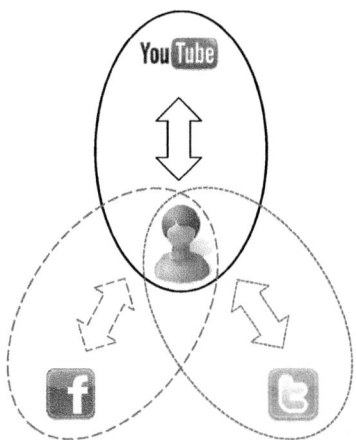

Fig. 1. The conceptual model of the proposed approach. Incorporation of user information into automatic cyberbullying detection both within a particular system and multiple systems.

Data Collection

One important shortage in this field of study is the lack of a standard labelled dataset. As in the state-of-the-art studies the personal information of the users such as age and gender is not taken into account, the currently available datasets are not suitable for our study. Not having a precise definition of the phenomena referred to as cyberbullying is also an obstacle for agreement on what could be a useful dataset for training.

Moreover, the ratio of bullying to non-bullying comments/posts is small, therefore, collecting training data to evaluate our approach is a challenging and time-consuming task. The most urgent need is a text dataset, such as comments or discussions, which contains a sufficient number of bullying posts. In order to study the effect of user's information on a better detection, the dataset authors should consist of different gender groups as well as age groups. Meanwhile, to train the classifier we need to have a labelled dataset. One avenue we are currently exploring to label a data set is the use of crowd sourcing. By using manual annotations obtained through platforms such as MTurk or Crowdflower, we will be able to label large amounts of data. For developing the dataset which will be used for cross-system analysis, first we will manually identify a set of users who are involved in conversations which are considered to be bullying, in a network such as YouTube. Then, using applications such as Google Social Graph API and Mypes[3], we will select those who also have a publicly accessible profile in another network, such as Twitter. By using the networks' streaming API, the selected users' posts and activities will be traced and collected for a period of 6 to 9 months, to make sure that all types of posts (bullying and non-bullying) are collected.

Auxiliary Information and Cross-System Analysis

We will use a supervised learning approach to train a classifier for detecting online bullying. We will employ a Support Vector Machines (SVM) model in WEKA [13] as classification tool. The target language for this experiment is English. To develop our model and train our classifier we will use several types of features. We will employ the TFIDF value of profane words in each post including their abbreviations and acronyms. The other feature is the TFIDF value for personal pronouns used in each post, grouped into second person and other pronouns. We will also make use of user features, such as age (adult versus teenager) and gender. The classifiers will be trained separately for each group. A plausible way to monitor post-harassing behaviour of the users and track their reactions in other systems is users' cross-system modelling. Aggregation of users' profiles from different systems can provide us with more information. Mypes is one of the available services that allow this [14]. By finding and tracing the users' behaviour over a period of time we can gain more accurate information about whether a user is a real bully or a victim. Over time it may become clear whether someone uses a vulgar behaviour towards everyone and in all conversations or has only targeted a specific person to bully. Similarly, by analysing a user's posts and activities, we can find how they react to other people's comments and behaviours and whether they are being victimized.

4   Discussion and Future Work

The main focus of the technical studies which have been conducted so far on cyber-bullying detection is mainly on textual content and there is one single system under

---

[3] http://mypes.groupme.org/mypes/

study at a time. We propose the incorporation of the users' profiles, such as age and gender, their characteristics, and post-harassing behaviour, to improve the accuracy of cyberbullying detection and to take into account the social networking graph. One of the main shortcomings in this field of study is the lack of access to suitable datasets, mostly because of privacy issues. The bullying comments and conversations that happen in public access networks, such as YouTube, are usually not continuous and their negative effects are softened by the support of other users who fight back the bullies and are against hate messages. There are also instances where the victim even becomes a hero for other users because of being a bullying target. Therefore, the real bullying incidents that are a match to our definition of bullying, mostly happen in private conversation such as chat logs and private messages on network profiles. So far we have investigated the gender-based approach for cyberbullying detection in a particular system, in which we observed improvements in classification. We are also going to investigate the age groups differences in cyberbullying. The next phase of our research aims to identify the victim of cyberbullying through monitoring the user's behaviour after a potential case of bullying across other systems. This means that after detecting a harassing post, we investigate the effect of the sentence on the targeted user and whether it had caused the user to feel bullied or threatened. To do so we can employ cross-system user modelling and make use of services that allow the aggregation of users' profiles.

References

1. Campbell, M.A.: Cyber bullying: An old problem in a new guise? Australian Journal of Guidance and Counselling 15, 68–76 (2005)
2. Espelage, D.L., Swearer, S.M.: Research on school bullying and victimization. School Psychology Review 32, 365–383 (2003)
3. Smith, P.K., Mahdavi, J., Carvalho, M., Fisher, S., Russell, S., Tippett, N.: Cyberbullying: its nature and impact in secondary school pupils. Journal of Child Psychology and Psychiatry 49, 376–385 (2008)
4. Kowalski, R.M., Limber, S.P., Agatston, P.W.: Cyber bullying: Bullying in the digital age, p. 224. Blackwell Publishing (2008)
5. Yin, D., Xue, Z., Hong, L., Davison, B.D., Kontostathis, A., Edwards, L.: Detection of harassment on Web 2.0. In: Proceedings of CAW2.0, Madrid, April 20-24 (2009)
6. Dinakar, K., Reichart, R., Lieberman, H.: Modelling the Detection of Textual Cyberbullying. In: ICWSM 2011, Barcelona, Spain, July 17-21 (2011)
7. Kontostathis, A.: ChatCoder: Toward the tracking and categorization of internet predators. In: Proceedings of SDM 2009, Sparks, NV, May 2 (2009)
8. Tan, P.N., Chen, F., Jain, A.: Information assurance: Detection of web spam attacks in social media. In: Proceedings of Army Science Conference, Orland, Florida (2010)
9. Chisholm, J.F.: Cyberspace violence against girls and adolescent females. Annals of the New York Academy of Sciences 1087, 74–89 (2006)
10. Carmagnola, F., Cena, F.: User identification for cross-system personalisation. Information Sciences 179, 16–32 (2009)

11. Abel, F., Araújo, S., Gao, Q., Houben, G.-J.: Analyzing Cross-System User Modeling on the Social Web. In: Auer, S., Díaz, O., Papadopoulos, G.A. (eds.) ICWE 2011. LNCS, vol. 6757, pp. 28–43. Springer, Heidelberg (2011)
12. Argamon, S., Koppel, M., Fine, J., Shimoni, A.R.: Gender, genre, and writing style in formal written texts. Text - Interdisciplinary Journal for the Study of Discourse 23, 321–346 (2003)
13. Hall, M., Frank, E., Holmes, G., Pfahringer, B., Reutemann, P., Witten, I.H.: The WEKA data mining software: an update. ACM SIGKDD Newsletter 11, 10–18 (2009)
14. Abel, F., Henze, N., Herder, E., Krause, D.: Linkage, aggregation, alignment and enrichment of public user profiles with Mypes. In: Proceedings of I-SEMANTICS, Graz, Austria, pp. 1–8 (September 2010)

Plag-Inn: Intrinsic Plagiarism Detection Using Grammar Trees

Michael Tschuggnall and Günther Specht

Databases and Information Systems,
Institute of Computer Science, University of Innsbruck
{michael.tschuggnall,guenther.specht}@uibk.ac.at

Abstract. Intrinsic plagiarism detection deals with the task of finding plagiarized sections of text documents without using a reference corpus. This paper describes a novel approach to this task by processing and analyzing the grammar of a suspicious document. The main idea is to split a text into single sentences and to calculate grammar trees. To find suspicious sentences, these grammar trees are compared in a distance matrix by using the pq-gram-distance, an alternative for the tree edit distance. Finally, significantly different sentences regarding their grammar and with respect to the Gaussian normal distribution are marked as suspicious.

Keywords: intrinsic plagiarism detection, grammar trees, stylistic inconsistencies, pq-gram distance, NLP applications.

1 Introduction

With the increasing amount of data available on the web it becomes more and more important to verify authorships. Especially on text documents like Master- or PhD theses written in an academic context, the identification of plagiarized passages is crucial. Currently, this problem is encountered using two main approaches [9]:

1. *External plagiarism detection* algorithms, which compare a suspicious document with a given set of source documents (e.g. the whole web), and
2. *Intrinsic plagiarism detection* algorithms, which try to find plagiarized sections by inspecting the suspicious document only.

The most successful algorithms currently existing in the intrinsic detection field use n-grams [12] or other features like the frequency of words from predefined word-classes [8] to find suspicious sections. The idea of the approach described in this paper is to use a syntactical feature, namely the grammar used by an author, to identify passages that might have been plagiarized. Due to the fact that an author has many different choices of how to formulate a sentence using the existing grammar rules of a natural language, the assumption is that the way of constructing sentences is significantly different for individual authors. For example, the sentence[1]

---

[1] example taken and modified from the Stanford Parser website [15]

G. Bouma et al. (Eds.): NLDB 2012, LNCS 7337, pp. 284–289, 2012.
© Springer-Verlag Berlin Heidelberg 2012

(1) *The strongest rain ever recorded in India shut down the financial hub of Mumbai, officials said today.*

could also be formulated as

(2) *Today, officials said that the strongest Indian rain which was ever recorded forced Mumbai's financial hub to shut down.*

which is semantically equivalent but differs significantly according to its syntax. The main idea of this approach is to quantify those differences and to find outstanding sentences which are assumed to have a different author and thus may be plagiarized.

The rest of this paper is organized as follows: Section 2 describes the algorithm in detail, while its evaluation is shown in Section 3. Finally, Section 4 summarizes related work and Section 5 recaps the main ideas and discusses future work.

2   The Plag-Inn Algorithm

The aim of the algorithm is to analyze a suspicious document and to find potentially plagiarized sections based on syntactical changes within that document. According to the task of intrinsic plagiarism detection, the algorithm inspects the given document only and needs no reference corpus. In detail, the algorithm consists of the following basic steps:

1. Parse the document and split it into single sentences using Sentence Boundary Detection (SBD) algorithms [14], which detect beginnings and endings of sentences, respectively. Currently, this is implemented by using *OpenNLP*[2], an open source tool that integrates SBD algorithms.

2. Analyze each sentence according to the grammar that was used to build the sentence. Figure 1 shows an examplary grammar tree which results from parsing sentence (2) in Section 1. The labels of the tree correspond to the tag definitions of the Penn Treebank [7], where e.g. *NP* corresponds to a noun phrase or *JJS* corresponds to a superlative adjective. This step can be done e.g. by using the *Stanford Parser* [6], an open source NLP tool that extracts grammatical features of a sentence. Besides information like part-of-speech tagging or dependencies between words [3], the parser also generates a grammar tree which is used in this case for further calculations.

3. The result of the last step is a set of grammar trees representing a document. Now these trees are evaluated by calculating the distance between each tree. I.e. every sentence is compared to every sentence in the document, and the distance is recorded. Because the general tree edit distance is very costly [2], the much more efficient *pq-gram-distance* [1] is used to identify the distance between two grammar trees. As it is shown by Augsten et al. the pq-gram-distance is a lower bound of the fanout weighted tree edit distance, i.e. sensitive to structure changes and thus very suitable. Although the usage of

---

[2] Apache OpenNLP, `http://incubator.apache.org/opennlp`, visited January 2012

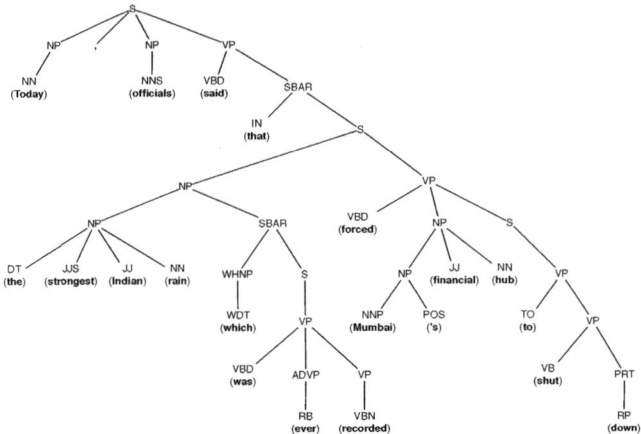

Fig. 1. Grammar Tree resulting from parsing sentence (2)

individual words may be informative as well for further improvements, this approach only takes the structure of a sentence into account and removes the words of each tree, i.e. the leafs.

Each distance is finally stored into a distance matrix D which is triangular and hence requires $\binom{n}{2} = \frac{n(n-1)}{2}$ distance computations, where n corresponds to the number of sentences in the document:

$$D_n = \begin{pmatrix} d_{1,1} & d_{1,2} & d_{1,3} & \cdots & d_{1,n} \\ d_{1,2} & d_{2,2} & d_{2,3} & \cdots & d_{2,n} \\ d_{1,3} & d_{2,3} & d_{3,3} & \cdots & d_{3,n} \\ \vdots & \vdots & \vdots & \ddots & \vdots \\ d_{1,n} & d_{2,n} & d_{3,n} & \cdots & d_{n,n} \end{pmatrix} = \begin{pmatrix} 0 & d_{1,2} & d_{1,3} & \cdots & d_{1,n} \\ * & 0 & d_{2,3} & \cdots & d_{2,n} \\ * & * & 0 & \cdots & d_{3,n} \\ \vdots & \vdots & \vdots & \ddots & \vdots \\ * & * & * & \cdots & 0 \end{pmatrix}$$

4. Having computed all distances between every sentence regarding its grammar tree, the last task is to identify sentences which have a significantly higher distance to most of the other sentences. To find these sentences, first the median distance value for each row in D is calculated, i.e. for each sentence. By applying the inverse Gaussian normal distribution over the resulting vector $(\bar{x}_1, \bar{x}_2, \bar{x}_3, \ldots, \bar{x}_n)$ the mean value μ and standard deviation σ are calculated in the second step. Every sentence i that has a higher mean distance than a predefined threshold δ is marked as plagiarized, i.e. where $\bar{x}_i > \delta$. As shown in Section 3, the best results could be achieved by choosing $\delta \approx \mu + 3\sigma$.

5. Finally, the algorithm tries to combine sets of plagiarized sentences into sections if marked sentences are close to each other. This is a very useful step, because e.g. sentences with few words do not have a significant grammar tree and are therefore not detected. However, if they are positioned in the middle of two marked sentences it is likely that these sentences are also plagiarized. The decision whether *clean* sentences should be grouped into marked sentences and thus also be marked depends on several thresholds

which will not be discussed in detail in this paper: for example the number of sentence-lookaheads or the minimum mean distance of a clean sentence. Moreover, marked sentences can also be unmarked if they are isolated and if they meet predefined conditions.

3   Experimental Results

The described algorithm was implemented and evaluated using seven randomly selected documents out of the PAN 2011 test corpus [10]. They are all written in English, consist of 170 up to 6200 sentences and contain up to 11 sections of plagiarism per document. Figure 2 shows a selected documents visualization of the distance matrix after having calculated the distances between the grammar trees of all sentences. Although the original matrix is triangular and computed as stated in Section 2, the visualization contains the mirrored values as well to provide better visibility. It can be seen easily that there are sentences for which the grammar tree differs significantly from all other trees.

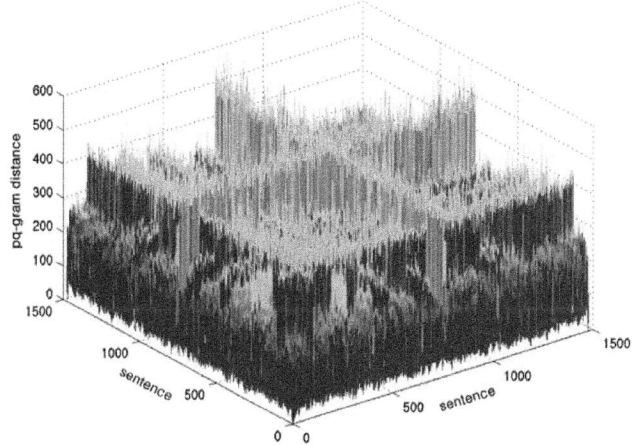

Fig. 2. Visualized Distance matrix of a selected English Document

According to the algorithm the next step computes the mean distance value for each sentence, i.e. traverses each row of the matrix shown in Figure 2. The resulting mean distance values for each sentence can be seen in Figure 3. All sentences with a mean distance exceeding the threshold δ are marked as suspicious. Evaluations showed that the best results could be gained with $\delta = \mu + 3\sigma - \frac{\sigma}{4}$. By inspecting various combinations of all other thresholds required, the best setting produced an average precision and recall value of about 32%, respectively[3]. In some test cases, an F measure of over 60% could be reached. Despite the fact that

---

[3] Precision and recall values are calculated as stated in the PAN workshop [10], i.e. based on detected sections rather than words.

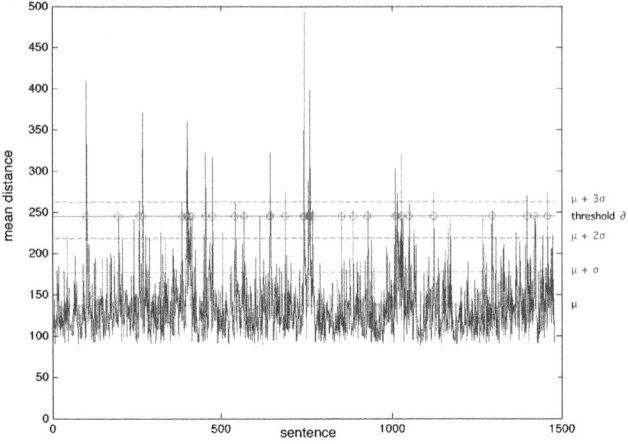

Fig. 3. Mean Distance per Sentence incl. Gaussian Mean and Standard deviation.

these values were adjusted to several documents only, this result is very promising[4]. By optimizing the algorithm including its thresholds the performance can surely be increased in the future.

4   Related Work

An often applied concept in the intrinsic plagiarism detection task is the usage of n-grams [12][5], where the document is split up into chunks of three or four letters, grouped and analyzed through sliding windows. Another approach also uses the sliding window technique but is based on word frequencies, i.e. the assumption that the set of words used by authors is significantly different [8].

An approach that uses the comparison of binary strings calculated from word groups like nouns or verbs using complexity analysis is described in [11]. Finally, approaches in the field of author detection and genre categorization use NLP tools to analyze documents based on syntactic annotations [13] or word- and text-based statistics like the average sentence length or the average parse tree depth [4].

5   Conclusion

In this paper a new approach for intrinsic plagiarism detection is presented which tries to find suspicious sentences by analyzing the grammar of an author. By comparing grammar trees and applying statistics the algorithm searches for significant different sentences and marks them as suspicious. First evaluations showed that this approach is feasible and reached promising results with an average F-measure of about 32%. Future work includes optimization and the

---

[4] The currently (2012) best intrinsic plagiarism detector achieves a recall and a precision value of about 33%, respectively, in average over thousands of documents [8].

evaluation of the algorithm on a bigger, representative test set as well as doing error analysis and research on a possible application on other languages. This approach could also be combined with other existing approaches to enhance the overall quality of an intrinsic and/or external plagiarism detection system. Moreover, the Plag-Inn algorithm could also be adjusted to attribute authors in multi-author documents.

References

1. Augsten, N., Böhlen, M., Gamper, J.: The pq-Gram Distance between Ordered Labeled Trees. ACM Transactions on Database Systems (2010)
2. Bille, P.: A survey on tree edit distance and related problems. Theoretical Computer Science 337, 217–239 (2005)
3. Catherine De Marneffe, M., Maccartney, B., Manning, C.D.: Generating typed dependency parses from phrase structure parses. In: LREC (2006)
4. Karlgren, J.: Stylistic Experiments For Information Retrieval. PhD thesis, Swedish Institute for Computer Science (2000)
5. Kestemont, M., Luyckx, K., Daelemans, W.: Intrinsic Plagiarism Detection Using Character Trigram Distance Scores. In: CLEF 2011 Labs and Workshop, Notebook Papers, Amsterdam, The Netherlands (2011)
6. Klein, D., Manning, C.D.: Accurate unlexicalized parsing. In: Proceedings of the 41st Annual Meeting on Association for Computational Linguistics, ACL 2003, Stroudsburg, PA, USA, vol. 1, pp. 423–430 (2003)
7. Marcus, M.P., Marcinkiewicz, M.A., Santorini, B.: Building a large annotated corpus of English: The Penn Treebank. Comp. Linguistics Linguistics (June 1993)
8. Oberreuter, G., L'Huillier, G., Ríos, S.A., Velásquez, J.D.: Approaches for Intrinsic and External Plagiarism Detection. In: CLEF 2011 Labs and Workshop, Notebook Papers, Amsterdam, The Netherlands (2011)
9. Potthast, M., Eiselt, A., Barrón-Cedeño, A., Stein, B., Rosso, P.: Overview of the 3rd International Competition on Plagiarism Detection. In: Petras, V., Forner, P., Clough, P. (eds.) Notebook Papers of CLEF 11 Labs and Workshops (2011)
10. Potthast, M., Stein, B., Barrón-Cedeño, A., Rosso, P.: An Evaluation Framework for Plagiarism Detection. In: Proceedings of the 23rd International Conference on Computational Linguistics (COLING 2010), Beijing, China (August 2010)
11. Seaward, L., Matwin, S.: Intrinsic Plagiarism Detection using Complexity Analysis. In: CLEF (Notebook Papers/Labs/Workshop) (2009)
12. Stamatatos, E.: Intrinsic Plagiarism Detection Using Character n-gram Profiles. In: CLEF (Notebook Papers/Labs/Workshop) (2009)
13. Stamatatos, E., Kokkinakis, G., Fakotakis, N.: Automatic text categorization in terms of genre and author. Comput. Linguist. 26, 471–495 (2000)
14. Stevenson, M., Gaizauskas, R.: Experiments on sentence boundary detection. In: Proc. of the 6th Conference on Applied Natural Language Processing, ANLC 2000, Stroudsburg, PA, USA, pp. 84–89 (2000)
15. The Stanford Parser, http://nlp.stanford.edu/software/lex-parser.shtml (visited January 2012)

On the Application of Spell Correction to Improve Plagiarism Detection

Daniel Micol[1], Óscar Ferrández[2], and Rafael Muñoz[1]

[1] Department of Software and Computing Systems,
University of Alicante,
San Vicente del Raspeig, Alicante, Spain
[2] Department of Biomedical Informatics,
University of Utah,
Salt Lake City, Utah, United States of America
{dmicol,rafael}@dlsi.ua.es, oscar.ferrandez@utah.edu

Abstract. In this paper we present the accuracy gains that spell corrector systems can provide to the plagiarism detection task when the appropriations contain spelling mistakes. These may have been introduced on purpose to avoid detection systems from finding the aforementioned appropriations, which could happen specially if such systems are based on lexical similarities. This document will detail the components that we have developed for both plagiarism detection and spell correction, and the significant gains that their combination produces.

1 Introduction

The task of finding plagiarisms becomes more complex if the person who does the appropriation inserts spelling mistakes or alterations in order to avoid being detected. This can be a serious obstacle for plagiarism recognition systems that are mostly based on lexical similarities. In this case, the application of a spell corrector system could lead to accuracy improvements in the detection process, and this is what we aim to prove in this paper. For this purpose, we have developed a baseline system for plagiarism detection that is mainly based on term occurrence frequencies. In such scenario, a misspelling in the processed documents could introduce a high level of noise, since they will have a lower document frequency, and therefore a higher inverse document frequency. Also, if a misspelling appears in a suspicious and a source document, these will be heavily linked by this term, and their similarity score may not be fair when comparing it with other documents. Therefore, we believe it would be beneficial to apply a spell corrector over the documents in our corpora to have more realistic term frequency comparisons.

2 Methods

Here we describe a system composed of two main modules: i) a shallow knowledge based plagiarism detection component; and ii) a spell corrector processor.

G. Bouma et al. (Eds.): NLDB 2012, LNCS 7337, pp. 290–295, 2012.

Our goal is to build a pipeline of processes that improves the detection of appropriations with the benefits of applying a spell corrector to detect plagiarisms.

2.1 A Shallow Knowledge Based Plagiarism Detector

The approach that we have implemented as a baseline plagiarism detection system [Micol et al., 2010] compares pairs of documents, namely source and suspicious, to determine the degree of similarity between them. For this purpose it first normalizes the contents of such files and then tries to find their largest common substring. A brute force solution to this problem would have a very high complexity and would present scalability limitations, specially for large documents.

In order to improve performance, we hash all overlapping n-grams of both the suspicious and source documents, storing, for every n-gram, the positions where they appear in the text. Next, we iterate over the contents of the suspicious document, extract its n-grams starting at every given offset, find them in the hash of n-grams of the aforementioned source document, and seek directly to the positions where the given n-gram appears, limiting unnecessary comparisons. From these points we will try to find the largest common substring to both documents.

In our system we wanted to choose an n-gram size that would maintain a passage's meaning, and that is also not too strict. Smaller n-grams have the problem that they might match parts of a single word and therefore the meaning would be lost. On the other hand, longer n-grams don't behave well when there is plagiarism obfuscation, so if the plagiarized sentence has been somehow altered, they won't work well. In our experiments we used an n-gram size of 30 characters.

After having extracted a set of identical substrings that appear in both the suspicious and source documents, it would be beneficial to make this comparison less restrictive by allowing additional words or characters to appear between matches, so that our system works well with certain kinds of low level obfuscations. This is common among plagiarisms when the corresponding person introduces additional words or rearranges part of the appropriated text. To overcome this issue, we perform a match merging operation that attempts to group matches and the text in-between if they are close enough in both source and suspicious documents. This is a greedy process that recursively attempts to merge matches if they satisfy the following two heuristics:

- The length of the text in-between two matches, m_i and m_j, is smaller than the length of m_i plus the length of m_j. This applies to the source and the suspicious part of the matches.
- The length of the merged match cannot be longer or equal than twice the length of m_i plus the length of m_j. This applies to the source and the suspicious part of the matches.

2.2   A Ranker Based Speller

We use the system described in [Gao et al., 2010] and [Sun et al., 2010], co-authored by the first author of this paper, in order to identify and correct spelling mistakes in the suspicious documents corpus. This speller system has two different phases: candidate generation and re-ranking. In the former of these, an input query is first tokenized into a sequence of terms. For each term q, we consult a lexicon to identify a list of spelling suggestions c whose score from q is lower than some threshold. This score is approximated by $-\log P(Q|C) \propto EditDist(Q, C)$.

These generated spelling suggestions are stored using a lattice data structure. We then use a decoder to identify the 20-best candidates by applying a standard two-pass algorithm. The first pass uses the Viterbi algorithm to find the best correction, and the second uses the A-star algorithm to find the 20-best, using the Viterbi scores computed at each state in the first pass as heuristics.

The core component in the second stage is a ranker, which re-ranks the 20-best candidate corrections using a set of features extracted from the query-correction pairs. If the top correction after re-ranking is different from the query it will be proposed to the user. These features can be grouped into the following categories:

- Surface-form similarity features, which check whether C and Q differ in certain patterns (e.g. whether C is transformed from Q by adding an apostrophe, or by adding a stop word at the beginning or end of Q).
- Phonetic-form similarity features, which check whether the edit distance between the metaphones [Philips, 1990] of a query term and its correction candidate is below certain thresholds.
- Entity features, which check whether the original query is likely to be a proper noun based on an in-house named entity recognizer.
- Dictionary features, which check whether a query term or a candidate correction are in one or more human-compiled dictionaries, such as the extracted Wiki, MSDN, and ODP dictionaries.
- Frequency features, which check whether the frequency of a query term or a candidate correction is above certain thresholds in different datasets, such as query logs and Web documents.

3   Experimentation and Results

We will now present the effect of using a spell correction system together with the plagiarism detector previously described. For this purpose we have used the corpora described in [Clough and Stevenson, 2011] and artificially added spelling mistakes. More concretely, we altered the suspicious documents by randomly inserting typos in 1%, 10%, 15% and 20% of the text's characters, respectively in the several experiments that we ran. This is because these corpora typically don't have spelling mistakes, however introducing typos could be a strategy to obfuscate appropriations.

The way in which we introduced spelling mistakes was completely random, by replacing characters with others randomly generated. We are aware that this is not very accurate given that in real life certain characters are more likely to be misspelled given the keyboard's key distribution, but we think that for a first set of tests it's valid to introduce these random misspellings.

As a baseline run we tested our plagiarism detection system over the original corpora. In addition, we ran the aforementioned speller system against these corpora, to analyze the noise that it can introduce due to false positives. Next, we simulated spelling mistakes by modifying 1% of the characters in every given suspicious document, and evaluated our plagiarism detection system on this modified corpus as well as after running the spell corrector. This percentage of characters was completely random, ignoring to which words they belonged. Afterwards, the same task was performed modifying 10%, 15% and 20% of the characters. Table 1 and Figure 1 show the results of these experiments.

Table 1. Results obtained by running our plagiarism detection system together with a spell corrector

Corpora	Accuracy	Precision	Recall	F-score
Original	0.916	0.930	0.930	0.930
Spell corrected	0.916	0.930	0.930	0.930
Typo (1%)	0.916	0.930	0.930	0.930
Typo (1%) + spell corrected	0.916	0.930	0.930	0.930
Typo (10%)	0.737	1.000	0.561	0.719
Typo (10%) + spell corrected	0.863	0.978	0.789	0.873
Typo (15%)	0.611	1.000	0.351	0.520
Typo (15%) + spell corrected	0.747	0.971	0.596	0.739
Typo (20%)	0.495	1.000	0.158	0.273
Typo (20%) + spell corrected	0.737	1.000	0.561	0.719

As we can see in the previous table and figure, adding a 1% of misspelled characters didn't alter the results. This is due to the fact that, although it was more difficult for the plagiarism detection system to find appropriations, documents were large enough to contain at least a few similar parts, and hence were recognized as plagiarized. However, this doesn't apply when the number of misspellings increases, since we can see that overall accuracy and f-score decrease. If we apply the spell corrector system we can see how both metrics improve considerably, showing that the application of such speller produces net gains by combining it with our plagiarism detection system.

We will now take a deeper look by analyzing the underlying text fragments that were found to be plagiarized. Table 2 shows the number of fragments that were detected to be appropriations, as well as the amount of characters in those fragments.

We can observe that even though having a 1% of misspellings didn't affect the overall results, it did decrease the number of characters being found to be copied. In addition, the number of fragments increased because those misspellings didn't allow our plagiarism detection system to detect large amounts of consecutive text, and therefore increased fragmentation. We can see that applying the spell corrector system does provide gains by increasing the number of characters that

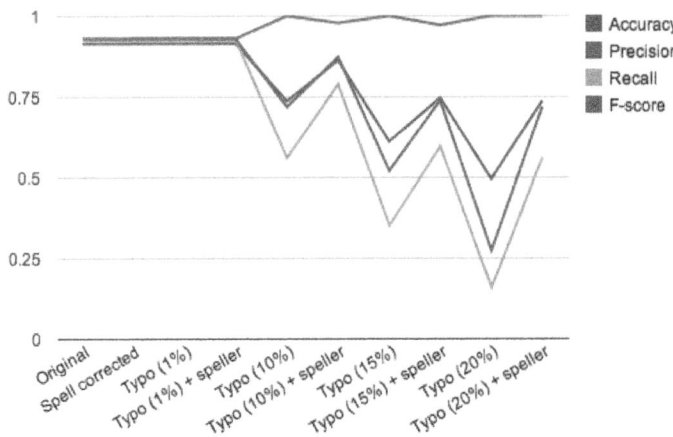

Fig. 1. Representation of the results obtained by running our plagiarism detection system together with a spell corrector

Table 2. Number of plagiarized fragments and characters obtained by running our plagiarism detection system together with a spell corrector

Corpora	Fragments	Characters
Original	292	29, 601
Spell corrected	310	29, 521
Typo (1%)	381	27, 163
Typo (1%) + spell corrected	372	28, 331
Typo (10%)	111	4, 443
Typo (10%) + spell corrected	233	10, 308
Typo (15%)	24	836
Typo (15%) + spell corrected	96	3, 603
Typo (20%)	10	347
Typo (20%) + spell corrected	56	2, 017

were plagiarized. The same applies to the case where we have a higher percentage of misspellings, as we can see that these typos do limit considerably the applicability of our plagiarism detector, which is in large mitigated by the use of the speller.

4   Conclusions and Future Work

In this paper we have proven the positive impact of using a spell corrector processor when the plagiarized documents are obfuscated with typos. In this sense, we are now able to improve the detection of text appropriations that were missed by shallow lexical comparisons and similarities. Furthermore, we have seen that the net gain produced by the application of spellers can be considerably high depending on the percentage of misspellings in the corresponding documents, hence making it very appropriate as a component of a plagiarism detection system. However, spell correctors are complex and expensive to run, and therefore using them will increase the run time of the whole system.

As future work we would like to explore with statistical models for automatic misspelling generation. These would take into account keyboard key positions as well as word length in order to predict which words are more likely to be misspelled, and in which way.

References

[Clough and Stevenson, 2011] Clough, P., Stevenson, M.: Developing a corpus of plagiarised short answers. Language Resources and Evaluation 45(1) (2011)

[Gao et al., 2010] Gao, J., Li, X., Micol, D., Quirk, C., Sun, X.: Learning Phrase-Based Spelling Error Models from Clickthrough Data. In: Proceedings of the 23rd International Conference on Computational Linguistics, Beijing, China (August 2010)

[Micol et al., 2010] Micol, D., Ferrández, Ó., Llopis, F., Muñoz, R.: A Lexical Similarity Approach for Efficient and Scalable External Plagiarism Detection. In: Proceedings of the SEPLN 2010 Workshop on Uncovering Plagiarism, Authorship and Social Software Misuse, Padua, Italy (2010)

[Philips, 1990] Philips, L.: Hanging on the metaphone. Computer Language Magazine 7(12), 38–44 (1990)

[Sun et al., 2010] Sun, X., Gao, J., Micol, D., Quirk, C.: A Large scale Ranker-Based System for Search Query Spelling Correction. In: Proceedings of the 48th Annual Meeting of the Association for Computational Linguistics, Uppsala, Sweden, pp. 266–274 (2010)

PIRPO: An Algorithm to Deal with Polarity in Portuguese Online Reviews from the Accommodation Sector

Marcirio Silveira Chaves[1], Larissa A. de Freitas[2],
Marlo Souza[2], and Renata Vieira[2]

[1] Business and Information Technology Research Center (BITREC),
Universidade Atlântica, Oeiras, Portugal
marcirio.chaves@uatlantica.pt
[2] Faculdade de Informática,
Pontifícia Universidade Católica do Rio Grande do Sul, Porto Alegre, Brazil
{larissa.freitas,marlo.souza}@acad.pucrs.br, renata.vieira@pucrs.br

Abstract. This paper presents the algorithm Polarity Recognizer in Portuguese (PIRPO) to classify sentiment in online reviews. PIRPO was constructed to identify polarity in Portuguese user generated accommodation reviews. Each review is analysed according to concepts from a domain ontology. We decompose the review in sentences in order to assign a polarity to each concept of the ontology in the sentence. Preliminary results indicate an average F-score of 0.32 for polarity recognition.

Keywords: Portuguese Online Reviews, Portuguese Sentiment Analysis, Portuguese Opinion Mining, Accommodation Sector.

1 Introduction

Portuguese user generated reviews can be a valuable source of information for customer orientated business such as restaurants, hotels and airlines. However, manually processing lots of user generated content (UGC), such as reviews or tweets, is a costly and time-consuming task.

Sentiment in user generated content can be an indicator of wider customer opinion, consequently it is necessary to automatically recognise the sentiment expressed by the reviewers. According to Liu [7], sentiment classification is an area that has had a great deal of attention in the community. It treats sentiment analysis as a text classification problem, in which document-level sentiment classification aims to find the general sentiment of the author in an opinionated text. However, as document level sentiment classification fails to detect sentiment about individual aspects of the topic [10], in this work we separate out review attributes (concepts or features) within review and treat them differently instead of seeing a review as a whole, as suggested by Tang et. al. [10].

The objectives of this paper are two fold: (a) present the algorithm Polarity Recognizer in Portuguese (PIRPO) based on specific sets of handcrafted

G. Bouma et al. (Eds.): NLDB 2012, LNCS 7337, pp. 296–301, 2012.

lexico-syntactic patterns to sentiment classification in Portuguese online reviews and (b) evaluate the proposed algorithm using comments annotated by a human as reference for evaluation. We adopt the definition to sentiment given by Wilson et. al. [12]: "a positive or negative emotion, evaluation, or stance".

2   Related Work

A popular Opinion Mining (OM) approach is a lexicon based strategy [9]. Lexicon-based approaches include the use of a list of nouns, verbs, adjectives and adverbs [4]. Ding et. al. [5] use all of these parts of speech in a holistic lexicon-based approach.

Chesley et. al. [4] use subjective and objective features, and polarity features to classify English opinionated blog texts. The first features include subjective (e.g. believing and suggesting) and objective (e.g. asking and explaining) verb classes, textual features (e.g. exclamation points and question marks) and Part-Of-Speech(POS)(i.e. first- and second-person pronouns and the number of adjectives and adverbs). Polarity features are composed by positive (e.g. approving and praising) and negative (e.g. abusing and doubting) verb classes, and positive and negative adjectives provided by Wikitionary. PIRPO uses: 1. subjective features, 2. objective features and 3. polarity features, but reviews are not annotated with a POS tagger.

Ding et. al. [5] determine the semantic orientations (i.e. polarities) of opinions expressed on product features in reviews. Their system processes reviews using a set of rules based on the polarity of the adjectives in each sentence. The algorithm uses the context of the previous or the next sentence (or clauses) to resolve opinion ambiguity. The idea is that people usually express the same opinion (positive or negative) across sentences unless there is an indication of opinion change using words such as "but" and "however". Experiments indicate a F-score of 91 outperforming state-of-the-art methods. Instead of English, PIRPO is designed to analyse the polarity of service features in reviews written in Portuguese.

Kar and Mandal [6] use lists of adverbs and adjectives and apply fuzzy logic to find opinion strength in English web reviews. They also use a POS tagger and a list of stop words before detect features and extract patterns based on the POS tags. The intensity finder is the module where the features are weighted and the review is scored based on the adverbs and adjectives found in the review. As Kar and Mandal, we want to quantify the sentiment expressed in customer reviews into a score. While their system assigns a score to a review, PIRPO assigns a score to each concept of an ontology expressed in a review.

SentiStrength [11] is an emotion detection algorithm which uses a sentiment word strength list, composed by positive and negative terms classified for either positive or negative sentiment strength. This algorithm also uses a set of rules to correction spellings and punctuation and two lists: negative words and emotions. SentiStrength was applied to process MySpace comments, which tend to have more spelling errors and atypical punctuation and emoticons. PIRPO uses list of adjectives with the corresponding polarity (positive, negative or neutral) associated to it. As we can find in the literature, a number of popular approaches for sentiment classification have been used to process English texts but only a small number have been adapted for Portuguese.

Recently, there have been a number of commercial sentiment tools available from software companies, for example: Attensity (`http://attensity.com`), Clarabridge   (`http://clarabridge.com`),   Sentimetrix   (`http://www.`sentimetrix.com `), and Synthesio (`http://synthesio.com`). Attensity and Synthesio explicitly claim to process Portuguese whilst Clarabridge and Sentimetrix are language independent, but there is no publicly available evaluation of these tools on Portuguese.

3   Our Approach to Recognise Polarity in Portuguese Online Reviews

Figure 1 presents the PIRPO information architecture. PIRPO receives as input a set of reviews which are pre-processed in order to extract their sentences and detect whose reviews that are split into positive and negative segments. Some information sources allow users to input positive and negative opinion in separately fields.

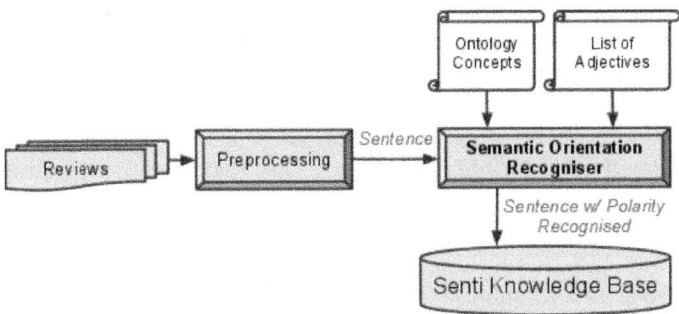

Fig. 1. PIRPO Information Architecture

PIRPO has a main module which recognise polarity of each sentence. This module receives as input sentences, a list of adjectives and concepts from an ontology. The output is a list of sentences with polarity recognised in each concept detected, which is stored in a knowledge base named Senti Knowledge Base.

The module Semantic Orientation Recognizer receives as input:

Reviews: The reviews are about Small and Medium Hotels in the Lisbon area, and it was provided by Chaves et. al. [2,3]. The information sources are Tripadvisor (`www.tripadvisor.com`) and Booking.com (`www.booking.com`). Full dataset is composed by 1500 reviews from January 2010 to April 2011 in Portuguese, English and Spanish, from which 180 in Portuguese.

Ontology Concepts: The concepts used to classify the reviews are provided by Hontology, which in its current version, has 110 concepts, 9 object properties and 30 data properties. To the best of our knowledge, Hontology is the first multilingual (i.e. Portuguese, English and French) ontology for the hotel domain. More detail about this ontology is given in Chaves et. al. [1].

List of adjectives: It is composed of sentiment-bearing words. This list of polar adjectives in Portuguese contains 30.322 entries provided by Souza et. al. [8]. This list is composed by the name of the adjective and a polarity which can assign one of three values: +1, -1 and 0. These values corresponding to the positive, negative and neutral senses of the adjective. PIRPO uses this list to calculate the semantic orientation of the concepts found in the sentences.

PIRPO assigns one out of three values to the polarity of a review segment: positive, negative, and mixed (when the segment contains one positive opinion and another negative about an object).

PIRPO output is a list of sentences with polarity that reflects the polarity of the words characterising the concepts of the ontology in the reviews. This approach eases the extension of PIRPO for other systems which use a different level of granularity.

```
Algorithm PIRPO(sentences, adjectives, concepts)
begin
     for each sentence si in sentences do
           sentence_polarity = 0
           for each concept cj in concepts do
               for each word wk in si do
                   if cj = wk then
                        let window as wk-n to wk+n, without wk
                        for each adjective ax in adjectives do
                            if ax is in window then
                                 sentence_polarity += polarity(ax)
                            endif
                        endfor
                   endif
               endfor
               save (si, cj, sentence_polarity)
           endfor
     endfor
end
```

PIRPO is implemented in Python. The algorithm above basically goes through each preprocessed sentence, each Hontology concept and each word in a pre-processed sentence. If the word in a pre-processed sentence is a concept of Hontology, we create a window of n words and we search these words in the list of adjectives. At least one word of the window should be in the list of adjectives for the sentence polarity to be calculated. The function *polarity* calculates the polarity based on the values in the list of adjectives. The function *save* saves

Table 1. Polarity assigned by a human evaluator

Concepts	Positive	Negative	Mixed
Hotel	3	2	22
Location	3	0	45
Room	3	5	28
Staff	2	2	35
Score	11	9	**130**

Table 2. Polarity classification by concept

Concepts	Precision			Recall			F-score		
	P	**N**	**M**	**P**	**N**	**M**	**P**	**N**	**M**
Hotel	0.20	0.00	1.00	0.33	0.00	0.22	0.25	0.00	0.36
Location	0.00	0.00	0.89	0.00	0.00	0.24	0.00	0.00	0.37
Room	0.11	0.00	1.00	1.00	0.00	0.19	0.20	0.00	0.32
Staff	0.08	0.00	1.00	1.00	0.00	0.12	0.15	0.00	0.21
Average	0.10	0.00	**0.97**	**0.58**	0.00	0.19	0.15	0.00	**0.32**

the results in comma separated values (CSV), which is to be stored in the Senti Knowledge Base.

4   Experiment, Evaluation and Discussion on the Results

The dataset selected to evaluate PIRPO is composed of reviews on Small and Medium Hotels in the Lisbon area, and it was provided by Chaves et al. [2]. We select only the 180 comments in Portuguese, which constitutes the subset used in our experiments. Each review sentence was evaluated as its polarity by a human. Polarity was classified as positive, negative or mixed. Table 1 presents the human evaluation.

According to Table 1, 86% of the human evaluations were classified as mixed. The concepts hotel, location, room and staff were the most mentioned in the online reviews. This fact indicates that these concepts were evaluated by hotel guests with positive and negative aspects. An in-depth analysis of the concepts classified as mixed is needed to find out more fine-grained information.

We evaluated PIRPO against the human evaluation. Table 2 presents the results for each concept. According to these results, PIRPO reached a better recall for concepts with positive polarity, while mixed polarity had a higher precision. We plan to verify the inter-annotator agreement using the kappa coefficient. Moreover, we also plan to evaluate PIRPO with other annotated corpus of online reviews.

5   Final Remarks

This paper introduced PIRPO, an algorithm to deal with polarity in Portuguese online reviews from accommodation sector and also described its preliminary results. PIRPO was designed to be domain-independent and can be applied to other domains such as tourism or books. To do this we need other domain ontologies. Domain specific lexicon might also improve the classification task.

As future work, we will work on the improvement of the results reached by PIRPO so far. Moreover, we are going to extend PIRPO to recognize the strenght of polarity underlying in each review. The final goal is to be able to integrate PIRPO in higher-level NLP systems, such as Information Extractors and Question Answer.

References

1. Chaves, M.S., Trojahn, C.: Towards a multilingual ontology for ontology-driven content mining in social web sites. In: 1st International Workshop on Cross-Cultural and Cross-Lingual Aspects of the Semantic Web (ISWC 2010), Shanghai, China (2010)
2. Chaves, M.S., Gomes, R., Pedron, C.: Analysing reviews in the Web 2.0: Small and Medium Hotels in Portugal. Tourism Management (2011)
3. Chaves, M. S., Gomes, R. e Pedron, C.: Decision-making based on Online Reviews: The Small and Medium Hotel Management. In: 20th European Conference on Information Systems, Barcelona, Spain, June 10-13 (to appear, 2012)
4. Chesley, P., Vincent, B., Xu, L., Srihari, R.: Using verbs and adjectives to automatically classify blog sentiment. In: AAAI Symposium on Computational Approaches to Analysing Weblogs (AAAI-CAAW), pp. 27–29 (2006)
5. Ding, X., Liu, B., Yu, P.S.: A holistic lexicon-based approach to opinion mining. In: Proceedings of First ACM International Conference on Web Search and Data Mining, WSDM 2008 (2008)
6. Kar, A., Mandal, D.P.: Finding opinion strength using fuzzy logic on web reviews. International Journal of Engineering and Industries 2(1) (2011)
7. Liu, B.: Sentiment analysis and subjectivity, 2nd edn. Handbook of Natural Language Processing (2010)
8. Souza, M., Vieira, R., Busetti, D., Chishman, R., Alves, I.M.: Construction of a Portuguese opinion lexicon from multiple resources. In: The 8th Brazilian Symposium in Information and Human Language Technology (STIL 2011), Cuiabá, Brazil (2011)
9. Taboada, M., Brooke, J., Tofiloski, M., Voll, K.D., Stede, M.: Lexicon-based methods for sentiment analysis. Computational Linguistics 37(2), 267–307 (2011)
10. Tang, H., Tan, S., Cheng, X.: A survey on sentiment detection of reviews. Expert Systems with Applications 36(7), 10760–10773 (2009)
11. Thelwall, M., Buckley, K., Paltoglou, G., Cai, D., Kappas, A.: Sentiment in short strength detection informal text. JASIST 61(12), 2544–2558 (2010)
12. Wilson, T., Wiebe, J., Hoffmann, P.: Recognizing contextual polarity: An exploration of features for phrase-level sentiment analysis. Computational Linguistics 35, 399–433 (2009)

Towards Interrogative Types
in Task-Oriented Dialogue Systems

Markus M. Berg[1,2], Antje Düsterhöft[2], and Bernhard Thalheim[1]

[1] University of Kiel, Germany
markus.berg@mail.uni-kiel.de, thalheim@is.informatik.uni-kiel.de
[2] University of Wismar, Germany
{markus.berg,antje.duesterhoeft}@hs-wismar.de

Abstract. The classification of questions and the identification of their respective answer types are crucial for different parts of a dialogue system and especially important for Rapid Application Development purposes. A common taxonomy of question types helps to connect parsers, grammars, pattern-based language generation methods and the natural language understanding module. Thus, in this paper we will present an overview of different question types and propose an abstract question description.

1 Introduction

As speech interfaces become more and more popular, methods for an easy and efficient design process have to be found. Today, we are only able to create general question-elements, but we cannot specify them any further. So we don't have a type-based linkage between questions and answers. Moreover, adaptability is an important step towards the acceptance of such systems. Depending on the user, we need to adapt the system with regard to formal and informal phrasing. The manual definition of prompts , as well as the formalisation of the answer domain, is a very time consuming task. Here rule- and template based approaches obviously facilitate the design process. In order to apply these methods, we need a set of different question types. Hence, in this paper we propose a list of different interrogative types and an abstract question description.

2 Question Types

According to Hirschman and Gaizauskas [4], questions can be distinguished by answer type (factual, opinion, summary) and by question type (yes/no, wh, indirect request, command). The last two categories describe utterances like *"I would like to know..."* and *"Name all persons that..."*. In order to explicitly include these and due to the fact that *questions* are often only seen as utterances that need to have a trailing question mark, we prefer the term *interrogatives*. An often encountered problem in connection with command-like interrogatives is the missing of the wh-word. So instead of saying *"Where do you want to go*

G. Bouma et al. (Eds.): NLDB 2012, LNCS 7337, pp. 302–307, 2012.

to?" a possible formulation would be *"Name your destination!"*. Thus, we need to find a general way of classifying interrogatives.

Classification by Category: Graesser [3] defines 16 question categories that mostly refer to tutoring and do not relate to task-oriented dialogue systems. Instead of asking informative questions like asking for the travel dates, they ask for expectations, judgements or ask questions about the content. Only five categories seem to be relevant for task-oriented dialogue systems. Concept completion refers to *wh-questions*, feature specification refers to questions that expect an adjective as answer, and quantification questions ask for numbers. *Disjunctive* questions don't qualify for an own category in this context as they only influence the surface and not the intention or answer domain of the question. Indeed, most questions can be formulated in a disjunctive way, e.g. *"Where do you want to go to?"* and *"Do you want to go to Paris or London"* both refer to a location.

Classification by Question Words: A very basic classification can be done by question words like *who, what, where, etc.* This leads to interpretation difficulties because there is no link to the answer type. A what-question can for example refer to a time (*"What's the time?"*) or a building (*"What is the highest building in the world?"*).

Classification by Question Words and Answer Type: Consequently, in order to specify concept completion, Srihari & Li [6] defined classes for factoid questions, i.e. named entities, temporal expressions, and number expressions. They extended the named entity classes from the Message Understanding Conferences by types like duration, weight, area, email, temperature and many more. Moreover, they introduced the *asking point* of a question, i.e. more or less the question word (sometimes in connection with an adjective) like *who, why* or *how long* that relates to a specific answer type (e.g. person, reason, length).

Also Moldovan et al. [5] realised that the question word is not sufficient for finding answers because question words like *what* are ambiguous. So they decided to use a combination of question class (question word), subclass and answer type, e.g. what/what-who/person (*"What costume designer decided that Michael Jackson should only wear one glove?"*) or what/what-where/location (*"What is the capital of Latvia?"*). Here the question-*focus*, i.e. the word the question is looking for, is the key for the correct identification of subclass and answer type (e.g. focus: capital, answer type: location).

Classification by Answer Type: We clearly see, that the answer type plays an important role when classifying questions. Hamblin confirms that when putting up the following postulates [7]: (a) An answer to a question is a sentence or statement, (b) the possible answers to a question form an exhaustive set of mutually exclusive possibilities, (c) to know the meaning of a question is to know what counts as an answer to that question. So, defining the answer helps defining the question. Existing work describes different levels of answer types. Whereas Hirschman and Gaizauskas [4] distinguish *factual, opinion* and *summary*, Moldovan [5] defines categories like *money, number, person, organization,* *distance* and *date*. Also Srihari & Li [6] introduce similar types that they call *named entity classes*.

Classification by Intention: Also Hovy et al. state that "there are many ways to ask the same thing" [2]. So they analysed 17,384 questions and answers in order to create a QA typology for Webclopedia. They did not focus on question words or the semantic type of the answer but instead attempted to represent the user's intention. So instead of classifying *"Who was Christoph Columbus?"* as who-question or *"What are the Black Hills known for?"* as what-question, they are both tagged as *Q-WHY-FAMOUS*.

Classification by Dialogue Act: In *DIT++* [1], questions are defined as *information seeking functions*, that are divided into *set questions, propositional questions*, and *choice questions*. A set question is described as wh-question and propositional questions are yes/no-questions. Choice questions can be answered with one element of a list of alternative choices.

3   Task-Oriented Interrogative Types

Questions are strongly connected to answers: A who-question results in an answer that contains a person. So the question word *who* and the answer type *person* are self-evidently closely related. Still, the answer type is not bijectively bound to the question word. Also the question word *what* can relate to the answer type *person*. So the classification of the following questions is not easy. They all have the same aim, yet the formulation is fundamentally different.

1. Who was the first man on the moon?
2. What's the name of the first man on moon?
3. I'd like to know the name of the first man on the moon.
4. Name the first man on moon!
5. Do you know who was the first man on the moon?

The first question can be classified as *factual question* (Hirschman and Gaizauskas), *concept-completion question* (Graesser), as a question that asks for a *named entity* (Srihari and Li), as a question that asks for a *person* (Moldovan), as a question with *focus "first man"* (Moldovan), or as a question with the *asking point "who"* (Srihari and Li). As you can see, there are many different possibilities how to classify these questions. Moreover, because a question is a specific type of dialogue act, we also need to differentiate between secondary and primary illocution. The last question looks like a yes/no question but in fact should be answered with a name. This also reinforces the problem of classifying a question only by structure or question word. Instead, for task-oriented systems, the user goal is most important, i.e. what kind of information the user expects.

We now propose an Abstract Question Description (AQD). We believe that the answer type is the most crucial information in order to choose the right grammar for the speech recogniser and for choosing the correct natural language understanding module. Furthermore, when using question generation methods, the answer type gives hints about how to formulate a question. Both, *city* and *country* refer to the question word *where*. Nevertheless, only the answer types enable us to create the correct formulation, e.g. *"In which city was Jim Parsons born?"* vs. *"In which country was Jim Parsons born?"*. The elements of the AQD will be described below.

Type: The expected answer type is the most important information, as this allows to classify questions depending on the information they ask for. The combination of different classification schemes leads to a taxonomy with increasing specificity, e.g. `fact > named entity > animated > person > astronaut`, `fact > named entity > non-animated > location > city`, or `decision > yesNo`. A full overview can be seen in figure 1.

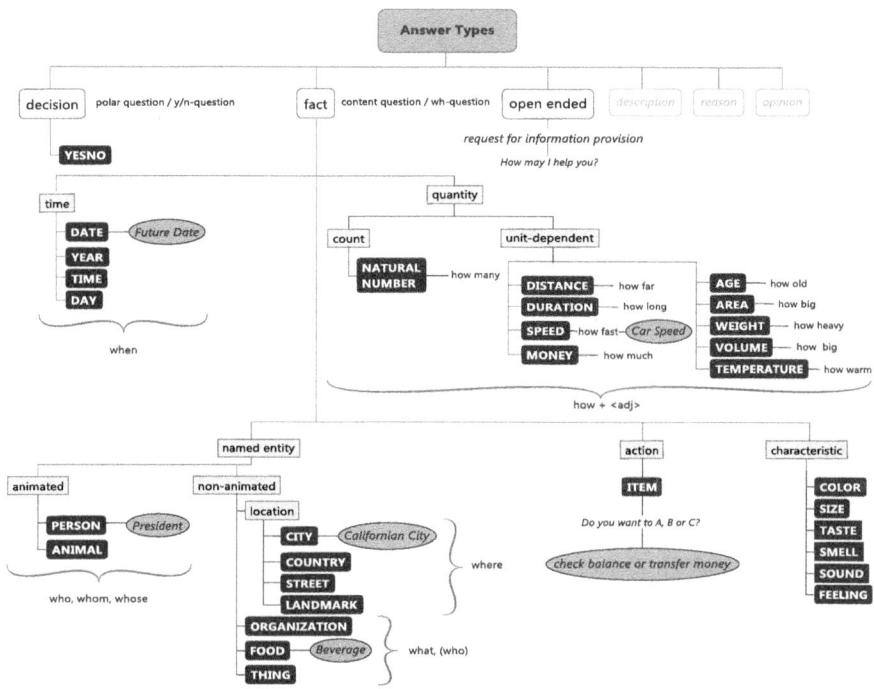

Fig. 1. Answer Types

When taking a closer look at the following formulations, that all have the same intention, we can see that the answer type `fact.namedEntity.nonAnimated.location.CITY` provides an abstract generalisation of the questions:

- Where do you want to go to?
- What's your destination?
- Do you already have a destination in mind?

- I need to know where you want to go to.
- Tell me about your travel plans!

When refining information-seeking acts, we describe what we ask for (user's aim), e.g. to ask for a location. By this means, we describe the answer type of a question, e.g. location or distance. In this connection we observe that a question may refer to different types at the same time: *"Do you want to go to London?"* obviously is a decision question that should be answered with yes or no. At the same time the question relates to a local fact (city). So *"No, to Paris"* would be a valid answer, too. Also *"Would you like a drink?"* can be answered with *"A coke please"*, i.e. without mentioning *yes* or *no* at all. This phenomenon is closely related to immediate correction and over-information and thus mostly occurs in connection with polar questions, like verifications or proposals. In order to address this very natural feature, we distinguish between *answer type* and *reference type*. The former refers to the type that is expected (direct answer type), whereas the latter refers to a category that is used within the question (indirect answer type).

Purpose: Furthermore, we need to describe the *purpose* of a question, i.e. *check, gather information, control the dialogue flow* or *social obligation*. The first category refers to questions that verify or clarify existing information, the second category aims at retrieving new information that is crucial for the processing of the task, the third one includes questions that ask for the next step in the dialogue (e.g. task selectors like *"Do you want to book a trip or do you need weather information?"*) and finally there exist social obligations like *"How are you?"*.

Surface Modifier: Some authors categorise questions by terms like *disjunctive* or *negative*. We consider this as a *formulation-modifier* because it doesn't change the message of a question. In fact, most questions can be reformulated to be disjunctive: *"Do you want to go to London?"* vs. *"Do you want to go to London or Paris?"*. The formulation itself does not change the answer type. However, it should be noted that there exist disjunctive questions that involve several answer types. In the request *"Please say your zip code or your city name!"* there are two answer types, which can be modelled by a set of answer types: {zip, city}.

Cardinality: In the Text Retrieval Conferences, questions are divided into *factoid, list* and *definition*. This classification is unfavourable for our issue as it mixes answer type (factoid, definition) and cardinality (list). Also, as you can see in the following examples, the expected cardinality of the answer does not always match the actual cardinality ($\to$ assumption/fact).

- *"Which players were in the German Soccer Team of 2012?"* $\to$ N/N
- *"Who wrote 'Do Androids Dream of Electric Sheep'?"* $\to$ 1/1
- *"Who produced 'Avatar'?"* $\to$ 1/N
- *"Name all international capitals beginning with Z!"* $\to$ N/1
- *"Name the editor of 'i, Robot'!"* $\to$ 1/N

This is due to the fact that the interrogator does not know the answer and makes an assumption in order to be able to formulate the question. Although the interrogator uses a plural in the fourth question (*capitals*), there is only one element in the answer. So in many cases it is not possible to infer the number of answers from the formulation of the question. Furthermore, we have to differentiate between the cardinality itself and the actual value of an answer, i.e. a *how many*-question only has a single answer (cardinality=1). In our opinion, the cardinality is negligible for the categorisation of questions. But as this feature is important for the question generation, we still use the assumed cardinality as part of our description.

These characteristics are now combined into an AQD. The question *"So you want to go to San Francisco?"* is formulated in a positive way and expects a single answer. The answer type is `Yes/No` and the reference type is `City`, or more specifically `Californian City`. This leads to the following AQD, a quintuple consisting of answer type, reference type, purpose, surface modifier and expected cardinality: $\epsilon = (decision.YN, fact.namedEntity.nonAnimated.local.$ $CITY.CalifornianCity, check, positive, 1)$. This definition is useful for both, processing and generation. When the system asks a question, it can be used to choose the correct grammar(s) for the speech recogniser and it defines which type of answer is expected. Moreover, it allows the Language Generation Module to create a suitable formulation for the question.

4   Conclusion and Future Work

In this paper we have presented an abstract question description for refining information seeking dialogue acts. This question description bases on answer types and on Hamblin's realisation, that the understanding of questions strongly relates to the identification of possible answers. We discovered, that when dealing with decision questions, we have to consider two different types: the *answer type* itself and the *reference type*. Besides the type, the question description also includes the expected cardinality and a surface modifier. The last element of the question description is the purpose. In our future work, we will evaluate the AQD with real data and implement it into a RAD tool.

References

1. Bunt, H., Alexandersson, J., et al.: Towards an ISO Standard for Dialogue Act Annotation. In: LREC (2010)
2. Eduard, H., Gerber, L.: et al.: Question Answering in Webclopedia. In: Proceedings of the Ninth Text Retrieval Conference, pp. 655–664 (2000)
3. Graesser, A., Rus, V., Cai, Z.: Question classification schemes. In: Workshop on the Question Generation Shared Task and Evaluation Challenge pp. 8–9 (2008)
4. Hirschman, L., Gaizauskas, R.: Natural language question answering: the view from here. Nat. Lang. Eng. 7, 275–300 (2001)
5. Moldovan, D.I., et al.: LASSO: A Tool for Surfing the Answer Net. In: Text Retrieval Conference (1999)
6. Srihari, R., Li, W.: Information Extraction Supported Question Answering. In: Proceedings of the Eighth Text Retrieval Conference (TREC-8), pp. 185–196 (1999)
7. Stokhof, M., Groenendijk, J.: Handbook of Logic and Language, ch.19, p. 21. Elsevier Science B.V (1994)

Teaching Business Systems to Agree[*]

Fred van Blommestein

University of Groningen

Abstract. In Business-to-Business (B2B) commercial relationships, the exchange of messages relies heavily on standards. However, standards are in practice seldom sufficient. Nearly all bilateral commercial contracts need human negotiations, followed by expensive software integration to realize automated B2B message exchange. Accordingly, business systems integrations are nearly always to some extent hard coded. This prohibits widespread adoption of automated support of commercial B2B relations. To solve this problem, this paper proposes a handshake protocol to agree on the semantical aspects of the interface between the computers of independent organizations.

1   Introduction

The present paradigm for B2B communication is based on rigid message and process standards. These standards are published by standardization- and sector-organizations such as GS1, BME, UBL and UN/CEFACT. Businesses then are expected to obey those standards. Often, however, the message structures and the process flow need to be adapted when establishing a specific B2B relationship. Programming or configuring an interface per trading partner appears only to be cost effective for a few stable partners with high message volumes.

Steel (1994) long ago identified a number of problem areas for the slow penetration of B2B system integration. Standards are produced too slowly, they are too generic and there exist too many of them. As a result in each local industry, implementation guidelines are produced that 'interpret' in their own way how to implement the 'standard'. At bilateral level again customizations are applied. The current standardization is not directed towards an Open-EDI environment where prior arrangements between trading partners do not occur.

In B2B literature to date, authors have proposed to let the standards describe trade relations in a more fundamental way, by standardizing models and ontologies instead of message structures (Huemer et al, 2006 and McCarthy & Geerts, 2002). They also have proposed to 'enrich' the standards with intentions of business partners (Jayaweera, 2004) and by structuring B2B messages conform formal logic (Kimbrough 2001). Yet, they all rely on standardization bodies to develop the standards for business communities. By not investigating how (systems of) individual businesses may develop or at least customize their own B2B communication, they did not address the problems Steel identified.

---

[*] Work for this paper was supported by the European Commission through the 7th FP project ADVANCE (http://www.advance-logistics.eu/) under grant No. 257398.

G. Bouma et al. (Eds.): NLDB 2012, LNCS 7337, pp. 308–313, 2012.

Lehmann (1995) presents an example dialog between two computer systems, resulting in the exchange of operational EDI messages. The computer systems, nor their users, have dealt with each other before. The dialog negotiates technical issues *and* semantics. In this paper, we take Lehmanns example as a challenge and attempt to describe how an infrastructure is to support such a dialog.

Here computer science and the theory of information systems, that traditionally regard computer communication as a highly structured process, meet natural language analysis that is used to the phenomenon that protocols are fuzzy. In computer communication protocols, usually each message structure is prescribed to the level of a bit pattern. In human communication, however, the utterance of facts and opinions is alternated with the introduction or instantiation of new objects and even with the definition of new concepts.

Businesses need to have the possibility to negotiate their inter-organizational processes. Innovation and new technology make process flows and semantics a dynamic part of business. Standards are too rigid to support these dynamics and are developed too slow. The main gap in theory and practice for dynamic and flexible B2B systems is the lack of a handshake protocol to agree on the semantical and process aspects of the interface between the computers of independent organizations

In this paper, a reference model is presented in which the aspects of a B2B relation are described and positioned in relation to each other. The reference model is based on general principles of human communication and independent from the technology used. The model is used as reference for the design of a dynamic B2B protocol.

2   The B2B Reference Model

Ogden & Richards (1923) created a model for human communication: the "Semiotic Triangle". One of the corners of the triangle is a perception of some real world phenomenon. The perception is symbolized in a sign that is sent to a recipient over a communication network. The recipient interprets the sign which induces a similar perception in his mind (see Figure 1).

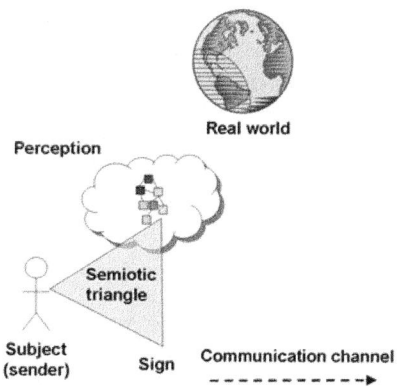

Fig. 1. Semiotic triangle

The recipient can only attach the right meaning to the sign, if he can observe (or at least imagine) the same part of the real world the sign refers to. The part of the real world the parties share perceptions on is called their 'universe of discourse'. Meaningful communication is only possible if both parties refer in it to the same universe and if at least part of their views can be aligned (figure 2).

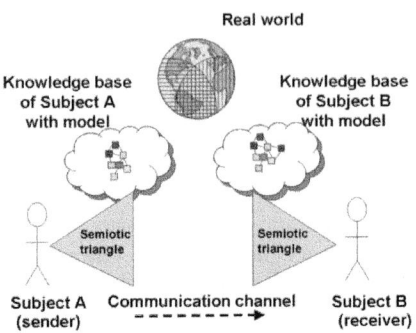

Fig. 2. B2B Reference model

The view, a party has on the real world is called the "knowledge base" of the party with respect to that world. A party's knowledge base consists of all his observations, inferences and decisions, shaping a model of the world. For B2B communication we assume that the knowledge base can be structured in some way. We also assume that only rational, organized parties can do business with each other. Only when the knowledge bases of business partners are aligned and contain similar views on the universe of discourse, they can co-ordinate their behavior. Thus, synergy can be reached in achieving their goals, e.g., completing a trade transaction. We can combine Figures 1 and 2, and extend those with the knowledge bases as in Figure 3.

A business conversation consists of a series of utterances, uttered by two business partners. Utterances are composed of signs. Utterances relate to real world events that a party has observed (such as the production of goods or the delivery of consignments) or to decisions the party has made (such as the decision to purchase products from the other party).

If the knowledge bases are implemented in the partners' automated information systems, the updates are made through those systems. Some mechanism should synchronizes the two systems, by exchanging encoded utterances. This process is illustrated in Figure 3.

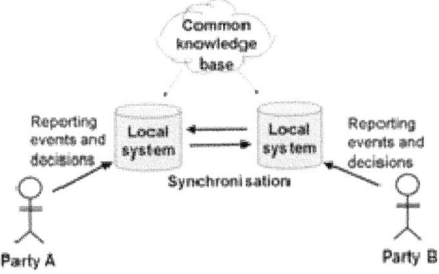

Fig. 3. B2B communication system

The knowledge base is filled with observations, inferences and decisions of business people. Business people, like all humans, use natural language for communication of their utterances.

Syntactically, a natural language assertion or utterance about the real world has the structure of a verb phrase, which contains a verb and may have several 'slots' that are filled with noun phrases. According to Sowa (1999), a verb represents an event, activity or process in the real world. Gruber (1965) introduced Thematic Roles (or Theta roles) of phrases as opposed to syntactical roles. In this paper, we are interested in semantics, rather than in syntax. Therefore, the focus will be on Thematic (or semantic) roles rather than on syntactic roles. Fillmore & Baker (2001) developed a theory in which each verb can be assigned a "frame" with a number of "slots", each to be filled by a semantic or thematic role. Verbnet has defined frames and identified role slots for some 3500 verbs. Verbnet lists 23 thematic roles.

In dynamic business communication, partners should be able to introduce new concepts. Metadata should be extensible and negotiable. It is necessary to be able to define new concepts, e.g., new products, services or conditions. Traditional (legacy) information systems are however not capable of reprogramming themselves and to accept new concept definitions on-the-fly. Therefore new concepts are to be based on existing concepts that can be handled by the information systems. New concepts are introduced by specializing existing ones, and narrowing their set of properties.

As an illustration, suppose the concept of Vehicle is already known in the common knowledge base of two communication partners. A Vehicle has a number of wheels as its property. A Bicycle may then be defined as a Vehicle with the number of wheels being '2'. An information system that may process data on Vehicles, thus, can also process data on Bicycles. The term Bicycle simply sets the number of wheels to have the value '2'.

3   Design of the Knowledge Base

In addition to observations (on existing objects) [1] and definitions [2] (of new concepts), it must be possible to instantiate[3] new objects as members of a set of objects that are defined as a concept. It must also be possible to define object states [4] and observation types or perceptions [5]. State definitions are crucial when negotiating process flows. Perceptions are mapped on transactions in the business information systems. Finally it should be possible to expand [6] and restrict [7] the set of properties on concepts.

Each of these seven utterance types has the same basic structure, which is derived from a natural language verb phrase. The core of the phrase is a verb together with two slots that may be filled with concepts, each fulfilling a thematic role. Each utterance must refer to a previously exchanged utterance on which it is based. This way instantiation may refer to defined concepts, and newly defined concepts may be based on more generic concepts. Observations should be based on earlier instantiations.

Identification				Intention			Core Proposition					Cardinality			Definition				
1	2	3	4	5	6	7	8	9	10	11	12	13	14	15	16	17	18	19	20
Utterance #	Based on Utterance #	Timestamp	Uttered by Party	Action	Stereotype	With Intention	Source Concept Name	Source Role Name	Verb	Target Role Name	Target Concept Name	Part of ID #	Min Repetition	Max Repetition	To utter by Party	With allowed Stereotype	With allowed Intention	Precondition #	Transaction with #

Fig. 4. B2B knowledge base structure

In addition to the basic utterance structure, slots should be available for quantification, intention and identification of the utterance itself. Figure 4 shows the full lay-out of a B2B knowledge base structure.

All in all, a knowledge base for B2B communication may be represented as a simple tabular structure with no more than 20 columns. The table structure may be implemented in middleware with the capability of automatically mapping newly defined concepts and transactions on the interfaces and databases of the business information systems it enables to communicate. If such middleware has access to the policies and capabilities of the business and its information systems it serves, it may independently negotiate with its peers on business communication processes.

4   Conclusion

By teaching business systems a little bit of natural language, business systems may get interconnected without the use of standardization committees and integration consultants. This technology may boost automated B2B communication, avoiding manual rekeying of information and speeding up inter-organizational business processes.

The mechanism as proposed in this paper is being implemented in a system that is to provide management information and decision support in networks of logistic companies. This system is developed in the ADVANCE project. An ADVANCE system extracts information for decision support to logistics personnel from all available resources, such as an ADVANCE system in another company, operational business systems and devices, such as on board computers and RFID readers.

References

1. ADVANCE, http://www.advance-logistics.eu/ (retrieved June 20, 2011)
2. Fillmore, Baker: Frame Semantics for Text Understanding. In: Proceedings of WordNet and Other Lexical Resources Workshop, NAACL, Pittsburgh (2001)
3. Gruber: Studies in Lexical Relations. MIT Working Papers in Linguistics, Boston (1965)
4. Huemer, et al.: UN/CEFACT's Modeling Methodology (UMM) UMM Meta Model – Foundation Module. United Nations Centre for Trade Facilitation and Electronic Business (2006)

5. Jayaweera.: A Unified Framework for e-Commerce Systems Development: Business Process Pattern Perspective. PhD thesis, Department of Computer and Systems Sciences Stockholm University and Royal Institute of Technology (2004)

6. Kimbrough: A Note on Getting Started with FLBC. ACM SIGGROUP Bulletin 22(2) (2001)

7. Lehmann: Machine-Negotiated, Ontology-Based EDI. In: Proceedings of CIKM 1994 Workshop on Electronic Commerce. Springer, Heidelberg (1995)

8. McCarthy, Geerts: An Ontological Analysis of the Primitives of the Extended REA Enterprise Information Architecture. The International Journal of Accounting Information Systems 3 (2002)

9. Ogden, Richards: The Meaning of Meaning: A Study of the Influence of Language Upon Thought and of the Science of Symbolism. Routledge & Kegan, London (1923)

10. Sowa: Knowledge Representation: Logical, Philosophical, and Computational Foundations. Brooks Cole Publishing Co., Pacific Grove (1999)

11. Steel: Another Approach to Standardising EDI. EM - Electronic Markets No. 12 (1994)

12. UN/CEFACT TBG, http://www.uncefactforum.org/TBG/TBG%20Home/tbg_home.htm (retrieved February 20, 2011)

13. Verbnet, http://verbs.colorado.edu/~mpalmer/projects/verbnet.html (retrieved February 20, 2011)

Improving Chinese Event Construction Extraction with Lexical Relation Pairs

Liou Chen[1,*], Qiang Zhou[2], and Hongxian Wang[1]

[1] Department of Computer Science and Technology, Tsinghua University, China
{chenlo,wanghx}@cslt.riit.tsinghua.edu.cn
[2] National Laboratory for Information Science and Technology, Tsinghua University, China
zq-lxd@mail.tsinghua.edu.cn

Abstract. Semantic Role Labeling (SRL) has proved to be a valuable tool for performing automatic analysis of natural language text. SRL requires additional information besides lexical and syntactical features to recognize the candidate arguments better. This paper investigates the use of Lexical Relation Pair (LRP) information for the task of Event Construction Extraction. Experimental results show that introducing LRP features can achieve some improvements. Further experiments indicate that using automatically extracted LRPs can obtain even better performance than manually annotated ones. This observation enables the large-scale usage of LRPs in the future.

Keywords: Event Construction extraction, Lexical Relation Pair.

1 Introduction

Semantic Role Labeling (SRL) is an important and challenging task in natural language processing. It mainly consists of two steps: recognize the arguments of target predicate and assign them with proper semantic role tags. Among them, argument identification has the greatest impact on the whole system.

At present, the main stream researches in argument recognition are focusing on feature engineering. The traditional full parsing based approaches (e.g., Xue, 2008) extracted features at syntactic tree level and proposed a wide range of lexical and structural features. Some shallow parsing based (Sun, et al. 2009; Sun, 2010) and semantic chunking based (Ding and Chang, 2009) methods attempted to substitute full structural features with shallow syntactic features. In addition, some higher-level features, like hierarchical semantic knowledge (Lin, 2010) and word sense information (Che, 2010) were also discussed.

This paper takes the novel Lexical Relation Pair (LRP) feature into consideration. We use the LRP information together with several other features into Chinese Event Construction (Zhou, 2010) extraction system to explore the power of LRPs in argument recognition. In addition, we extract LRPs automatically and try to replace manual pure LRPs with automatic noisy ones.

---

* Corresponding author.

G. Bouma et al. (Eds.): NLDB 2012, LNCS 7337, pp. 314–319, 2012.

Experimental results indicate that the application of new LRP features can improve the performance of argument recognition. Another valuable observation is that the system that applies automatically extracted LPRs can achieve better performance.

The remainder of this paper is organized as follows. In Section 2 we provide a framework of the EC extraction system and present a baseline paradigm. Section 3, 4 and 5 describe our proposed EC extraction system in detail. Finally we draw conclusions and discuss future directions in Section 6.

2   Chinese Event Construction Extraction

Unlike the "Predicate-Arguments" structure in SRL system, this paper plans to extract a new structure named "Event Construction (EC)" in input Chinese sentence. The new structure was first proposed by (Zhou, 2010). It is a sequence of event content chunks as line *EC* in Fig. 1 shows. The event content chunks are at syntactic level and labeled with grammatical function tag and syntactic constituent tag.

EC extraction aims to find all event chunks and label them with syntactic tags, like line *EC* in Fig. 1. Similar as SRL, we divide EC extraction into two key modules, one is event content chunk recognition (simplified as "ECCR") and the other is syntactic tag labeling (abbreviated as "STL"). ECCR aims to determine the boundaries of each event chunk and STL means assigning each recognized event chunk with its syntactic tags.

Existing studies have shown that the shallow parsing based methods can obtain better performance in recognizing the boundaries of arguments (Sun et al., 2009; Sun, 2010) compared with full parsing and semantic chunking based approaches. For this reason, we realize our system on the basis of base chunks (as line *BC* in Fig. 1 shows).

Two different paradigms are compared in this paper: one is the baseline paradigm that only uses basic lexical and shallow syntactic features in ECCR sub task. The other one is our proposed approach that introduces LRP information into extraction system.

The feature templates applied in the baseline system are "standard", which have been used in previous Chinese SRL researches (Sun et al., 2009; Ding and Chang, 2009). We give a brief description of these features:

Features for ECCR: We denote a token base chunk $[c_k w_i...w_h...w_j]$, where w_h is the head word of this chunk and c_k denotes the constituent tag of this chunk. The complete set of features is listed here:

POS tag and word of predicate, $w_h\_c_k$, the nearest verb before and after chunk, head word context within a window of size -2/2, constituent context within a window of size -1/1, position, distance, constituent path between predicate and chunk, the conjunction of target predicate and w_h,   the conjunction of predicate and c_k.

Features for STL: Denote 1) a given event chunk as $w_{i-1}[w_i w_{i+1}...w_{j-1} w_j]w_{j+1}$, and 2) a given predicate w_v. The features for syntactic tag labeling are listed as follows:

w_i, w_j, POS tag of w_i, POS tag of w_j, w_i_POS of w_i, w_j_POS of w_j, w_{i-1}_POS of w_{i-1}, w_{j+1}_POS of w_{j+1}, chunk length, $w_{i+1}...w_{j-1}$, POS of w_{i+1}...POS of w_{j-1}.

WORD:	以	乐观	态度	对待	疾病	的	学生
POS:	p	a	n	v	n	uJDE	n
BC:	_	[np	]	[vp]	[np]	_	[np]
EC:	[D-pp		]	[P-vp]	[O-np]	_	[H-np]

The student who treats the disease with optimistic attitude

Fig. 1. An example for EC Extraction

3   Extraction with LRP Information

A LRP takes the form of $t= (w_i, w_j, r_{ij})$, where w_i and w_j denote two words, and r_{ij} is a string meant to denote a relationship between them. The possible relations between two words are pre-defined as: 1) inter-chunk relations, such as *"Subject-Predicate"* relation (ZW), *"Predicate-Object"* relation (PO), etc.; 2) intra-chunk relations, including *"Coordination"* relation (LH), *"Modifier-Head"* relation (DZ), etc.

Denote 1) a token base chunk c_k and its left adjacent chunk as $[c_{k-1}...w_{h-1}...w_{i-1}][c_k w_i...w_h...w_j]$, where w_{h-1} and w_h indicate the headword of the two chunks; and 2) a given predicate w_v. We propose two basic assumptions by analyzing the relationships between event structures and relation pairs as follows:

1. Long-distance dependency relationship: if w_h and w_v constitute a LRP, c_k is likely to be inside one event content chunk, with w_h as the head word of this event chunk.
2. Adjacent dependency relationship: if w_{h-1} and w_h constitute a LRP, c_{k-1} and c_k are usually act as an entirety inside or outside event chunks.

Based on our assumptions, the LRP information is used in the form of two features:

1. The relative position of w_h with w_v + LRP type of w_h and w_v.
2. LRP type of w_{h-1} and w_h + LRP type of w_{h-1} and w_v + LRP type of w_h and w_v.

We apply "NULL" as a special relation type when there are no possible LRPs between two words.

4   Experimental Results

Data Preparation: The corpus used in this paper is EventBank, which is automatically built from Tsinghua Chinese Treebank (TCT) (Zhou et al., 1997). The data set consists of 49,042 training sentences, 11,792 development sentences and 24,361 test sentences. Each sample is as line *EC* in Fig. 1 shows.

The manually annotated LRP knowledge base used in our experiments is from four data sources, including TCT, Chinese collocation dictionary (Qi, 2003), People's Daily annotated corpus (Yu et al., 2001) and PKU grammatical dictionary (Yu et al., 1998). The final manual knowledge base contains 953,828 different LRPs.

As we implement our extractor on the basis of base chunks, a base chunk parser (Yu, 2009) trained on ParsEval-2009 base chunk bank (Zhou and Li, 2009) whose F-score is 91.31% is applied. For an input sentence assigned with gold standard word segmentation and POS tags, the parser can provide base chunks as line *BC* in Fig. 1.

Classifier: For ECCR subtask, we use the CRF++ toolkit[1] because of its outstanding performance on sequence labeling. STL module chooses the maximum entropy tool-box[2] as the multi-class classifier due to its training efficiency.

Evaluation Measures: As the Event Construction Extraction system consists of two sub-tasks, we evaluate them separately. The ECCR sub task is evaluated by Chunk-Recall (R), Chunk-Precision (P), and Chunk-F-score. For STL sub task, we use the Chunk-Micro-Accuracy (A) to evaluate the performance of classifier.

The performances of whole system are evaluated at two levels: chunk level and event level. The Chunk-P, Chunk-R and Chunk-F are used to evaluate the chunk level performances and the event level applies the Event-Micro-A measure.

System Performance: Table 1 lists the performances of the baseline and our proposed system on the test set.

Comparing the two performances of ECCR in Table 1, we get the conclusion that introducing LRP features can enhance the baseline system. From the chunk-level and event-level evaluation results of whole system, we find that to understand the "whole" meaning of a sentence is very difficult.

Table 1. Performances of baseline and our proposed EC Extraction systems

	ECCR			STL	Whole System			
	P (%)	R (%)	F (%)	A (%)	P (%)	R (%)	F (%)	A (%)
Baseline	87.93	85.43	86.66	95.61	85.08	82.66	83.85	62.16
Ours	88.29	86.35	87.31	95.61	85.54	83.66	84.59	63.38

Although our proposed system does not achieve too much increase on the overall dataset compared with the baseline method, we further analyze the two systems along two key dimensions: chunk length and number of training samples.

Event Chunk Length: Fig. 2 displays the performance curves of the baseline and proposed approaches under different chunk lengths. We observe that LRP features perform better on longer chunks. The reason is that the LRP features can provide both the long-distance "Predicate-Token" dependency information and the adjacent dependency relations, which can help to distinguish the complex structure better.

Number of Training Samples: We divide our test data set into three parts according to the number of training samples of each predicate as 1) high-frequency predicates that have more than 100 samples in training set; 2) middle-frequency predicates that

---

[1] http://sourceforge.net/projects/crfpp/
[2] http://homepages.inf.ed.ac.uk/lzhang10/maxent_toolkit.html

occur between 10 and 100 times and 3) low-frequency predicates whose occurrences are lower than 10.

Fig. 3 reports the Event-Accuracy of two systems on predicates with different training samples. One encouraging observation from the histogram is that the LRP features can obtain more increases on the predicates that are insufficiently trained. The reason is that high-frequency predicates already have large amounts of training samples that the lexical and syntactic features are enough to represent a sentence. So the LRP information will perform better on middle and low frequency predicates.

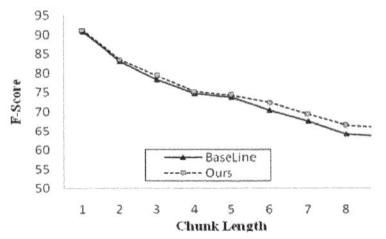 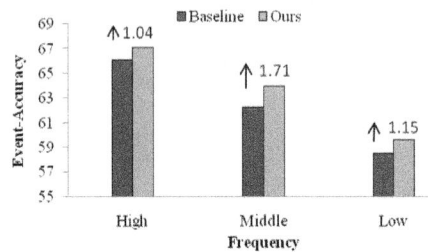

Fig. 2. F-score of two systems under different event chunk lengths

Fig. 3. Event-accuracy of two systems on predicates with different frequencie

5   Extraction System with Automatic LRPs

Previous explorations have shown that introducing manual LRP information can improve the performance of EC extraction system. We then think about applying automatically extracted LRPs to clear whether the system can attain similar improvements.

LRP extraction can be treated as the "inverse" of LRP application. Therefore, we use the baseline system firstly to get ECs from unlabeled sentence and then extract LRPs from the event chunks base on the two assumptions: 1) the head word of an event chunk can form an inter-chunk LRP with predicate; and 2) the head words of base chunks inside an event chunk are likely to form an intra-chunk LRP.

We use 868,974 sentences from the manually annotated people's daily (1998 and 2000) corpus as unlabeled data. Then 997,367 LRPs are extracted.

Table 2 displays the performances of EC Extraction system with automatic and manual LRPs. It is surprising that when fully automatic LRP features are used, our extraction system can achieve even better performance than the one applies manual ones, though the gap is not evident. One reason is the test set is also extracted from People's Daily data. So they have the same text sources with automatic LRPs.

Table 2. Performance of four different EC extraction systems

	KB-Size	Chunk-F	Event-A
Manual	953,828	84.59	63.38
Automatic	997,367	84.65	63.59

6   Conclusion

The contributions of this paper are: 1) explores the special use of LRP information in EC extraction system and improves the performance compared with the baseline system, especially on the predicates which are insufficiently trained. 2) Puts forward an extractor to automatically extract a number of reliable LRPs. These automatic LRPs can obtain even better performance than manual ones.

Two conclusions can be drawn. First, LRP features are effective in EC extraction. Second, the size and quality of LRP knowledge base can affect the performance. Therefore in future, we will continue to explore the usage of LRP features and focus on how to extract more high-quality LRPs automatically.

Acknowledgments. This work was supported by National Natural Science Foundation of China (Granted No. 60873173), Tsinghua-Intel Joint Research Project.

References

1. Ding, W., Chang, B.: Fast Semantic Role Labeling for Chinese Based on Semantic Chunking. In: Li, W., Mollá-Aliod, D. (eds.) ICCPOL 2009. LNCS, vol. 5459, pp. 79–90. Springer, Heidelberg (2009)
2. Lin, X., Zhang, M., Wu, X.: Chinese semantic role labeling with hierarchical semantic knowledge. In: Electrical and Control Engineering International Conference, pp. 583–586 (2010)
3. Qi, Y.: Chinese collocation dictionary. Guangming Daily Press, Beijing (2003)
4. Sun, W., Sui, Z., Wang, M., Wang, X.: Chinese semantic role labeling with shallow parsing. In: Proceedings of the 2009 Conference on Empirical Methods in Natural Language Processing, pp. 1475–1483 (2009)
5. Sun, W.: Semantics-driven shallow parsing for Chinese semantic role labeling. In: Proceedings of the ACL 2010 Conference Short Papers, pp. 102–108 (2010)
6. Xue, N.: Labeling Chinese predicates with semantic roles. Computational Linguistics 34(2), 225–255 (2008)
7. Yang, S., Chen, J.: Research on structural ambiguity in Chinese automatic syntactic parsing. Journal of Kunming University (Science & Technology) 30(2) (2005)
8. Yu, H., Zhou, Q.: Intra-chunk relationship analyses for Chinese base chunk labeling system. Journal of Tsinghua (Science & Technology) 49(10) (2009)
9. Yu, S., Zhu, X., Wang, H., Zhang, Y.: PKU grammatical dictionary. Tsinghua University Press, Beijing (1998)
10. Yu S., et al.: Guideline of People's Daily Corpus Annotation. Technical report, Beijing University. (2001)
11. Zhou, Q., Li, Y.: The Design of Chinese Chunk Parsing Task. In: The Tenth Chinese National Conference on Computational Linguistics, pp. 130–135 (2009)
12. Zhou, Q., Zhang, W., Yu, S.: Chinese Treebank Construction. Journal of Chinese Information Processing, 42–51 (1997)
13. Zhou, Q.: SemEval-2010 Task 11: Event detection in Chinese news sentences. In: ACE 2010 Workshop, p. 86 (2010)
14. People's Daily (1998, 2000) corpus,
   http://icl.pku.edu.cn/icl_res/default_en.asp

A Classifier Based Approach
to Emotion Lexicon Construction

Dipankar Das, Soujanya Poria, and Sivaji Bandyopadhyay

Department of Computer Science and Engineering, Jadavpur University,
188, Raja S.C. Mullick Road, Kolkata 700 032, India
{dipankar.dipnil2005,soujanya.poria}@gmail.com,
sivaji_cse_ju@yahoo.com

Abstract. The present task of developing an emotion lexicon shows the differences from the existing solutions by considering the definite as well as fuzzy connotation of the emotional words into account. A weighted lexical network has been developed on the freely available ISEAR dataset using the co-occurrence threshold. Two methods were applied on the network, a supervised method that predicts the definite emotion orientations of the words which received close or equal membership values from the first method, Fuzzy c-means clustering. The kernel functions of the two methods were modified based on the similarity based edge weights, Point wise Mutual Information (PMI) and universal Law of Gravitation (LG_r) between the word pairs. The system achieves the accuracy of 85.92% in identifying emotion orientations of the words from the WordNet Affect based lexical network.

Keywords: Emotion orientations, ISEAR, Fuzzy Clustering, SVM, PMI, Law of Gravitation, WordNet Affect.

1 Introduction

Several researchers have contributed their efforts regarding the semantic orientation of words [2] but most of the works are in English and are used in coarse grained sentiment analysis (e.g., *positive*, *negative* or *neutral*). There are some examples like the word, "*succumb*" which triggers mix of multiple emotions (*fear* as well as *sad*) to a reader. Thus, considering the problem of word emotion identification as a multi-label text classification problem, the present attempt of generating emotion lexicon by incorporating the fuzzy nature of emotion is different from the existing solutions in this field. Therefore, the Fuzzy c-means clustering [4] has been employed followed by Support Vector Machine (*SVM*) [5] based classification to accomplish the goals. The objective function of Fuzzy c-means clustering and kernel function of SVM have been modified by adding some more functions like Point wise Mutual Information (*PMI*), universal Law of Gravitation (*LG_r*) and similarity based edge weights between the word pairs. It has been observed that SVM helps to predict the definite emotion orientations of the words that receive close or equal membership values from the

G. Bouma et al. (Eds.): NLDB 2012, LNCS 7337, pp. 320–326, 2012.

Fuzzy c-means clustering method. The proposed method extracts emotion orientations with high accuracy on the weighted lexical networks.

2   Construction of Lexical Network

A generalized lexical network (G) has been constructed from the psychological statements of the ISEAR (International Survey of Emotion Antecedents and Reactions) dataset [6]. The psychological statements contain about 3~4 sentences pre-classified into seven categories of emotion (*anger, disgust, fear, guilt, joy, sadness* and *shame*). A set of standard preprocessing techniques was carried out, *viz.*, *tokenizing, stemming* and *stop word removal* using *Rapidminer's text plugin* [1]. The Stanford Part-of-Speech (POS) Tagger[2] and the *WordNet* stemmer [7] have been used for (*POS*) tagging and lemmatization of the statements. A total of 320 stop words (quite frequent words such as "*any~thing/body*" and "*Wh~en/ile*" etc.) are removed. Negation words include the 6 words ("*not*", "*no*", "*never*" etc.). In addition to the usual negation words, the words and phrases which mean negations in a general sense have also been included (e.g., "*free from*" and "*lack of*"). Finally, to construct the lexical network, two words are linked if one word appears with the other word in a single statement. A total of 449060 words were found.

Co-occurrence Based Network (G_{Co}): Co-occurrence identifies the chances of frequent occurrence of two terms in a text corpus alongside each other in a certain order as an indicator of semantic proximity. In contrast to collocation, co-occurrence assumes interdependency of the two terms. A co-occurrence network (G_{Co}) has been developed from the generalized network (G) based on the co-occurrence *threshold* > 1 in the whole corpus. The co-occurrence network (G_{Co}) consists of 82,457 words.

WordNet Affect Based Network (G_{WA}): The generalized lexical network (G) is transformed into another lexicalized network based on the words present in the *Wordnet Affect* [1]. Each word of a statement and its stem form is searched in any of the six *WordNet Affect* lists. If any match is found, the word is tagged as the *Emotion Word <EW>*. Two words are linked if one word is an *Emotion Word <EW>* and appears in the context of the other word in a statement. This type of *WordNet Affect* based network (G_{WA}) consists of 63,280 words.

Similarity Based Weighted Networks: The weights (ρ) are assigned to the edges of the word pairs in the networks based on two types of similarity measure, WordNet distance based similarity (W_{Sim}) and Corpus distance based similarity (C_{Sim}).

WordNet Distance Based Similarity (W_{Sim}). English WordNet 2.1[3] has been used to measure the semantic distance between two words. WordNet::Similarity is an open-source package for calculating the lexical similarity between word (or sense) pairs based on variety of other similarity measures.

---

[1] http://rapid-i.com/content/blogcategory/38/69/
[2] http://nlp.stanford.edu/software/tagger.shtml
[3] http://www.d.umn.edu/tpederse/similarity.html

Corpus Distance Based Similarity (C_{Sim}). In case of Corpus distance based similarity (C_{Sim}), we averaged the lexical distances (L_d) of the co-occurred word pairs in the statements without considering their ordering. We consider the inverse values of the W_d and L_d by assuming that the larger distance between two words implies lower similarity and vice versa. The weights are calculated based on the summation of two similarity scores as shown below. The corresponding weighted networks are denoted as G ' G_{Co}' and G_{WA}'.

3   Fuzzy Clustering Cum SVM Based Classification

The Fuzzy c-means clustering algorithm [4] followed by a SVM based supervised classification [5] has been employed by modifying the objective function and kernel functions using three more functions like Point of Mutual Information (*PMI*) (α), universal Law of Gravitation (LG_r) (β) and similarity based edge weights (ρ). The PMI identifies the association between two words (w_1 and w_2) while the universal Law of gravitation (LG_r) incorporates the lexical affinity between two words for belonging to an emotion cluster. Considering the lexical network as the universe of words, the word pairs as the two masses and the root mean square of their lexical distances (L_d) as the distance between the masses, the affinity or attraction between the word masses has been calculated using the notion of Gravitational Force (F). The Gravitational constant G has been calculated based on the average degree of the nodes present in that network.

$$PMI, \alpha = \log \frac{p(w_1, w_2)}{p(w_1).p(w_2)}, \quad F, \beta = G.\frac{m_1 m_2}{r^2},$$

where $m_1 = f_{count}(w_1, w_2)$, number of times the word w_1 appears before the word w_2 in the corpus and $m_2 = f_{count}(w_2, w_1)$, number of times the word w_2 appears before the word w_1 in the corpus and the distance r, root mean square of the lexical distances (L_d) between an word pair is,

$$r = \sqrt{\frac{\sum_{i=1}^{n} L^2_{d_i}}{n_{co}}}, \quad \text{where } n_{co} \text{ is the number of Co-occurrences.}$$

The modified objective functions (J) of the Fuzzy c-means clustering is as follows,

$$J = \sum_{i=1}^{c}\sum_{k=1}^{N} \mu_{ik}^{p} (\| x_k - v_i \|)^2 + (\alpha + \beta + \rho)\sum_{i=1}^{c}\sum_{k=1}^{N} \mu_{ik}^{p} (\sum_{x_r \in N_k} (\| x_k - v_i \|)^2), p > 1$$

where, p, the exponential weight influences the degree of fuzziness of the membership function. x_k is the co-ordinate in an n-dimensional space of kth point where the values of n features are mapped into n-dimensional space such as $x_k = (p_1, p_2, p_3, ..., p_n)$ where p_i is the value of the ith feature. The other functions are as follows, To acquire the optimality, the modified objective function is differentiated with respect to v_i (for fixed u_{ij}, $i = 1, \ldots, k$, $j = 1, \ldots, n$) and to u_{ij} (for fixed v_i, $i = 1, \ldots, k$), $\frac{\partial J}{\partial v_i} = 0$. If we assume $\| x_k - v_i \|^2 = D_{ik}$ and $(\sum_{x_r \in N_k} (\| x_r - v_i \|)^2 = y_i$, the modified objective function is turned into the following equation where the Lagrange multiplier (λ) is introduced for solving it.

$$F_m = \sum_{i=1}^{c} \sum_{k=1}^{N} (\mu_{ik}{}^p D_{ik} + (\alpha + \beta + \rho)\mu_{ik}{}^p y_i) + \lambda(1 - \sum_{i=1}^{c} \mu_{ik})$$

If the modified objective function is differentiated with respect to the membership variable μ_{ik}, the membership value is obtained as follows,

$$\frac{\partial F_m}{\partial \mu_{ik}} = p\mu_{ik}{}^{p-1} D_{ik} + (\alpha + \beta + \rho)\mu_{ik}{}^{p-1} y_i - \lambda = 0 \text{ and } \frac{\partial F_m}{\partial \mu_{ik}} = 0$$

$$\mu_{ik} = \frac{\lambda}{p(D_{ik} + (\alpha + \beta + \rho)y_i)}, \text{ where } \sum_{i=1}^{c} \mu_{ik} = 1,$$

$$\lambda = \frac{p}{(\sum_{j=1}^{c}(\frac{1}{(D_{jk} + (\alpha + \beta + \rho)y_j)})^{1/p-1})^{p-1}}, \mu_{ik} = \frac{1}{\sum_{j=1}^{c}(\frac{D_{ik} + (\alpha + \beta + \rho)y_i}{D_{jk} + (\alpha + \beta + \rho)y_j})^{1/p-1}}$$

The centre of the cluster is identified using the following equation.

$$\frac{\partial F_m}{\partial v_i} = 0, \ v_i = \frac{\sum_{k=1}^{N} \mu_{ik}{}^p (x_k + (\alpha + \beta + \rho) \sum_{x_r \in N_k} x_r)}{(1 + \alpha + \beta + \rho) \sum_{k=1}^{N} \mu_{ik}{}^p}$$

The hyper parameter was selected by using the universal law of gravitation and the Co-occurrence *threshold* >1. The previous results show that the clustering technique can help to decrease the complexity of SVM training [8]. Thus, the *SVM* based classifier has been employed to identify the definite class for an emotion word by modifying its kernel function (using α, β and ρ, the similar functions used in Fuzzy clustering), dividing the training data into sections and excluding the set of clusters with minor probability for support vectors. For the clusters of mixed and uniform category, the support vectors by SVM are extracted and formed into reduced clusters. The best feature set for the classifier has been identified based on the performance of the classifier in terms of accuracy of the classifier. Information Gain Based Pruning (*IGBP*) was carried out to remove the words (e.g. *game, gather, seem* etc.) that do not play any contributory role in the classification. Features with high Information Gain reduce the uncertainty about the class to the maximum. The input vectors have been generated from the concept template of the statements. Each of the concept template contains the *lexical* pattern *p* using a context window that contains the words around the left and the right of the *Emotion Word, <EW>*, e.g.,

$$p = [l\text{-}i\ldots l\text{-}3l\text{-}2\ l\text{-}1 <EW> \ldots</EW>\ l+1\ l+2\ l+3\ldots l+i],$$

where, *l±i* are considered as the *context* of *p* as the contexts are good predictors for emotion in a corpus. Equivalent emotional seed words are used to learn the context rules that contain information about the equality of the words in context.

4   Experiments

The proposed method is evaluated in terms of precision for the words that are classified with high confidence. In case of SVM, it was found that the top 100 words of each emotion classes achieved 82.78% accuracy. Therefore, the membership value (μ_{ik}) of each cluster as identified during fuzzy clustering was regarded as a confidence measure and incorporated into the SVM. Then, the words with the highest membership values are evaluated with respect to a cluster and the top 100 words for seven emotion classes achieved more than 95% accuracy. The results (in Table 1) show that the membership value of each cluster can work as a confidence measure of the classification as well as it enhances the precision on several lexical networks. It has been observed that the system performs better on the weighted lexical networks in comparison with their ordinary versions. It implies that the similarity based scoring influences the fuzzy clustering as well as the SVM to predict their emotion orientation. Additionally, further experiments conducted using the SVM alone on these networks reveal that the fuzzy clustering helps the SVM to clarify the lexical ambiguities. In most cases, the co-occurrence and the similarity based scoring information from the corpus improve accuracy.

Table 1. Precision of different Classifiers (in %) for top 100 words on various networks

Approaches	G	G'	G_{WA}	G_{WA}'	G_{Co}	G_{Co}'
SVM	83.22	85.19	88.23	91.67	84.10	86.77
Fuzzy c-means +SVM	88.01	90.78	92.56	95.02	87.44	91.60

Impact of Psychological Features: Feature plays a crucial role in any machine-learning framework. By reviewing the ISEAR dataset, the following features related to emotion and psychology have been selected to accomplish the classification task. Each of the features is assigned with different numeric values as supplied by the ISEAR dataset. Each word associated with the following features is represented as the feature vector. The detail impact of the psychological features is described in Table 2.

A. Background Variables (*Age, Gender, Religion, Occupation, Country*)
B. General Variables (*Intensity, Timing, Longevity*)
C. Physiological Variables (*Ergotropic and Trophotropic Arousals and Felt temperature*)
D. Expressive Behavior (*Movement, Non-verbal and Paralinguistic activity*)

The individual results based on the general and physiological variables show various interesting insights of the variables from the perspective of emotion (e.g., low *intensity* for emotion classes of *shame* and *guilt* and high for *joy, fear* and *sadness*) [3] in addition with the less contributing features, background variables and expressive behaviour. Finally, the system achieves 85.92% accuracy on the *WordNet Affect* based weighted lexical network.

Error Analysis: Mainly four types of errors have been investigated. The identification of emotion words present in a statement but in two separate sentences does not always carry the similarity between their lexical patterns and therefore performs poor on the weighted networks. The second error type is that some of the ambiguous words are still present (*"faint"*, *"sick"*, *"humble"* etc.) even after including the membership value of the Fuzzy clustering algorithm into SVM. The third error type is the presence of implicit emotions (*"Dreams are in the eyes of the children"* do not contain any direct emotion word but contain an emotional sense). The fourth error type is related to the idiomatic expressions. Idiomatic expressions often do not detect the emotion orientation of the words even after including the present text based psychological knowledge. The current model cannot deal with these types of errors. We leave their solutions as future work.

Table 2. Classification accuracies (%) for different feature combinations

Features	Fuzzy c-means	SVM	Fuzzy c-means +SVM
PMI	47.45	49.59	51.02
LGr	33.27	46.07	50.23
PMI + LGr	53.55	57.82	59.12
PMI + LGr + A	69.78	70.44	75.72
PMI + LGr + B	61.23	64.88	67.21
PMI + LGr + C	56.87	58.22	60.78
PMI + LGr + D	54.90	57.98	59.14
PMI + LGr + A+B	72.09	74.80	76.06
PMI + LGr + A+B+C	76.22	79.78	81.77
PMI + LGr + A+B+D	73.34	75.80	78.25
PMI + LGr + A+B+C+D	79.77	82.10	84.55

5   Conclusion and Future Work

A method has been proposed for extracting the emotion orientations of words with high accuracy on different types of lexical networks and their weighted versions. There are a number of directions for future work. One is the incorporation of syntactic information and other is to identify solutions for the above mentioned error cases. The textual clues related to psychology may be included to improve the performance.

326 D. Das, S. Poria, and S. Bandyopadhyay

Acknowledgments. The work reported in this paper was supported by a grant from the India-Japan Cooperative Programme (DSTJST) 2009 Research project entitled "Sentiment Analysis where AI meets Psychology" funded by Department of Science and Technology (DST), Government of India.

References

1. Strapparava, C., Valitutti, A.: Wordnet affect: an affective extension of wordnet. Language Resource and Evaluation (2004)
2. Turney, P.D., Littman, M.L.: Measuring praise and criticism: Inference of semantic orientation from association. ACM TIS 21(4), 315–346 (2003)
3. Das, D., Bandyopadhyay, S.: Analyzing Emotional Statements – Roles of General and Physiological Variables. In: The SAAIP Workshop, 5th IJCNLP, pp. 59–67 (2011)
4. Bezdek, J.C.: Pattern Recognition with Fuzzy Objective Function Algoritms. Plenum Press, New York (1981)
5. Joachims, T.: Text Categorization with Support Machines: Learning with Many Relevant Features. In: Nédellec, C., Rouveirol, C. (eds.) ECML 1998. LNCS, vol. 1398, pp. 137–142. Springer, Heidelberg (1998)
6. Scherer, K.R.: What are emotions? And how can they be measured? Social Science Information 44(4), 693–727 (2005)
7. Miller, A.G.: WordNet: a lexical database for English. Communications of the ACM 38(11), 39–41 (1995)
8. Cervantes, J., Li, X., Yu, W.: Support Vector Machine Classification Based on Fuzzy Clustering for Large Data Sets. In: Gelbukh, A., Reyes-Garcia, C.A. (eds.) MICAI 2006. LNCS (LNAI), vol. 4293, pp. 572–582. Springer, Heidelberg (2006)

Technical Term Recognition
with Semi-supervised Learning
Using Hierarchical Bayesian Language Models

Ryo Fujii and Akito Sakurai

Keio University,
3-14-1 Hiyoshi, Kohoku-ku,
Yokohama, Kanagawa, Japan
{roy,sakurai}@ae.keio.ac.jp

Abstract. To recognize technical term, term dictionaries or tagged corpora are required, but it will take much cost to compile them. Moreover, the terms may have several representations and new terms may be developed, which complicates the problem further, that is, a simple dictionary building can't solve the problem. In this research, to reduce the cost of creating dictionaries, we aimed at building a system that learns to recognize terminology from small tagged corpus using semi-supervised learning. We solved the problem by combining a tag level language model and a character level language model based on HPYLM.

We performed experiments on recognition of biomedical terms. In supervised learning, we achived 65% F-measure which is 8% points behind the best existing system that utilizes many domain specific heuristics. In semi-supervised learning, we could keep the accuracy against reduction of supervised data better than exisiting methods.

1 Introduction

Term recognition or term extraction is not only one of the major research fields in natural language processing, but also a fundamental technology for improving accuracy of syntax analysis. Technical term recognition is a specific field where technical term dictionaries could be a solution, but new technical terms are invented on a daily basis, thus it is impossible to deal with such new technical terms only by simply preparing dictionaries in real applications.

An automatic technical term recognition is a solution. Assuming that we have a tagged corpus of the target technical terms, we solve the problem as we solve general tagging problems. For conventional methods, we need to prepare sufficent amount of domain specific corpus, which is costly, and we need some heuristics to obtain useful features, which are again domain specific.

A problem which seems to be specific but actually general in technical term recognition is that the terms are very often compound words, sometimes comprising of three or more words. The words in compound words may be new spelling or commonly used words. In the latter case, we have to consult context

G. Bouma et al. (Eds.): NLDB 2012, LNCS 7337, pp. 327–332, 2012.

to recognize the commonly used words as part of technical terms. To view such words as part of technical terms correctly, not only the term, we need to consider also a wide range of context around the term. However, the existing methods [1][2] basically assume the 2-gram model which is insufficient when it comes to solving the above mentioned problems.

In this paper, we propose to combine two types of **Variable-order Hierarchical Pitman-Yor Language Model** (VPYLM)[3], one of which is for tag sequences that will find an appropriate context to recognize common words in a technical term, and the other is for character sequences that will find useful character strings (prefixes or suffixes) to recognize unknown technical terms.

2   HPYLM and VPYLM

The Hierarchical Pitman-Yor Language Model (HPYLM)[4] is an n-gram language model based on Pitman-Yor language process.

In HPYLM, the depth of the hierarchy is common and fixed for all the contexts. As a result, where n becomes greater, the size of the CRP tree of HPYLM becomes greater in the order of Σ^n . This fact has virtually made the computation of a high dimensional n-gram HPYLM impossible. VPYLM[3] eases this restriction by considering one more probability process on branches.

To explain VPYLM we use a Chinese restaurant process. When a new customer comes to a VPYLM, the customer goes down as in HPYLM but may land at the node earlier. Where to stop is determined according to $p(v)$ defined below.

The probability of passing the node $q(h)$ and that of arriving at the node $1 - q(h)$ can naturally be estimated from the number of passes of each node $t(h)$ and the number of arrivals $s(h)$. With the assumption that all the number of passes of the node and the number of arrivals at the node have as a prior distribution the beta distribution, the above mentioned probabilities are estimated as $q(h) = \frac{t(h)+\alpha_h}{s(h)+t(h)+\alpha_h+\beta_h}$. Therefore the probability $p(h_l)$ that w is observed at the node h is given by $p(h_l) = (1 - q(h_l)) \prod_{i=0}^{l-1} q(h_i)$ where h_l is a lower level context or subcontext of h with the length $l(\leq L)$, where the entire length of h is L.

In addition, by introducing a certain small probability threshold ϵ, we could stop the growth of tree, i.e., stop branching, at the node such that $\epsilon > p(h_l)$ by forcing the customer to arrive at the node. By doing so, we could control the growth of the CRP tree to obtain a reasonable size tree with good accuracy.

3   Combining Language Models

We consider the problem of technical term recognition in the framework POS tagging. We suppose that a technical term has a special POS tag, e.g. ⟨protein⟩ for protein names. Under the framework, we could use context which is a sequence of tags before a target tag to properly recognize compound words.

A simple way of applying HPYLM to POS tagging including tagging technical terms is to use it in place of HMM to predict the next tag probabilities based on its preceding tag n-gram probability. It would be expected that the HPYLM predict tag probabilies better than HMM in a sense that HPYLM smoothes infrequent tag n-grams. This model is to be referred to as **tag-HPYLM** (tH).

We think technical terms have their features in their spelling and are reliably predicted based on the features. To incorporate the knowledge of features by machine learning, we propose to build a character-level HPYLM which predicts the probability for word to have a POS or to be a technical term based on their spellings. By combining the tag-HPYLM and the character-level HPYLM, for predicting tag-sequence probability and for predicting word emission probability would get better prediction of probability. The combined model is to be referred to as **nested tag-HPYLM** (ntH). The character-level HPYLM was proposed, combined with word-level HPLYM, and is called NPYLM in [5].

The difference of our proposal with NPYLM is that we introduce some specific tokens in character-level HPYLM, we use VPYLM for tag-HPYLM, and consequently we need to invent new forward filtering-backward sampling scheme.

3.1 Character-Level HPYLM

Since we could not foresee the appropriate length of characters to be used for character-level HPYLM and the length could be long (maybe longer than five characters), we use VPYLM instead of HPYLM as in [5].

Using the character-level HPYLM, probability of word emission $Q(w) = \prod_{i=1}^{L} P_{char}(c_i|S_{i-1})$ where w spelled $\langle c_1, c_2, \ldots, c_L \rangle$ and S_i is $\langle c_1, c_2, \ldots, c_i \rangle$, $P_{char}(c|S)$ is probability for character c with context S in character-level HPYLM.

3.2 Introduction of Tokens

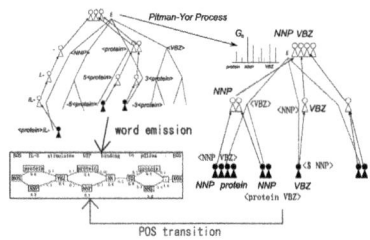

Fig. 1. Nested Tag-HPYLM

The purpose of character-level VPYLM is to automatically learn characteristic spelling specific in technical terms. The specific spellings are mainly observed in prefixes and suffixes and we need to somehow specify some words are technical terms and some words are not. To solve the problem, we propose to insert some tokens specifying POS or that the following word is a specific technical term and that the preceding word is a specific one.

A token representing "protein" is inserted before and after every word or word strings that mean protein. Now word emission probability from tag t $P_{emit}(w|t)$ is replaced by $Q(w_t)$, where w_t is $\langle t, c_1, \ldots, c_L, t \rangle$. Assume "IL-5" is a protein name and our character-level VPYLM will learn that the trigram $\langle protein, \mathrm{I}, \mathrm{L} \rangle$ is a frequent trigram. When the VPYLM is asked $Q(\langle protein, \mathrm{I}, \mathrm{L}, -, 2 \rangle)$, it would return far larger probability to $\langle protein, \mathrm{I}, \mathrm{L}, -, 2 \rangle$ than to $\langle noun, \mathrm{I}, \mathrm{L}, -, 2 \rangle$. Then we would say that "IL-2" would be a protein name rather than a commonly used name.

Additinally, the base distribution in POS-VPYLM is estimated from character-VPYLM, For character-VPYLM, every string starts with a POS token, so distribution after "beginning of string" is equal to tag unigram distribution.

4   Distant-Sampling

We use common forward filtering-backward sampling algorithm for tagging. Let $\alpha[i][j]$ be the forward probability that the j-th tag t_j is assigned the i-th word w_i, which is calculated by:

$$\alpha[i+1][j] = P_{\text{emit}}(w_{i+1}|t_j) \sum_{H \in \Gamma_i} P_{\text{tran}}(t_j|H) \cdot \alpha[i][\text{tail}(H)] \tag{1}$$

where H runs in the context Γ_i of a word w_{i+1}, and $\text{tail}(H)$ is the rightmost tag of H. It is difficult to compute this bacause Γ_i is a set of all the possible tag n-grams and its size is of exponentiaition of n.

The trick we adopted is to sample one context H_{ij} by Algorithm 1 for the word w_i at position i and a possible t_j, and not consider all possible context in H in the expression (1). Then (2) will be used for the forward probability.

Algorithm 1. forward context sampler for distant-sampling

---

input: i, j, $\alpha_k^*[l][n]$ (for $0 \leq l < i$, $0 \leq n \leq m$, $0 \leq k \leq m$)
output: H_{ij}

    $H \Leftarrow \{t_{ij}\}, k \Leftarrow j$
    for $l = i-1, i-2, \ldots, 1$ **do**
        sample $j^* \sim P_{\text{tran}}(t_k|H_{ij}) \cdot \alpha^*[i][j]$
        $H^* \Leftarrow \{t_{lj^*}\} + H$
        if exists tag-VPYLM node of H^* **then**
            $H \Leftarrow H^*, k \Leftarrow j^*$
        else
            return H
        end if
    end for

$$\alpha^*[i+1][j] = P_{\text{emit}}(w_{i+1}|t_j) \sum_{k=1}^{m} P_{\text{tran}}(t_j|H_{ik}) \cdot \alpha^*[i][k] \tag{2}$$

There are two reasons that this sampling works well. First, since backoff smoothing is implied in VPYLM, even if an inappropriate tag is sampled in H, its effect is mitigated by the probability estimates of those shorter contexts.

Second, the depth of CRP tree is controlled by ϵ the stopping condition of branching as is described in 2. Therefore the context which will be sampled is among those with relatively high probabilities.

5   Experiments

5.1   Supervised Technical Term Recognition

With the same assumption of the competition held by Natural Language Processing in Biomedical and its Applications (NLPBA)[1], we conducted learning by using all data given as supervised data in order to compare the accuracy of the proposed method with the existing methods. We added POS tags to learning data except technical terms, trained the infinite-ntH with test data as they are, and evaluated the tagging results by applying the evaluation script distributed.

NLPBA dataset uses two kinds of tags, one is with prefix "B-", which means the beginning of a term, and other is with prefix "I-". But this expression is not appropriate for our model because the model will treat middle and tail words of a term. To avoid this problem, we defined a trivial mapping from these tags to four kinds of tags, "L-/R-" for left/right boundary, "I-" for middle words, and "S-" for just a single word expression of a term.

Table 1 shows the results of this system. Our result is better than the competition baseline and [7], almost equal to [6] but worse than [8]. [6] uses plenty of abstracts from MEDLINE to compute word similarity. [8] uses Penn Treebank and domain specific techniques. We would argue that our proposed method outperforms a method with domain specific techniques but could not compete well with the well-engineered domain specific techniques and supplementary corpus.

Table 1. Result for NLPBA test

model	complete match			right boundary match			left boundary match		
	precision	recall	F-score	precision	recall	F-score	precision	recall	F-score
Zho04[1]	0.6942	0.7599	0.7255	0.7527	0.8239	0.7867	0.7253	0.7939	0.7580
Zha04[6]	0.6098	0.6907	0.6478	0.6889	0.7803	0.7318	0.6570	0.7442	0.6979
Lee04[7]	0.4759	0.5080	0.4914	0.5485	0.5854	0.5664	0.5389	0.5752	0.5564
infinite-ntH	0.6092	0.6881	0.6462	0.6952	0.7852	0.7374	0.6554	0.7402	0.6953

5.2   Technical Term Recognition by Using Semi-supervised Learning

We examined to what extent the accuracy could be maintained when unsupervised data is supplied instead of supervised data in NLPBA. From the data set used in the previous experiments, without adding any change in test data, we randomly separated it supervised and unsupervised data.

Table 2 shows the result of the test with semi-supervised learning. This shows us that for the reduced supervised data the accuracy is kept relatively well. The tagging experiment on the GENIA corpus by CRF that conducted semi-supervised learning [9], shows only precision higher than our results which may indicate their learning is biased to higher precision. In our experiments F-score does not reduce much when even the supervised data decreases from 50% to 20% which suggests our proposed model is robust to reduction of supervised data.

---

[1] http://research.nii.ac.jp/~collier/workshops/JNLPBA04st.htm

Table 2. Result of NLPBA test with semi-supervised learning

	complete match			right boundary match			left boundary match		
supervised	precision	recall	F-score	precision	recall	F-score	precision	recall	F-score
50%	0.5565	0.6538	0.6012	0.6355	0.7466	0.6866	0.6153	0.7228	0.6647
40%	0.5457	0.6421	0.5900	0.6273	0.7382	0.6782	0.6037	0.7105	0.6528
30%	0.5331	0.6159	0.5715	0.6141	0.7094	0.6583	0.5986	0.6915	0.6417
20%	0.5120	0.6041	0.5543	0.5953	0.7025	0.6445	0.5778	0.6818	0.6255
10%	0.4704	0.5683	0.5148	0.5613	0.6781	0.6142	0.5376	0.6495	0.5883

Therefore, as we expected, infinite-ntH was confirmed to be a model that gives accurate tags from a reduced amount of supervised learning data, by using semi-supervised learning.

6   Conclusion

We successfully combined a tag-level and a character-level Bayesian language model to acheive high performance in technical term recognition without any domain speicific knowledge and tools. One of the ideas for success is to insert tokens into character-level corpus to extract automatically prefixes and suffixes that are specific for technical terms. The other is an introduction of computationally efficient but accurate forward filtering-backward sampling method.

References

1. Zhou, G.D., Su, J.: Exploring deep knowledge resources in biomedical name recognition. In: JNLPBA 2004, pp. 96–99. ACL (2004)
2. Song, Y., Kim, E., Lee, G.G., Yi, B.-K.: Posbiotm-ner in the shared task of bionlp/nlpba 2004. In: JNLPBA 2004, pp. 100–103. ACL (2004)
3. Mochihashi, D., Sumita, E.: The infinite markov model. In: NIPS 2007, pp. 1–2 (2007)
4. Teh, Y.W.: A hierarchical bayesian language model based on pitman-yor processes. In: ACL, pp. 985–992 (2006)
5. Mochihashi, D., Yamada, T., Ueda, N.: Bayesian unsupervised word segmentation with hierachical language modeling. ACL 1(36), 49 (2009)
6. Zhao, S.: Name entity recognition in biomedical text using a hmm model. In: JNLPBA 2004, pp. 84–87. ACL (2004)
7. Lee, C., Hou, W.J., Chen, H.-H.: Annotating multiple types of biomedical entities: a single word classification approach. In: JNLPBA 2004, pp. 80–83. ACL (2004)
8. Park, K.-M., Kim, S.-H., Lee, K.-J., Lee, D.-G., Rim, H.-C.: Incorporating lexical knowledge into biomedical ne recognition. In: JNLPBA 20, pp. 76–79. ACL (2004)
9. Jiao, F., Wang, S., Lee, C.-H., Greiner, R., Schuurmans, D.: Semi-supervised conditional random fields for improved sequence segmentation and labeling. In: ACL, pp. 209–216 (2006)

An Experience Developing a Semantic Annotation System in a Media Group[*]

Angel L. Garrido[1], Oscar Gómez[1], Sergio Ilarri[2], and Eduardo Mena[2]

[1] Grupo Heraldo - Grupo La Información, Zaragoza - Pamplona, Spain
{algarrido,ogomez}@heraldo.es
[2] IIS Department, University of Zaragoza, Zaragoza, Spain
{silarri,emena}@unizar.es

Abstract. Nowadays media companies have difficulties for managing large amounts of news from agencies and self-made articles. Journalists and documentalists must face categorization tasks every day. There is also an additional trouble due to the usual large size of the list of words in a thesaurus, the typical tool used to tag news in the media.

In this paper, we present a new method to tackle the problem of information extraction over a set of texts where the annotation must be composed by thesaurus elements. The method consists of applying lemmatization, obtaining keywords, and finally using a combination of Support Vector Machines (SVM), ontologies and heuristics to deduce appropriate tags for the annotation. We have evaluated it with a real set of changing news and we compared our tagging with the annotation performed by a real documentation department, obtaining very good results.

Keywords: Semantic tagging and classification, Information Extraction, NLP, SVM, Ontologies, Text classification, Media, News.

1 Introduction

In almost every company in the media industry, activities related to categorization can be found: news production systems must filter and sort the news at their entry points, documentalists must classify all the news, and even journalists themselves need to organize the vast amount of information they receive. With the appearance of the Internet, mechanisms for automatic news classification often become indispensable in order to enable their inclusion in web pages and their distribution to mobile devices like phones and tablets.

To do this job, medium and big media companies have documentation departments. They label the published news, and the typical way to do that is by using thesauri. A thesaurus [1] is a set of items (words or phrases) used to classify things. These items may be interrelated, and it has usually the structure of a hierarchical list of unique terms.

---

[*] This research work has been supported by the CICYT project TIN2010-21387-C02-02.

G. Bouma et al. (Eds.): NLDB 2012, LNCS 7337, pp. 333–338, 2012.

In this paper, we focus on the inner workings of the NASS system (*News Annotation Semantic System*), which provides a new method to obtain thesaurus tags using semantic tools and information extraction technologies. The seminal ideas of this project have been recently presented in [2] and with this paper we want to delve deeper into the operation of the system.

In our system we propose to obtain the main keywords from the article by using text mining techniques. Then, using Natural Language Processing (NLP) [3], the system retrieves other type of keywords called *named entities*. After, NASS applies Support Vector Machines (SVM) [4] text classification in order to filter articles. The system uses the keywords and the named entities of the filtered texts to query an ontology about the topic these texts are talking about, using the answers to those queries to increase the probability of obtaining correct thesaurus elements for each text, and it updates that matching score in a table. Finally, NASS looks at this table and selects the terms with a score higher than a given threshold, and then it labels the text with the corresponding tags. We have tested this method over a set of thousands of real news published by a leading Spanish media. As we had the chance to compare our results with the real tagging performed by the documentation departments, we have benefited from this real-world experience to evaluate our method.

This paper is structured in the following way. Section 2 explains our method. Section 3 discusses the results of our experiments. Section 4 cites some related work about news categorization. Finally, Section 5 provides our conclusions.

2   NASS Methodology

The outline of our method is as follow. First, NASS obtains from the text of every piece of news the names of the main characters, places and institutions, and then it guesses which the key ideas and major themes are. Second, the system uses this information in order to find related thesaurus terms. Finally, NASS assigns corresponding thesaurus terms to the text. The architecture of our solution which is shown in Figure 1, fits the general schema of an Ontology-Based Information Extraction (OBIE), as presented by Wimalasuriya and Dou [5].

Before obtaining keywords, NASS lemmatizes all the words in the text. Lemmatization can be defined as the process of obtaining the canonical form representing one word, called *lemma*. This procedure simplifies the task of obtaining keywords and reduces the number of words the system has to consider later. It can also help to obviate *stop words*: prepositions, conjunctions, articles, numbers and other meaningless words. Moreover, it also provides us clues about the kinds of words appearing in the text: nouns, adjectives, verbs, and so on. For this purpose, we have used Freeling [6]. Freeling is an open source suite of language analyzers developed at TALP Research Center, at BarcelonaTech (Polytechnic University of Catalunya). After this process, the system has a list of significant words, which is the input to the next process: obtaining keywords.

We propose merging two methods to obtain a set of significant keywords from a given text. The first method that NASS applies is a simple term frequency

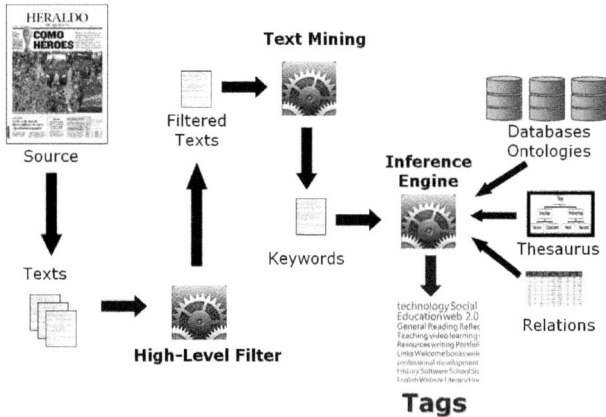

Fig. 1. General architecture of the NASS System

algorithm, but with some improvements. We call it *TF-WP* (*Term Frequency – Word Position*), which is obtained by multiplying the frequency of a term with a position score that decreases as the term appears for the first time towards the end of the document. This heuristic is very useful for long documents, as more informative terms tend to appear towards the beginning of the document. The TF-WP keyword extraction formula is as follows:

$$TF - WP = (\frac{1}{2} + \frac{1}{2} * \frac{nrWords - pos}{nrWords}) * TF$$

$$TF = \frac{nrRepetitions}{nrWords}$$

where *nrWords* is the total number of terms in the document, *pos* is the position of the first appearance of the term, *TF* is the frequency of each term in the document, and *nrRepetitions* is the number of occurrences of that term.

The second method is the well-known *TF-IDF* (*Term Frequency – Inverse Document Frequency*), based on the word frequency in the text but also taking into account the whole set of documents, not only the text considered. We have merged both methods by adding the two values TF-WP and TF-IDF after applying them a weight α and β, respectively. At the end of this task NASS obtains a list of keywords with their number of repetitions and weights.

In news, it is very common to find names of people, places or companies. These names are usually called *named entities* [7]. For example, if the text contains the words "Dalai Lama" both words could be considered as a single named entity to capture its actual meaning. This is a better option than considering the words "Dalai" and "Lama" independently. To do this task we have used Freeling too, which uses this identification method and provides a confidence threshold to decide whether to accept a named entity or not. In our system, this threshold is set at 75% in order to ensure a good result. NASS retrieves the more relevant named entities identified, replaces whitespaces by underscores (for example, "Dalai_Lama"), and finally adds them to the same collection of keywords obtained before.

At this point, NASS has a list of the most important keywords and named entities in the input article. It could tag the text with this information, but we want to take a step forward by using the thesaurus for tagging. The question now is: how could NASS obtain thesaurus terms from a list of keywords, when the words that best summarize the text are not present in it? There are a lot of interesting ways to *infer and deduce* terms according to their meaning and their relationships with other terms. We have chosen SVM text categorization because it is a powerful and reliable tool for text categorization [4]. Regarding the type of SVM used, we have used a modified version of the Cornell SVM-Light implementation with a Gaussian radial basis function kernel and the term frequency of the keywords as features [8].

However, we have discovered some limitations as soon as we apply SVM over real news sets. SVM has a strong dependence on the data used for training. While it works very well with texts dealing with highly general topics, it is not the case when we need to classify texts on very specific topics not included in the training stage, or when the main keywords change over time. So, we have improved the SVM results by using techniques from Ontological Engineering [9]. We advocate the use of knowledge management tools (ontologies, semantic data models, inference engines) due to the benefits they can provide in this context.

The first step is to design an ontology that describes the items we want to tag, but this is not an easy task in media business. The reason is that media cover many different themes, and therefore it is a disproportionate task to try to develop an ontology about all the publishable topics. We think that a better approximation is to design an ontology for every interesting subject we want to tag, whenever SVM results are not able to get good results. Our ontologies are not only composed by named entities, as we have also introduced relation-ships and actions that join concepts providing semantic information, which is a substantial difference that brings advantages not contemplated by SVM.

NASS tries to match each keyword with one of the words in the ontology. For this, we have prepared in advance a table with the help of the documenta-tion department and based on experimental and statistical analysis of the tags introduced manually. This table has two columns. In the first column we put ontology concepts. In the second column we put the probability of talking about a topic usually labeled with a term of the thesaurus when the system detects that concept in a text. Then, NASS submits SPARQL queries against the on-tology and it uses Jena as a framework and Pellet as an inference engine. As soon as it finds a keyword that matches a term of the ontology, it looks at its associated concept and then the system uses the previous table to retrieve the corresponding probability. At this point, it is important to mention that some keywords could be related with one or more thesaurus tags, and also a thesaurus tag could be related to one or more keywords. NASS increases the probability of tagging the article with a term each time it accesses a row of this table. A high number of accesses to the same thesaurus term guarantees that it can be used as a tag on the article. Through an extensive experimental evaluation we found useful to use 60% as a threshold to accept a term. Finally, NASS returns the thesaurus tags obtained by this method in order to label the text.

3 Experimental Evaluation

In our experiments we have used a corpus of 1755 articles tagged with thesaurus terms manually assigned by the documentation department of the Spanish company *Grupo Heraldo*. We found that the highest number of well-labeled articles occurs when we suitably populate the ontology, and if NASS fails it is due to a lack of semantic information. When the ontology has fewer elements, then the recall drops significantly (73% vs. 95%) but the precision is the same (98%). Summing up, the results obtained were really good. Furthermore, our system was able to detect additional labels that are relevant (even though they were not selected in the manual annotation) and avoid labels that were wrongly chosen by the documentation department.

4 Related Work

As commented along the paper, among other techniques, we have used a method commonly used for text categorization: the Support Vector Machine, or SVM [10]. This method has some appropriate features to face text classification problems, for example its capability to manage a huge number of attributes and its ability to discover which of them are important to predict the category of the text after a training stage. It is based on solid mathematical principles related to statistical learning theory and there are plenty of articles and books based on such mechanisms, not only for text classification but for any work related to cataloging all kinds of elements and entities.

As an example of project combining SVM and ontologies we can reference [11], a recent work where SVM is used to categorize economic articles using multi-label categorization. The big difference is that they use ontologies to create labels prior to the categorization process, and then use different types of SVM only in that process, which does not let them obtain neither a high degree of accuracy nor a high number of categories, whereas our proposal avoids these problems. We would also like to mention the work performed by Wu et al. [12], which faced this kind of problems using a quite interesting unsupervised method based on a Naive-Bayes classifier and natural language processing techniques.

5 Conclusions and Further Work

In this paper, we have presented a tagging system that helps media companies in their daily labeling labor when they use a thesaurus as the annotation tool.

We have performed experiments on real news previously tagged manually by the staff of the documentation department and our experimental results show that we are able to get a reasonable number of correct tags using methods like SVM, but the accuracy improves with the combined use of NLP and semantic tools when the training set of the SVM must be updated frequently. Instead of having to label news each year to create a reliable set for training, we propose that documentalists fill the instances of classes in a predefined ontology. Then, our system has enough information to label the news automatically by using semantic tools. We found this simpler and more intuitive for end users, and it helps to get better results. Besides, the accuracy of the automatic assignment of tags with our system is very good, obtaining 99% of correct labels. In fact, the current version of NASS is already being successfully used in several companies. In the future, we plan to introduce new and more powerful methods to enrich the system providing it with greater speed, wider scope and better accuracy.

References

1. Gilchrist, A.: Thesauri, taxonomies and ontologies: an etymological note. Journal of Documentation 59(1), 7–18 (2003)
2. Garrido, A.L., Gomez, O., Ilarri, S., Mena, E.: NASS: news annotation semantic system. In: Proceedings of ICTAI 2011, International Conference on Tools with Artificial Intelligence, pp. 904–905. IEEE (2011)
3. Smeaton, A.F.: Using NLP or NLP resources for information retrieval tasks. Natural Language Information Retrieval. Kluwer Academic Publishers (1997)
4. Joachims, T.: Text Categorization with Support Vector Machines: Learning with Many Relevant Features. In: Nédellec, C., Rouveirol, C. (eds.) ECML 1998. LNCS, vol. 1398, pp. 137–142. Springer, Heidelberg (1998)
5. Wimalasuriya, D.C., Dou, D.: Ontology-Based Information Extraction: an introduction and a survey of current approaches. Journal of Information Science 36(3), 306–323 (2010)
6. Carreras, X., Chao, I., Padró, L., Padró, M.: Freeling: an open-source suite of language analyzers. In: Proceedings of the 4th International Conference on Language Resources and Evaluation, pp. 239–242. European Language Resources Association (2004)
7. Sekine, S., Ranchhod, E.: Named entities: recognition, classification and use. John Benjamins (2009)
8. Leopold, E., Kindermann, J.: Text categorization with support vector machines. How to Represent Texts in Input Space? Machine Learning 46, 423–444 (2002)
9. Fernandez-Lopez, M., Corcho, O.: Ontological engineering. Springer (2004)
10. Cortes, C., Vapnik, V.N.: Support-vector networks. Machine Learning 20(3), 273–297 (1995)
11. Vogrincic, S., Bosnic, Z.: Ontology-based multi-label classification of economic articles. Computer Science and Information Systems 8(1), 101–119 (2011), ComSIS Consortium
12. Wu, X., Xie, F., Wu, G., Ding, W.: Personalized news filtering and summarization on the web. In: Proceedings of ICTAI 2011, International Conference on Tools with Artificial Intelligence, pp. 414–421. IEEE (2011)

From Requirements to Code: Syntax-Based Requirements Analysis for Data-Driven Application Development

Goran Glavaš, Krešimir Fertalj, and Jan Šnajder

Faculty of Electrical Engineering and Computing,
University of Zagreb, Unska 3, 10000 Zagreb, Croatia
{goran.glavas,kresimir.fertalj,jan.snajder}@fer.hr

Abstract. Requirements analysis phase of information system development is still predominantly human activity. Software requirements are commonly written in natural language, at least during the early stages of the development process. In this paper we present a simple method for automated analysis of requirements specifications for data-driven applications. Our approach is rule-based and uses dependency syntax parsing for the extraction of domain entities, attributes, and relationships. The results obtained from several test cases show that hand-crafted rules applied on the dependency parse of the requirements sentences might offer a feasible approach for the task. Finally, we discuss applicability and limitations of the presented approach.

1 Introduction

Majority of software engineering projects start with requirements analysis. These requirements are often expressed in natural language (NL) since it is very often the only mean of communication understood both by the customer and the system analyst. Requirements analysis is a procedure that translates NL representation of the system into some formal representation (i.e., model of a system) used for further system development. Requirements written in NL, however, often suffer from inaccuracy, incompleteness, and ambiguity. This can be remedied to a certain extent by constraining language constructs used in writing requirements [3,9]. However, some domain knowledge might not be expressible due to extensive language reduction. Formal requirements (e.g., first-order logic or UML) are unambiguous, but customers rarely have the technical knowledge to create them.

In this paper, we focus on processing NL requirements for small-scaled data-driven software applications. We define three different concepts crucial for any data-driven application as items to be recognized in NL requirements: domain entities, their attributes, and relationships. Our approach relies on dependency parsing of NL. Dependency parsing allows us to define extraction patterns for entities, attributes, and relations in a straightforward way and we consider it especially suitable for the task. Our target model is a simple, data-centered

G. Bouma et al. (Eds.): NLDB 2012, LNCS 7337, pp. 339–344, 2012.

domain model, which can be easily translated to program code. The procedure itself is language-independent, but it relies on language-specific natural language processing (NLP) tools.

The rest of the paper is structured as follows. Section 2 presents the related work. Rule-based extraction relying on dependency parsing is explained in Section 3. In Section 4 we report the evaluation results on several test cases. The conclusions and ideas for future work are outlined in Section 5.

2   Related Work

In an attempt to automate the translation of natural language requirements into formal models, researchers have dominantly been introducing intermediate models containing both linguistic characteristics and elements of the targeted formal model. Because the complexity of translation of NL specification into any formal model is considerable, many consider only the specifications written in a constrained NL. In [7] NL constructs are effectively limited to context-free grammar. Parsing ambiguities are resolved using a corpora of frequency-weighted parts of speech. The authors propose a formalism referred to as Two-Level Grammar (TLG) to be used as an intermediate model. Methodology described in [8] considers semi-automated approach to identifying conceptual object-oriented constructs. However, the recognition of concepts is based purely on POS-tagging which might be over-simplistic. In [4] authors consider requirements written in unconstrained natural language. POS-tagging and chunking are used in order to build triplets of the form *(subject, predicate, object)*. Verbal phrases are chosen as predicates, and the closest noun phrases as subjects and objects. Triplets are then combined into intermediate models, which are ultimately translated into three formal target models: domain, activity, and use case model.

Our idea was to exploit current state-of-the-art NLP tools, primarily statistical dependency parsing. Unlike many related work, we consider unconstrained natural language and our method does not require any precompiled lexicons nor vocabularies. Our method is designed primarily for data-driven applications development which, admittedly, simplifies the process of automated specification analysis by narrowing down the number of constructs to be recognized. While many of the approaches in the related work focus on fully automating the analysis process, we propose the approach that allows both fully automated and semi-automated execution. We believe that human intervention between automated steps can significantly improve the accuracy of the results while the overall human effort is still being rather low.

3   Syntax-Based Requirements Analysis for Data-Driven Applications

The overall workflow of our syntax-based method is depicted in Fig. 1. The input of the process is the requirements specification document of a data-driven system,

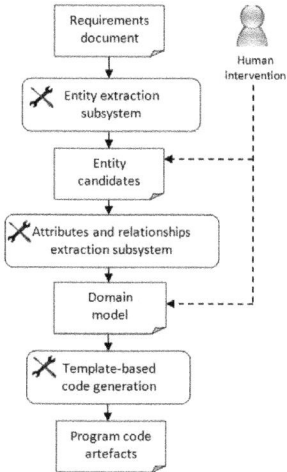

Fig. 1. Syntax-based requirements analysis workflow

written in unconstrained natural language. The automated entity extraction step generates a collection of entity candidates. In the semi-automated setting the list of automatically generated entity candidates can be modified by the domain expert (e.g., corrected by manual elimination of spurious entities). The collection of entities then serves as an additional input for the second step in which attributes and relationships are extracted. In the fully automated mode, each candidate is assigned a confidence score. The idea is to give false positive candidates low confidence scores and have them automatically removed from the list using the confidence thresholding.

3.1 Entity Extraction

In order to extract entity candidates from textual specification we define a set of rules (i.e., patterns) by which words or phrases are identified as possible domain entities. All extraction patterns were defined using dependency parses of requirements sentences. Each dependency relation is a triplet of the form (*relation, governor, dependent*) and it denotes the type of grammatical relation between the governing word and the dependent word [6,2]. It is presumed that the domain expert will recognize and eliminate false positives easily, whereas she might have difficulties noticing that some domain entities are missing. In order to decrease the number of false negatives, we consider every noun phrase that appears in the role of a subject in at least one of the sentences in the specification to be an entity candidate. If the word or phrase is an important domain concept (i.e., domain entity), it should be well described, which means it will probably appear as a subject of more than one sentence. In the fully automated approach, however, we cannot count on human help with false positives elimination. For this purpose we define heuristic rules based on which we estimate the confidence

Table 1. Rules for entity extraction

Rule name	Confidence	Example
Non-verbal predicate	decrease	The *name* is mandatory.
Determination	increase	Each *author*; Every *gallery*.
Existence of a direct object	increase	Each *author* has a collection of paintings.
Possession	decrease	The *name* of the theme.

Table 2. Rules for attribute and relationship extraction

Rule name	Example
Subject-object relationship	The *painting* has a unique **name**.
Clausal complement with internal subject	The **year** in which the *painting* was created.
Possession	*Author's* **name**; The **name** of the *painting*.

of some entity candidate being a genuine domain entity. Each rule is defined as a specific pattern of dependency relations. We then search for these patterns in specification sentences. If satisfied, each rule increases or decreases the confidence estimate for the candidate being the true entity. The exact amount of increase or decrease in confidence were determined experimentally, individually for each rule. The rules are shown in Table 1.

3.2 Attributes and Relationships Extraction

Attributes and relationships extraction uses the list of previously extracted entities and the original requirements document as input. Having identified a list of entities facilitates extraction of their attributes and relationships. Attribute candidates are limited to nouns and noun phrases that stand in a specific syntactic relation to previously extracted entities. Relationship candidates are limited to constructs of requirements document sentences in which two previously extracted entities co-occur. This approach also propagates errors from entity extraction level. Attributes that we extract are the data properties of entities (e.g., the *name* of a person). Relationships that we extract are limited to binary conceptual (i.e., entity-relationship) data relationships. In addition to identifying existence of a relationship between two entities, the cardinality of each relationship is extracted as well (e.g., 1:N or N:N). We created several patterns by which we identify attribute and relationship candidates. The patterns for attributes recognition are shown in Table 2.

We used the same set of rules for relationship extraction, only imposing additional requirement of both noun phrases occurring in the collection of previously extracted entities. Cardinalities of relationship ends were determined based on multiplicity tags provided by the POS-tagger. Finally, all pairs of 1:N relationships of opposite directions between the same entities were replaced with a single N:N relationship.

Table 3. Performance results of automated extraction

	Task	Precision (%)	Recall (%)	F1 (%)
	Entity extraction	86.7	96.9	91.3
Fully automated	Attribute extraction	85.2	82.4	83.2
Fully automated	Relationship extraction	82.6	83.6	82.2
Semi-automated	Attribute extraction	89.7	88.1	88.3
Semi-automated	Relationship extraction	91.8	93.0	92.0

4 Experimental Results

We tested the performance of our system on five different requirements documents, created by the graduate students of software engineering, each document describing a small-scale data-driven software system. The models extracted automatically by the system were evaluated against the gold standard domain models created by the students. The two-step nature of our method allows us to evaluate the performance of the attribute and relationship extraction step separately from the entity extraction step which is not possible for the fully automated procedure. We used Stanford dependency parser and POS-tagger for English [1,10]. In the semi-supervised approach, the domain expert was allowed to remove false positives from the entity candidates list produced after the first step. This allows us to evaluate the performance of attributes and relationships extraction separately, without propagating the errors from the entities extraction step. In Table 3 we report the performance of the system on all tasks.

The results indicate that the errors made by the system in the phase of entity extraction are propagated to the phase of attribute and relationship extraction. Expectedly, the precision and recall of the attributes and relationships extraction both increase once the correct set of entities is provided in the semi-automated approach. These results seem promising and indicate that the syntactic parsing is a feasible approach for the analysis of the requirements for data-driven applications. Most of the extraction errors are caused by one of the following reasons: incorrect syntax parse (the Stanford parser we used has the reported accuracy of 86.96% [5]), limited coverage of the hand-crafted extraction rules, and the non-articulation of common domain knowledge in the requirements document. Common domain knowledge was dominantly affecting the performance of relationship extraction because many of the relationships between common entities (e.g., *person* was born in one *city*) are considered to be common knowledge.

5 Conclusions and Future Work

In this paper we presented a simple two-step method for automated translation of natural language requirements to program code. We consider the requirements for data-driven applications and extract entities, their attributes, and relationships. We have shown that the hand-crafted rules applied on the dependency parse of the requirements sentences might offer a feasible approach for the task. Concepts of the domain model seem to appear in syntactically distinguishable patterns. Our research is limited to data-driven applications and binary relationships between entities, but this does not necessarily limit the suitability of the approach outside this scope. The performance on larger requirements specification remains to be tested as well. The set of hand-crafted rules we created seems somewhat limited although adequate for covering most common formulations of domain entities, attributes and relationships.

Replacing the hand-crafted rules with the supervised machine learning techniques is a possible step for future work. Future research will also focus on the extraction of the dynamic aspects of the software systems from the requirements documents, such as recognizing functions of domain entities and the interactions between them.

References

1. De Marneffe, M., Manning, C.: The Stanford typed dependencies representation. In: Proceedings of the Workshop on Cross-Framework and Cross-Domain Parser Evaluation, pp. 1–8. Association for Computational Linguistics (2008)
2. De Marneffe, M., Manning, C.: Stanford typed dependencies manual (2009)
3. Denger, C., Berry, D., Kamsties, E.: Higher quality requirements specifications through natural language patterns. In: Proceedings of the IEEE International Conference on Software, Science, Technology, and Engineering, pp. 80–90. IEEE (2003)
4. Ilieva, M., Ormandjieva, O.: Models derived from automatically analyzed textual user requirements. In: Proceedings of the Fourth International Conference on Software Engineering Research, Management and Applications, pp. 13–21. IEEE (2006)
5. Klein, D., Manning, C.: Accurate unlexicalized parsing. In: Proceedings of the 41st Annual Meeting on Association for Computational Linguistics, vol. 1, pp. 423–430. Association for Computational Linguistics (2003)
6. Kübler, S., McDonald, R., Nivre, J.: Dependency parsing. Synthesis Lectures on Human Language Technologies 1(1), 1–127 (2009)
7. Lee, B., Bryant, B.: Contextual natural language processing and DAML for understanding software requirements specifications. In: Proceedings of the 19th International Conference on Computational Linguistics, vol. 1, pp. 1–7. Association for Computational Linguistics (2002)
8. Overmyer, S., Lavoie, B., Rambow, O.: Conceptual modeling through linguistic analysis using LIDA. In: Proceedings of the 23rd International Conference on Software Engineering, pp. 401–410. IEEE Computer Society (2001)
9. Tjong, S., Hallam, N., Hartley, M.: Improving the quality of natural language requirements specifications through natural language requirements patterns. In: Proceedings of the Sixth IEEE International Conference on Computer and Information Technology. IEEE (2006)
10. Toutanova, K., Klein, D., Manning, C., Singer, Y.: Feature-rich part-of-speech tagging with a cyclic dependency network. In: Proceedings of the 2003 Conference of the North American Chapter of the Association for Computational Linguistics on Human Language Technology, vol. 1, pp. 173–180. Association for Computational Linguistics (2003)

Structuring Political Documents
for Importance Ranking

Alexander Hogenboom, Maarten Jongmans, and Flavius Frasincar

Erasmus University Rotterdam
P.O. Box 1738
NL-3000 DR Rotterdam
The Netherlands
{hogenboom,frasincar}@ese.eur.nl,
maarten.jongmans@gmail.com

Abstract. Today's parliamentary information systems make political data available to the public in an effective and efficient way by moving from the classical document-centric model to a rich information-centric model for political data. We propose a novel approach to exploiting the rich information sources available through such parliamentary information systems for ranking results of a typical query for debates in accordance with their importance, for which we have developed several proxies. Our initial evaluation indicates that debate intensity and key players have an important role in signaling the importance of a debate.

1 Introduction

Since the beginning of this century, people's on-line hunger for information related to politics has increased dramatically. While 18% of all adults in the United States was estimated to consume political news on-line in 2000, 44% of all American adults searched the Web for political news in 2009 [9]. This phenomenon has been taking place in the Netherlands as well, which appears to have caused Dutch political parties to start embracing this new era in which information technology plays an increasingly important role in the political space.

Given the increasing importance of the Web in the political space, one of the main problems is that anyone, e.g., citizens, politicians, political (pressure) groups, or qualified journalists, can publish anything at any time, in such a way that the published information is accessible to anyone. This can result in vast amounts of on-line "news" in which opinions may be represented as facts. As such, it is crucial to have a reliable information source on the Web allowing easy access to information produced in parliament, e.g., meeting notes or voting records. This is of paramount importance, as existing parliamentary information platforms are mainly visited by the public (44%) and businesses (24%) [8]. Such information sources can thus help bridging the gap between the public and the government, while forming a concrete foundation of democracy and providing a starting point for a true e-democracy.

G. Bouma et al. (Eds.): NLDB 2012, LNCS 7337, pp. 345–350, 2012.

Existing systems like PoliDocs [4] have already taken promising steps towards making political information easily accessible to the public. The nature of data published through such systems introduces new challenges for ranking results of queries on political data. As political documents form a natural collection for search tasks in which answers do not typically consist of documents but rather of information hidden in one or more documents [6], recent developments exhibit a tendency of moving from a document-centric to a structured, information-centric model for political data. Such a model enables linking concepts in various political documents, yet the typical incompleteness of these references thwarts the applicability of well-known query result ranking methods such as PageRank [1] and HITS [7]. In our current endeavors, we propose a way of using the information obtained from structured political documents in order to rank debates in accordance with their importance as perceived by political experts.

The remainder of this paper is structured as follows. First, we discuss some related work on political information systems in Sect. 2. Then, Sect. 3 demonstrates how well-structured political information can be used in order to rank debates in accordance with their importance, which is evaluated in Sect. 4. Last, in Sect. 5, we draw conclusions and propose directions for future work.

2   Parliamentary Information Systems

In order for parliamentary information systems to effectively and efficiently disclose political information to the public, political data must be available through an information-centric data model. Unfortunately, most political data has until now been published as unstructured natural language text. Governments have only fairly recently begun to become aware of the need for political data in a more structured data format like XML. Therefore, in order to be able to disclose the wealth of information hidden in past, present, and future parliamentary publications in natural language text, a principal method for converting natural language text into structured political information is of paramount importance.

One of the first attempts of structuring political data is conceptually surprisingly simple. Gielissen and Marx [3] take a collection of political documents in PDF format and convert these documents into an XML format in which each line of the original text is annotated with the properties of its bounding box. By applying some information retrieval heuristics on the text, the resulting XML file is subsequently enriched with additional annotations for concepts of interest in the political domain, e.g., political parties, members of parliament, etcetera.

With parliamentary data being available in a structured format, the possibilities are numerous. For instance, some researchers have explored structured political data for traces of sentiment, by automatically determining subjectivity in parliamentary publications and by subsequently determining the semantic orientation of the identified subjective parts [5]. The potential of structured data in parliamentary information systems has also been demonstrated by the utilization of such data in order to facilitate faceted search [4].

As political data is predominantly queried for information that does not typically consist of full documents but rather of information hidden in one or more documents [6], ranking original political documents for relevance with respect to a query is a far from trivial task. Well-established ranking techniques like PageRank [1] and HITS [7] are suitable for documents or concepts with many interconnections, but the incompleteness of the links in today's structured political data requires a different ranking approach for search results in political data. Ontology-based approaches [2], in which ranking is based on the (lack of) appearance of domain concepts, may seem a suitable alternative. Yet, in political recordings, it may not so much be the concepts that are discussed that make a document relevant, but rather the way in which these concepts are discussed. A ranking method that takes into account this phenomenon is yet to be proposed.

3   Importance Ranking of Debates

In order to make political documents accessible to the public through parliamentary information systems, the information contained by such documents needs to be processed first. We propose to follow a state-of-the-art method [3] by converting political documents into an XML format containing a structure which primarily models the documents' layout and subsequently using heuristics in order to add information into the structure of the XML such that a query mechanism like XQuery can be used to retrieve information to answer specific questions.

The data can be queried by means of techniques exploiting the newly introduced structure. An interesting application is to retrieve information on debates that are possibly interesting, given a user's query. Common methods of ranking query results rely on graphs of interlinked items. Although debating report documents have some references to documents containing, e.g., motions or amendments, these references are typically incomplete, thus rendering such methods less applicable to ranking debates. Therefore, we propose a novel method for ranking debates in accordance with their importance.

We define importance as the probability that the public finds a debate of great significance. We model importance as a three-dimensional construct. The first dimension is the intensity of a debate. In intense debates, people argue and interrupt one another a lot. When this happens, a debate often receives a lot of attention from the media, which suggests its importance to the public.

The second dimension of our importance measure is constituted by the quantity and quality of key players in a debate. Here, quantity refers to the number of participants in a debate, whereas quality is constituted by the people participating in a debate. For instance, the presence of the (deputy) prime minister or the number of participating floor leaders may signal the importance of a debate.

Last, the third dimension of our construct is formed by the debate length. This length refers to how much time is required for a debate. A debate can take very long when parties do not agree with each other or with the executive branch. Alternatively, a debate may consume a lot of time when a topic is so complex and important that it requires a lot of time to be discussed properly.

We propose to operationalize the three dimensions of importance by means of several attributes that can be extracted from the structured parliamentary data by means of a query mechanism like XQuery. We assume the importance I_d of a debate d to be a function of its n attributes $x_{d1}, \ldots, x_{dn}$ and their associated weights $\beta_1, \ldots, \beta_n$, i.e.,

$$I_d = \sum_{i=1}^{n} \beta_i x_{di}, \quad 0 \leq I_d \leq 1, \quad 0 \leq x_{di} \leq 1. \tag{1}$$

The attributes in (1) are distributed over the three dimensions of our importance construct. The intensity of a debate is operationalized as the amount of switches between speakers in a debate. The attributes constituting the dimension of key players are the percentage of attending members of parliament, the presence of the (deputy) prime minister, the percentage of speaking floor leaders, the number of members of parliament speaking in a debate, and the amount of members of the executive branch of the government speaking in a debate. The debate length is operationalized by means of the total number of words spoken in a debate, the time a debate closes, the number of blocks in a debate, and the number of times the executive branch refers to a possible second term.

The range of each variable is limited to the interval $[0, 1]$. We apply min-max normalization to the number of interruptions, such that its minimum is mapped to 0, and its maximum is mapped to 1. The percentage of members of parliament attending a debate is a value between 0 and 1 and as such does not need further normalization. The presence of the (deputy) prime minister is encoded as either 0 (not present) or 1 (present). Furthermore, the percentage of floor leaders speaking is recoded as 1 (percentages higher than 0.9), 0.3 (percentages between 0.6 and 0.9) or 0 (all other cases). If there are less than nine speakers, this recoded value is multiplied with 0.5. Additionally, we apply min-max normalization to the number of members of parliament speaking in a debate. The number of speaking members of the executive branch is recoded to 1 (three or more speaking members), 0.35 (two speaking members), or 0 (all other cases). Furthermore, we apply min-max normalization to the word count as well as to the number of minutes a debate ends after midnight. The number of blocks is encoded as either 0 (ten blocks or less) or 1 (more than ten blocks). Last, second term references are encoded as 0 (no references) or 1 (one or more references).

4   Evaluation

In order to evaluate how our proxies contribute to a debate's importance, we consider five different methods for distributing the attribute weights. First, we consider giving each attribute an equal weight of 0.091. Alternatively, we consider distributing all weight over attributes related to either the intensity dimension (a weight of 1.000 for the single attribute related to this dimension), the key players (where the six attributes are assigned a weight of 0.166), or the debate length (resulting in four attributes with a weight of 0.250). Last, we perform a Multivariate Linear Regression (MLR) analysis in order to optimize the weights.

Our evaluation is performed on a set of parliamentary recordings of 100 debates held in the Netherlands between January 1, 2009, and March 31, 2010, retrieved through PoliDocs [4]. These debates have been structured by means of a state-of-the-art approach [3]. Through a survey, 9 out of 17 approached Dutch political experts (analysts, scientists, etcetera) have assigned each of our 100 selected debates to the *Top 10*, *Top 11–20*, or *Top 21–30* debates, or an *Unordered* category for the remaining 70 debates. We have aggregated the expert rankings by first distributing 100 points over all debates for each survey. Each *Top 10*, *Top 11–20*, and *Top 21–30* debate received four, three, and two points, respectively, and the remaining points were equally distributed over the unranked debates. We have then averaged each debate's score over all nine surveys.

Our data is split into a training set for the MLR analysis (60%) and a test set (40%) for comparing the ranking of debates as produced by our five methods with the ranking made by political experts. This comparison is done by means of a P@k test, where we assess the percentage of the top k debates of our methods occurring in the experts' top k debates, for $k \in \{10, 20, 30\}$.

The weights of our models are reported in Table 1. The number of speaker switches has a relatively high positive correlation with a debate's importance. The presence of the (deputy) prime minister and parliament members, the number of floor leaders speaking, and the debate closing time exhibit a rather positive correlation too. Interestingly, the number of executive branch members speaking exhibits a negative correlation with a debate's importance in our data set.

Table 2 reports the precision of the top 10, 20, and 30 documents returned by our considered models. Of our four dimension-based models, both the model focusing on the intensity dimension and the model focusing on the dimension of key players perform comparably well. This is in line with the observed weight distribution found by our MLR analysis, which emphasizes attributes related to both of these dimensions. However, our MLR model appears to have a more stable overall performance. It outperforms all other models in terms of precision on the top 10 and 20 documents, while exhibiting a performance that is comparable with the other models on the top 30 documents.

Table 1. Attribute weight configurations per model

Attribute	Equal	Intensity	Players	Length	MLR
Speaker switches	0.091	1.000	0.000	0.000	0.195
Parliament members present	0.091	0.000	0.166	0.000	0.133
Prime minister	0.091	0.000	0.166	0.000	0.181
Deputy prime minister	0.091	0.000	0.166	0.000	0.144
Floor leaders speaking	0.091	0.000	0.166	0.000	0.135
Parliament members speaking	0.091	0.000	0.166	0.000	0.046
Executive branch speaking	0.091	0.000	0.166	0.000	-0.132
Word count	0.091	0.000	0.000	0.250	0.107
Closing time	0.091	0.000	0.000	0.250	0.149
Block count	0.091	0.000	0.000	0.250	-0.106
Second term	0.091	0.000	0.000	0.250	-0.050

Table 2. Importance ranking performance per model

Precision	Equal	Intensity	Players	Length	MLR
P@10	20%	30%	30%	10%	40%
P@20	55%	50%	55%	55%	65%
P@30	70%	70%	73%	77%	73%

5 Conclusions and Future Work

We have proposed a novel way of exploiting an information-centric model for political data in order to rank results of a typical query for parliamentary debates in accordance with their importance. To this end, we have developed several proxies for a debate's importance. Our results indicate that debate intensity and key players have an important role in signaling the importance of a debate.

A more extensive survey, in which political experts provide a more detailed ranking of more debates, may bring additional insights into what constitutes an important debate. Another direction for future work is to investigate the possibility of a non-linear relation between our proxies and a debate's importance.

Acknowledgments. The authors are partially supported by the Dutch national program COMMIT.

References

1. Brin, S., Page, L.: The Anatomy of a Large-Scale Hypertextual Web Search Engine. In: 7th International World-Wide Web Conference (WWW 1998), pp. 107–117. Elsevier (1998)
2. Frasincar, F., Borsje, J., Levering, L.: A Semantic Web-Based Approach for Building Personalized News Services. International Journal of E-Business Research 5(3), 35–53 (2009)
3. Gielissen, T., Marx, M.: Exemelification of Parliamentary Debates. In: Ninth Dutch-Belgian Workshop on Information Retrieval (DIR 2009), pp. 19–25 (2009)
4. Gielissen, T., Marx, M.: The design of PoliDocs: a Web Information System for the Disclosure of Dutch Parliamentary Publications. In: Sixth International Workshop on Web Information Systems Modeling (WISM 2009), vol. 461. CEUR-WS (2009)
5. Grijzenhout, S., Jijkoun, V., Marx, M.: Sentiment Analysis in Parliamentary Proceedings. In: Workshop From Text to Political Positions, T2PP 2010 (2010)
6. Kaptein, R., Marx, M.: Focused Retrieval and Result Aggregation with Political Data. Information Retrieval 13(5), 412–433 (2010)
7. Kleinberg, J., Kumar, R., Raghavan, P., Rajagopalan, S., Tomkin, A.: The Web as a Graph: Measurements, Models, and Methods. Springer (1999)
8. Marcella, R., Baxter, G., Moor, N.: The Effectiveness of Parliamentary Information Services in the United Kingdom. Government Information Quarterly 20(1), 29–46 (2003)
9. Smith, A.: The Internet's Role in Campaign 2008. Tech. rep., Pew Internet and American Life Project (2009)

Author Disambiguation
Using Wikipedia-Based Explicit Semantic Analysis

In-Su Kang

School of Computer Science & Engineering, Kyungsung University
Pusan. 608-736 South Korea
dbaisk@ks.ac.kr

Abstract. Author disambiguation suffers from the shortage of topical terms to identify authors. This study attempts to augment term-based topical representation of authors with the concept-based one obtained from Wikipedia-based explicit semantic analysis (ESA). Experiments showed that the use of additional ESA concepts improves author-resolving performance by 13.5%.

Keywords: Author Disambiguation, Explicit Semantic Analysis, Wikipedia, Topical Representation.

1 Introduction

Author disambiguation is to cluster author name occurrences mainly found in bibliography into groups of real-world authors. Its primary difficulties come from the fact that bibliographic records in which author names occur are not enough to provide author-identifying features. Among others, article titles are too short to discover topical association between same-name authors under the assumption that researchers normally attack the same or similar topics during a certain period of time [2].

Recently proposed Wikipedia-based explicit semantic analysis (ESA) enables one to explicitly represent text of any size in the form of a vector of weighted Wikipedia concepts [1]. To date, it has been successfully applied to many applications including text categorization [1], multilingual retrieval [6], etc. Thus, this study attempts to apply ESA to article titles to get additional concept-based topical representation of authors. Evaluations showed that a careful combination of term-based and concept-based topical representations helps to improve author disambiguation performance.

1.1 Related Work

To get better topical representation of authors, Song et al. [7] used the first page of the article to predict author's hidden topics by applying latent Dirichlet allocation. Kang and Na [4] applied automatic feedback to expand the article title with its topically-related titles retrieved from Google Scholar. Unlike these approaches, this study seeks to enrich author's topical description using natural concepts from massive amounts of open encyclopedic knowledge like Wikipedia.

G. Bouma et al. (Eds.): NLDB 2012, LNCS 7337, pp. 351–354, 2012.

2   ESA-Based Author Disambiguation

Let r be a bibliographic record that is normally comprised of co-author names $a_1,\ldots,$ $a_i,\ldots, a_n$, paper-title t, and publication-title p. Suppose that a_i is the name instance to be disambiguated. Author representation R for a_i from r is then defined to be the set of biographic representation $B=\{a_1,\ldots, a_n\}-\{a_i\}$ and topical representation $T_r=\{t, p\}$.

The goal of this study is to enrich short and insufficient topical description T_r with new concepts related to t and p using ESA. When title t is given to ESA as its input, it produces a vector of Wikipedia concepts (documents) each of which quantifies the similarity between the respective concept and t. Let such a vector be T_{ESA}, which can be viewed as another topical representation for a_i.

However, T_r and T_{ESA} have different characteristics. For example, T_{ESA} would be much larger than T_r. In addition, T_{ESA} is considered to be composed of concepts rather than terms T_r is composed of. Considering that the aforementioned topical enrichment should be targeted at improving author disambiguation, certain schemes would be needed to effectively combine such heterogeneous topical representations in calculating author similarities. This study suggests some of such combination schemes using the following author similarity formula [4], where $fsim_k$ is the feature similarity between k-th features of author representations R^i and R^j, and θ_k is a threshold value of k-th feature, and $\delta(p)$ returns 1 if p is true, 0 otherwise,.

$$Sim(R^i,R^j)= \delta\left(\left(\sum_{k=1}^{K}\delta\big(fsim_k\big(R_k^i,R_k^j\big)\geq \theta_k\big)\right)\geq \pi\right) \tag{1}$$

First, BLEND combination scheme simply unify topical representations T_r and T_{ESA} into a single representation $T=T_r \cup T_{ESA}$, ignoring the differences between T_r and T_{ESA}. In this case, the above author similarity equation would be like the following eq.(2), where $fsim_B$ indicates the similarity between biographic representations, and $fsim_T$ calculates the similarity between topical representations.

$$Sim(R^i,R^j)= \delta\big(\big(\delta\big(fsim_B\big(B^i,B^j\big)\geq \theta_B\big)+ \delta\big(fsim_T\big(T^i,T^j\big)\geq \theta_T\big)\big)\geq \pi\big) \tag{2}$$

Next, OR combination scheme separates two topical representations and deals with their similarity contribution equally as in eq.(3).

$$Sim(R^i,R^j)=$$
$$\delta\big(\big(\delta\big(fsim_B\big(B^i,B^j\big)\geq \theta_B\big)+ \delta\big(fsim_T\big(T_r^i,T_r^j\big)\geq \theta_{Tr}\big)+ \delta\big(fsim_T\big(T_{ESA}^i,T_{ESA}^j\big)\geq \theta_{TESA}\big)\big)\geq \pi\big) \tag{3}$$

Finally, SUM combination scheme accumulates independent similarity contributions from two topical representations to get a single topical similarity as in eq.(4).

$$Sim(R^i,R^j)=$$
$$\delta\big(\big(\delta\big(fsim_B\big(B^i,B^j\big)\geq \theta_B\big)+ \delta\big(\big(fsim_T\big(T_r^i,T_r^j\big)+ fsim_T\big(T_{ESA}^i,T_{ESA}^j\big)\big)\geq \theta_T\big)\big)\geq \pi\big) \tag{4}$$

3   Evaluation

To evaluate our method, PSU-CiteSeer-14 dataset[1] was used. It consists of 8,453 bibliographic records for 14 ambiguous English person names. All 8,453 name instances were manually disambiguated into 479 real-world author identifiers. Following the evaluation scheme of Pereira et al. [5], the dataset is randomly divided into train and test sets in halves, and the best parameter values tuned from the train set are used to evaluate the test set. This process is iterated ten times and their averaged F1 scores are reported. For biographic-feature similarity $fsim_B$, the number of common co-author names was computed. For topical-feature similarity $fsim_T$, tf×idf-based cosine similarity was used after Porter-stemming and stop-word removal. To get ESA-based topical representation, Research-ESA[2], an open source implementation of ESA, was utilized with BM25 as ESA retrieval model and fixed-size option as its pruning function. For each bibliographic record of which topical representation is to be enriched, its title with no preprocessing is submitted to Research-ESA, and then the system returns a list of top Wikipedia concepts or categories which corresponds to T_{ESA}.

Table 1. Performance of author disambiguation ($\pi=\theta_B=1$)

Method	Features	Rec.	Pre.	F1	Parameters
HAC	B,TP	0.4583	0.1348	**0.2083**	θ_T=0.61
BFC	B,TP	0.2841	0.5030	**0.3631** (baseline)	θ_T=0.3
BFC	B,ESA(Art-50)	0.2851	0.4663	0.3538	θ_T=0.23
BFC	B,ESA(Cat-50)	0.2975	0.4493	0.3580	θ_T=0.33
BLEND	B,TP,ESA(Art-50)	0.2441	0.5304	0.3344 (-7.9%)	θ_T=0.33
BLEND	B,TP,ESA(Cat-50)	0.2708	0.5058	0.3527 (-2.9%)	θ_T=0.31
OR	B,TP,ESA(Art-50)	0.2966	0.4852	0.3681 (+1.4%)	θ_{T_r}=0.31, θ_{TESA}=0.34
OR	B,TP,ESA(Cat-50)	0.3062	0.4985	0.3794 (+4.5%)	θ_{T_r}=0.31, θ_{TESA}=0.35
SUM	B,TP,ESA(Art-50)	0.3098	0.5017	0.3830 (+5.5%)	θ_T=0.35
SUM	B,TP,ESA(Cat-50)	0.3481	0.5054	**0.4132** (+13.5%)	θ_T=0.36

Table 1 shows evaluation results. B and TP means B and $T_r=\{t, p\}$ in Section 2. ESA(Art-50) and ESA(Cat-50) indicate respectively top 50 Article concepts and top 50 Category concepts obtained from Research-ESA. As the baseline, single-linkage HAC (Hierarchical Agglomerative Clustering) was not good compared to BFC (Biggest-First Clustering) [3]. Thus, all later evaluations were performed using BFC, which, given a graph, iterates the process that identifies a sub-graph g of the largest-degree node and its adjacent nodes from the input graph and assigns a new cluster identifier to each node of g or merges g into one of existing clusters until all nodes in the input graph belongs to a cluster.

Replacing TP with ESA concepts decreased the performance, indicating that ESA concepts should be secondary. Unifying or blending TP terms with ESA concepts was

---

[1] http://clgiles.ist.psu.edu/data/nameset_author-disamb.tar.zip
[2] http://www.multipla-project.org/
research_esa_ui/configurator/index/

not a good strategy, decreasing the baseline performance. A little improvement was obtained from not mixing TP terms from ESA concepts but considering their similarity contribution equally importantly using OR scheme. A further improvement was made by SUM scheme, which adds independent similarity contributions from TP terms and ESA concepts.

Interestingly, Wikipedia category space was better than its article space in representing author topics for all schemes combining ESA concepts with TP terms. This may be due to the fact that the category space is much smaller and more generalized than the article space. In other words, the category space are better suited to be used as a controlled list of research areas authors are interested in, enabling the higher likelihood that paper titles of two same-name authors are matched even with low similarity between title terms. This also explains that the use of ESA concepts increases recall rather than precision as observed in Table 1.

4   Conclusion

This study suggested some techniques to combine term-based and ESA concept-based topical representations for author disambiguation. Empirical evaluations showed that Wikipedia category concepts are more helpful than its article concepts in augmenting author's term-based topics. In the future, it would be needed to enhance author's topical representation through multiple ESAs from several knowledge sources.

Acknowledgments. This research was supported by Kyungsung University Research Grants in 2012.

References

1. Gabrilovich, E., Markovitch, S.: Overcoming the brittleness bottleneck using wikipedia: enhancing text categorization with encyclopedic knowledge. In: Proceedings of the 21st National Conference on Artificial intelligence, pp. 1301–1306 (2006)
2. Han, H., Giles, C.L., Zha, H.: A model-based k-means algorithm for name disambiguation. In: Proceedings of Semantic Web Technologies for Searching and Retrieving Scientific Data (2003)
3. Kang, I.S.: Disambiguation of author names using co-citation. Journal of Information Management 42(3), 167–186 (2011) (in Korean)
4. Kang, I.-S., Na, S.-H.: Disambiguating Author Names Using Automatic Relevance Feedback. In: Kim, T.-h., Gelogo, Y. (eds.) UNESST 2011. CCIS, vol. 264, pp. 239–244. Springer, Heidelberg (2011)
5. Pereira, D., Ribeiro-Neto, B., Ziviani, N., Laender, A.: Using Web information for author name disambiguation. In: Proceedings of the ACM/IEEE Joint Conference on Digital Libraries (JCDL), pp. 49–58 (2009)
6. Potthast, M., Stein, B., Anderka, M.: A Wikipedia-Based Multilingual Retrieval Model. In: Macdonald, C., Ounis, I., Plachouras, V., Ruthven, I., White, R.W. (eds.) ECIR 2008. LNCS, vol. 4956, pp. 522–530. Springer, Heidelberg (2008)
7. Song, Y., Huang, J., Councill, I., Li, J., Giles, C.L.: Efficient topic-based unsupervised name disambiguation. In: Proceedings of the ACM IEEE Joint Conference on Digital Libraries (JCDL), pp. 18–23 (2007)

Automatic Population of Korean Information in Linking Open Data

Shin-Jae Kang[1] and In-Su Kang[2]

[1] School of Computer and Information Technology, Daegu University
Gyeonsan, Gyeongbuk, 712-714 South Korea
[2] School of Computer Science & Engineering, Kyungsung University
Pusan, 608-736 South Korea
sjkang@daegu.ac.kr, dbaisk@ks.ac.kr

Abstract. This paper presents an automatic populating method which adds non-English factual information into Linking Open Data (LOD). Unlike previous approaches, we extract semantic data from non-structured information, i.e. sentences, and use WordNet to link resources in the semantic data to the ones in DBpedia. The techniques of cross-lingual link discovery and hedge detection are used to select factual information to be added into LOD.

Keywords: Hedge Detection, Cross-lingual Link Discovery, Linking Open Data, QA System.

1 Introduction

Linking Open Data (LOD)[1] interlinks vast amounts of structured information on the Web, and uses Uniform Resource Identifiers (URIs)[2] to identify all resources existing on the Web. Due to these characteristics, LOD can be employed as a knowledge base for QA. Some multilingual datasets exist in LOD, but LOD has been constructed mainly focusing on English, and contains sparse Korean knowledge compared to English.

There has been some research to extend LOD with the information of other languages. In order to populate Chinese semantic data extracted from Chinese Wikipedia, Niu *et al.* [1] used structured content from infobox templates as well as their instances in Wikipedia. Most of LOD is DBpedia[3]-centric linked structure obtained from exact matching of URI information. DBpedia's resource names are just taken from the URIs of Wikipedia articles, so resources in DBpedia and Chinese semantic data extracted can be connected based on the transitive properties of <owl:sameAs>. Choi *et al.* [2] also suggested that infobox schema management and multilingual infobox alignment with multilingual thesauri are needed in order to acquire multilingual

---

[1] http://www.w3.org/wiki/SweoIG/TaskForces/CommunityProjects/
LinkingOpenData
[2] http://www.w3.org/TR/uri-clarification/
[3] http://dbpedia.org/About

G. Bouma et al. (Eds.): NLDB 2012, LNCS 7337, pp. 355–359, 2012.

LOD. As the research of utilizing LOD, Herzig et al [3] used the URIs as interlingual document representations to implement multilingual retrieval, since there can be multiple labels in different languages for one URI and the URI itself can be seen as an interlingual representation for the resource it identifies.

This article presents an automatic populating method which adds non-English factual information into LOD. Unlike previous approaches, we extract semantic data in the form of RDF[4] triples from non-structured information, i.e. sentences, as well as structured information such as tables in Wikipedia pages, and use the synsets of WordNet and URIs to link resources in semantic data extracted to the ones in DBpedia. By doing this, we can get great help to create a LOD-based multilingual QA system.

2   Extending LOD with Korean Information

Five steps for extending LOD with Korean factual information in the form of sentences are illustrated in Fig. 1.

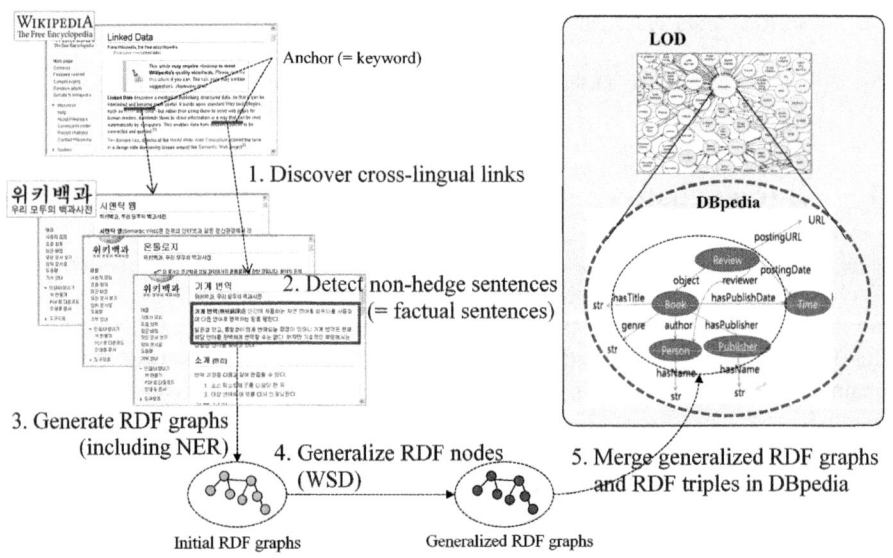

Fig. 1. Five steps to extend LOD with non-structural information

Fig. 2. shows an example of extending DBpedia according to these five steps.

---

[4] http://www.w3.org/RDF/

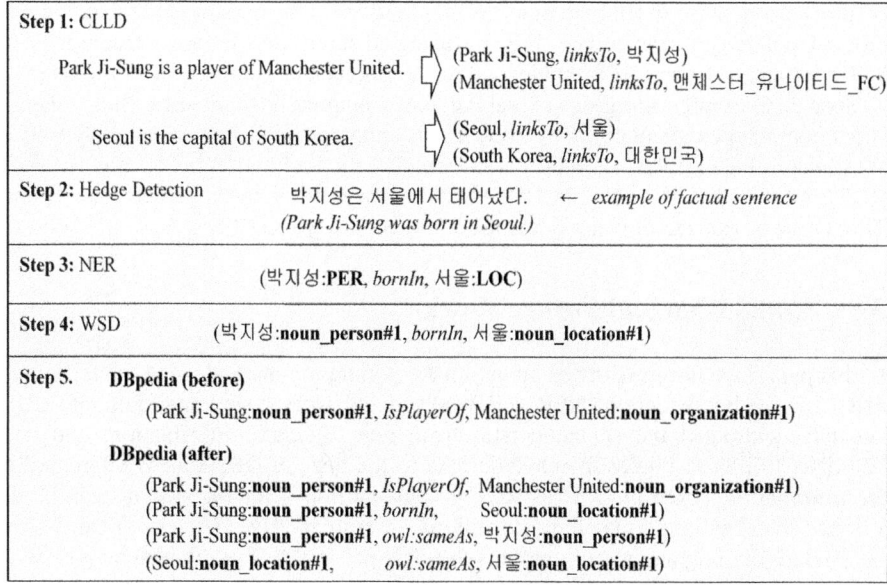

Fig. 2. Example of extending DBpedia according to the five steps

Step 1. Cross-lingual link discovery (CLLD) is a way of automatically finding potential links between documents in different languages [4]. Cross-lingual links between English-Korean Wikipedia documents are used as hints to locate the correct URIs in DBpedia for the identified Korean facts. The synsets of WordNet play the same role in Step 4. We've developed a CLLD system to discover English-to-Korean cross-lingual links by using resources such as link probability, title lists of Wikipedia articles and an English-Korean bilingual dictionary [5].

Step 2. In order to distinguish facts from unreliable or uncertain information, linguistic devices such as hedges [6] have to be identified. Examples of English hedges include *may*, *probably*, *it appears that*, etc. Hedge detection is applied to identify fact sentences from web documents, and the facts are regarded as information to be added into LOD. We evaluated various machine learning techniques for detecting hedges and defined the feature set to achieve the best performance [7].

Step 3. Named entities (NE) and their relations should be extracted from fact sentences. To do this, we first constructed NE recognizer using the SVM learning algorithm [8]. Korean morphological analysis and syntactic analysis were used to extract feature information, and then terms in sentences are classified into one of several types of NE defined. In addition to these results, their NE relations are needed to generate RDF triples, which can be viewed as initial RDF graphs. This work is ongoing.

Step 4. In order to map between the RDF graphs generated in Step 3 and DBpedia, nodes in the RDF graph need to be generalized to WordNet synsets, because resources in DBpedia include relevant WordNet synsets, which enable to automatically identify the correct URI for NEs in RDF triples Step-3 produces. Thus, this step requires word sense disambiguation (WSD), and we developed a generalization method of ontology instances, i.e. NE, by using unsupervised learning techniques for WSD, which uses open APIs and lexical resources such as Google and WordNet [9].

Step 5. Generally, merging several databases is quite difficult since their schemas should be mapped manually. However, in our approach, since the common synsets of WordNet in RDF triples bring the effect of automatic integration between generalized RDF triples and RDF triples in DBpedia, we just need to put together these triples. This is one advantage of using Semantic Web technologies [10].

3   Conclusion and Future Work

In this paper, we have presented an automatic populating method using the techniques of CLLD, hedge detection, NER and WSD to add Korean semantic data into LOD. We have extracted the semantic data from non-structured information, and used WordNet to link resources in semantic data to the ones in DBpedia. When regarding the information stored in LOD as RDF graphs, various semantic systems can be developed by using graph matching algorithms. In similar way, QA system can be implemented. Finding an answer can be viewed as processes of graph matching between a query graph and RDF graphs.

In the future, we are planning to complete NER system, and integrate our previous works such as CLLD, hedge detection, NER, and WSD to implement our idea described in this paper. We are performing a 5-year long-term project targeting the LOD-based multilingual QA system, which consists of three modules of *question analysis*, *LOD retrieval*, and *answer extraction* including answer fusion and generation. So we will do further research to develop efficient graph search and matching algorithms on huge graphs to implement the QA system.

Acknowledgments. This research was supported by Basic Science Research Program through the National Research Foundation of Korea (NRF) funded by the Ministry of Education, Science and Technology (2011-0007025).

References

1. Niu, X., Sun, X., Wang, H., Rong, S., Qi, G., Yu, Y.: Zhishi.me - Weaving Chinese Linking Open Data. In: Aroyo, L., Welty, C., Alani, H., Taylor, J., Bernstein, A., Kagal, L., Noy, N., Blomqvist, E. (eds.) ISWC 2011, Part II. LNCS, vol. 7032, pp. 205–220. Springer, Heidelberg (2011)
2. Choi, K.-S., Kim, E.-K., Choi, D.-H.: Multilingual Linking Open Data. In: Proceedings of the 3rd European Language Resources and Technologies Forum, Venezia, Italy (2011)
3. Herzig, D., Taneva, H.: Multilingual expert search using linked open data as interlingual representation. Notebook Papers of the CLEF 2010 Labs and Workshops, Padua, Italy (2010)
4. Tang, L.X., Geva, S., Trotman, A., Xu, Y., Itakura, K.: Overview of the NTCIR-9 Crosslink Task: Cross-lingual Link Discovery. In: Proceedings of the 9th NTCIR Workshop, Tokyo, Japan, pp. 437–463 (2011)

5. Kang, S.J.: Cross-lingual Link Discovery by Using Link Probability and Bilingual Dictionary. In: Proceedings of the 9th NTCIR Workshop, Tokyo, Japan, pp. 484–486 (2011)
6. Hyland, K.: Persuasion, interaction and the construction of knowledge: representing self and others in research writing. International Journal of English Studies 8(2), 8–18 (2008)
7. Kang, S.-J., Kang, I.-S., Na, S.-H.: A Comparison of Classifiers for Detecting Hedges. In: Kim, T.-h., Gelogo, Y. (eds.) UNESST 2011. CCIS, vol. 264, pp. 251–257. Springer, Heidelberg (2011)
8. Hwang, W.H., Kang, S.J.: Automatic semantic annotation of web documents by SVM machine learning. Journal of Korea Industrial Information Systems Society 12(2), 49–59 (2007) (in Korean)
9. Kang, S.J., Kang, I.S.: Generalization of Ontology Instances Based on WordNet and Google. Journal of Korean Institute of Intelligent Systems 19(3), 363–370 (2009) (in Korean)
10. Allemang, D., Hendler, J.: Semantic Web for the Working Ontologist. Elsevier (2008)

Initial Results from a Study on Personal Semantics of Conceptual Modeling Languages

Dirk van der Linden[1,2], Khaled Gaaloul[1], and Wolfgang Molnar[1]

[1] Public Research Centre Henri Tudor, Luxembourg, Luxembourg
{dirk.vanderlinden,khaled.gaaloul,wolfgang.molnar}@tudor.lu
[2] Radboud University Nijmegen, The Netherlands

Abstract. In this paper we present the initial results from our longitudinal study into the personal semantics of common meta-concepts used in conceptual modeling. People have an implicit understanding of many of the meta-concepts used for modeling purposes, although these are rarely ever made explicit. We argue that a proper understanding of how modelers personally interpret the meta-concepts they use in nearly all of their (domain) models can aid in several things, e.g. explicating a modeler's (proto)typical concept usage, finding communities that share a conceptual understanding and matching individual modelers to each other. Our initial results include the analysis of data resulting from our study so far and a discussion what hypotheses they support.

Keywords: conceptual modeling, personal semantics.

1 Introduction

A number of core (*meta-*)concepts (e.g., ACTORS, GOALS, RESOURCES[1]) used in Enterprise Modeling are common to most modeling languages (e.g., BPMN, e3Value, i*, ITML, UML). Because of the different focuses (e.g., processes, value-exchanges, goals) of such languages, these concepts are often used in subtly different ways. For instance, ACTORS typically being humans or machines, GOALS being mandatory things to achieve or just things to work toward. When models from such different languages are integrated or otherwise linked (e.g., in Enterprise Modeling [8]) it is important to be aware of these differences. Thus, it is important to treat different models carefully when analyzing them to be linked or integrated. If they, or parts of them are treated as equivalent because they are superficially similar (e.g. different concepts are referred to by the same words) the resulting integrated model could be semantically inconsistent and not represent the intended semantics of the original model and modelers (cf. [4]).

To deal with these differences in semantics it is not enough to look solely at the semantics of the modeling language itself. Modeling languages can have multiple specifications with no official standard (e.g., i* [3]) or have inconsistently or underspecified semantics (e.g., BPMN [5,10], UML [15]). Furthermore, even

---

[1] To differentiate between concepts and words referring to them we print concepts in SMALL CAPS.

G. Bouma et al. (Eds.): NLDB 2012, LNCS 7337, pp. 360–365, 2012.
© Springer-Verlag Berlin Heidelberg 2012

when a well-defined specification is available there is no guarantee modelers will use it exactly as specified, either because it clashes with their intuitions [12] or simply because of unintended misuse. Knowing how a modeler will use the semantics related to a concept can therefore not come only from knowledge of the modeling language itself, it is necessary to investigate their typical understanding of that concept.

It would be far too impractical to investigate each modeler involved in a modeling project on an individual basis. Instead it should be possible to achieve a generalized understanding of '(stereo)types' of modelers, which could be used to infer with a certain probability what kind of understanding a modeler has of a given concept. The first step towards achieving this is investigating whether there are discretely identifiable groups of people with a shared understanding of modeling concepts. Given that language itself is inherently bound by community [11,6], we propose that there are identifiably different communities that share to some degree an understanding of modeling language concepts (*hypothesis 1*). One would expect the shared understanding to be stronger for some specific concepts than others (e.g., two modelers might agree on common concepts they both use, yet diverge on those concepts outside of their area of expertise), meaning that to find communities one should look at the individual conceptual understandings rather than just overall scores (*hypothesis 2*).

To the extent of our knowledge, there is little other work on explicating personal semantics of modeling concepts. A focus on modelers over models has received some attention [1,7], but tends to focus on (automated) corpus analysis or defining formal data structures, whereas our work is focused on empirical acquisition of content. Moreover, personal semantics have been reasoned to be of valuable use in improving the quality of shared conceptualizations [2,13].

To test these hypotheses we have set up a longitudinal study using psychometrics to measure the understandings of computing and information science students during their undergraduate studies. The use of these students should ensure that we can clearly observe the formation and change of communities related to their understanding as they receive instruction into conceptual modeling theory, methods and practice. This setup will allow us to study the stability of understandings and speed of their change. Given that the potential participants, depending on their exposure to conceptual modeling (e.g., possible prior education) are mostly inexperienced we expect the majority of participants to express a neutral understanding of most concepts (*hypothesis 3*), while those who do have prior exposure should exhibit more polarized understandings (*hypothesis 4*).

2   Our Method for Eliciting Conceptual Understandings

We have used Semantic Differentials to elicit the conceptual understandings, keeping in mind the design criteria set out by [14]. Our interest was to monitor semantic understanding related to conceptual modeling regardless of aspect focus, so we opted for a wide-ranging set of categories (ACTOR, EVENT, GOAL, PROCESS, RESOURCE, RESTRICTION, RESULT), resulting from our earlier work on categorization of conceptual modeling languages and methods [9]. Participants were primed to conceptual modeling context by inclusion of a priming task. The words used for this task were the words used for the relevant concepts by the modeling languages and methods we have previously analyzed. We investigated the dimensions *natural, human, composed, necessary, material, intentional* and *vague*, also originating from our previous work.

With the necessary data available we constructed a semantic differential in LimeSurvey. The test itself consisted of a priming part and adjective scoring part for each concept, and was prefaced with a number of background questions (e.g., age, prior experience and education). Participants were actively approached in the first week of their curriculum, all of them following either computing science or information systems science tracks. We chose to follow a group of students from the same university in order to ensure a greater homogeneity in the (educational) stimuli they undergo during the duration of this longitudinal study. The test itself was constructed and administered in Dutch.

3   Results of the First Observation Moment

A total of 19 participants filled out the test completely and agreed to participate in subsequent tests during their studies. The majority of these (16) were computing science students, the remaining 3 being students of information systems science. The majority of the participants were native Dutch speakers. Only 2 of the group had a different primary language (German and Serbian) but were fluent in written and spoken Dutch.

The adjective scores for a given dimension were averaged and used to produce the individual score matrices that contain a score between 1 (strong agreement) and 5 (strong disagreement) for each concept-dimension combination. We then calculated a matrix with the median of all these scores and used this to calculate the overall standard deviation in concept-dimension combinations and participant's own deviation from the median. We then clustered the individual score matrices using repeated bisection clustering into the smallest amount of clusters where total membership was still ≥ 1 (see Fig. 1). Finally, to investigate concept-specific communities and provide an overview of the approximate semantic distance between these results we performed a principal component analysis (PCA) shown in Fig. 2.

4   General Discussion

Most importantly, the results support our *1st hypothesis*: the existence of communities with different typical understandings of modeling concepts. Figure 1 shows the three major clusters (i.e. communities) we found in the overall (i.e. all categories taken together) scores. As expected, the largest cluster (see Fig. 1(a)) exhibits a mostly neutral score (most scores being > 2 and < 4), supporting our *3rd hypothesis*. This is understandable given that the first phase of this study coincided with the very first week of academic education for these students. As

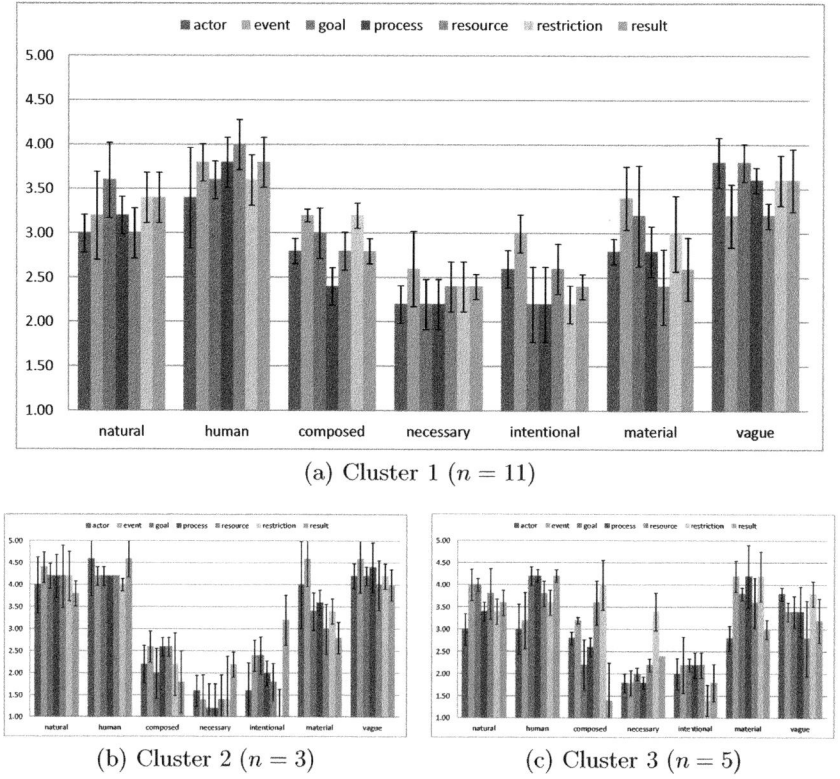

(a) Cluster 1 $(n = 11)$

(b) Cluster 2 $(n = 3)$ (c) Cluster 3 $(n = 5)$

Fig. 1. Clusters found in the overall results. The subfigures visualize the results of each concept-dimension combination and related standard deviation grouped per dimension. To illustrate, Fig. 1(a) first shows in what respect all the individual categories (ACTOR,. . . ,RESULT) are considered *natural* things, followed by in what respect the categories are considered *human* things, and so on. The higher a score (and thus bar), the more a given dimension is considered not fitting for a certain concept. The lower a score (and thus bar), the more a given dimension is considered fitting for a certain concept. The standard deviation is visualized on each bar and represents the range to which participants differ on how much a concept-dimension combination 'fits'.

such it is to be expected that a large group of participants has no strong feelings so far towards the concepts they are yet to be educated about.

Two smaller clusters (Fig. 1(b) and 1(c)) both exhibit stronger polarized responses (both negative and positive). The most common strong reactions that seem to hold for all investigated concepts in these are to the properties of being *necessary* and *intentional*. Furthermore, the properties of being *human, natural, material* and *vague* gave strongly positive responses for a number of concepts. There was no clear link between these specific clusters to any properties of the participants we were aware of such as sex, age, educational level and type, prior experience with modeling notations, programming languages, etc.

The manual clustering we performed of the participants into those with and without experience showed little significant overall difference, even though both groups were similar in size and composition. Furthermore, when we investigated the specific participants we found that there were counterexamples for *hypothesis 4* from both sides: those with prior experience scoring decidedly neutral, and those without prior experience having strongly polarized scores. On further investigation of specific concepts we also found no clusters that separated those with and without experience, causing us to reject our 4th *hypothesis*.

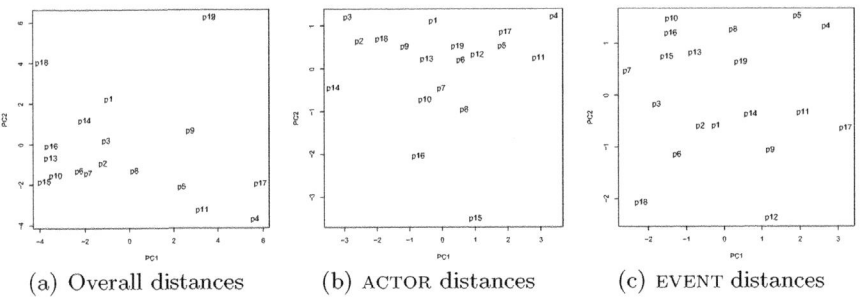

(a) Overall distances   (b) ACTOR distances   (c) EVENT distances

Fig. 2. Principal components found in the results of both overall and some concept-specific scores. These visualizations represent (roughly) the distance between understandings individual participants have. Shown are respectively the distances between overall understanding (taking all concepts into account) and two specific concepts.

An example of the differences between participants on a per-concept basis can be seen in Fig. 2. Here it can be seen clearly that a per-concept view offers greater discriminatory possibilities compared to overall results and that the distance between participants differs per concept, thus supporting our assumption that participants have specific understandings for each concept. The amount of different understandings between participants also differs between concepts. Some concepts contain a number of participants with quite similar understandings and some outliers, whereas some concepts like EVENT (Fig. 2(c)) have a far greater amount of variance. This can be explained to some degree by the conceptual width of a concept, as things like EVENTS can allow for far more possible instantiations than something like ACTORS. Over time we expect to also find more strongly delineated clusters in these wide concepts that correlate with the formation of communities with similar understandings. Nonetheless, these results support our 2nd *hypothesis* that shared understandings are more useful to investigate on a per-concept basis.

5   Conclusion and Future Work

We have shown the results of the first observation of our longitudinal study and found support for *hypotheses 1,2* and *3* (respectively the existence of communities with shared understanding of modeling concepts, per-concept more clearly identifying shared understandings and a tendency for the largest group in this phase to have a neutral understanding) while we had to reject *hypothesis 4* (people with prior experience having more polarized understandings). With additional data gained from subsequent observations we hope to investigate the evolution of participants' understandings, their transitions between communities and to what degree shifts in their understanding can be correlated to real-world phenomena.

References

1. Aimé, X., Furst, F., Kuntz, P., Trichet, F.: Ontology Personalization: An Approach Based on Conceptual Prototypicality. In: Chen, L., Liu, C., Zhang, X., Wang, S., Strasunskas, D., Tomassen, S.L., Rao, J., Li, W.-S., Candan, K.S., Chiu, D.K.W., Zhuang, Y., Ellis, C.A., Kim, K.-H. (eds.) WCMT 2009. LNCS, vol. 5731, pp. 198–209. Springer, Heidelberg (2009)
2. Almeida, M.B.: A proposal to evaluate ontology content. Applied Ontology 4, 245–265 (2009)
3. Ayala, et al.: A comparative analysis of i*-based agent-oriented modeling languages. In: SEKE 2005, Taipei, Taiwan, pp. 43–50 (2005)
4. van Buuren, R., Gordijn, J., Janssen, W.: Business case modelling for e-services. In: 18 th Bled eConference eIntegration in Action (2005)
5. Dijkman, R.M., Quartel, D.A.C., van Sinderen, M.J.: Consistency in multi-viewpoint design of enterprise information systems. Inf. Softw. Technol. 50(7-8), 737–752 (2008)
6. Hoppenbrouwers, S.J.B.A.: Freezing language: conceptualisation processes across ICT-supported organisations. Ph.D. thesis, Radboud University Nijmegen (2003)
7. Katifori, V., Poggi, A., Scannapieco, M., Catarci, T., Ioannidis, Y.: Ontopim: how to rely on a personal ontology for personal information management. In: Proc. of the 1st Workshop on The Semantic Desktop. Citeseer (2005)
8. Lankhorst, M.M.: Enterprise architecture modelling–the issue of integration. Advanced Engineering Informatics 18(4), 205–216 (2004)
9. van der Linden, D.J.T., Hoppenbrouwers, S.J.B.A., Lartseva, A., Proper, H.A(E.): Towards an Investigation of the Conceptual Landscape of Enterprise Architecture. In: Halpin, T., Nurcan, S., Krogstie, J., Soffer, P., Proper, E., Schmidt, R., Bider, I. (eds.) BPMDS 2011 and EMMSAD 2011. LNBIP, vol. 81, pp. 526–535. Springer, Heidelberg (2011)
10. Van Nuffel, D., Mulder, H., Van Kervel, S.: Enhancing the Formal Foundations of BPMN by Enterprise Ontology. In: Albani, A., Barjis, J., Dietz, J.L.G. (eds.) CIAO!/EOMAS 2009. LNBIP, vol. 34, pp. 115–129. Springer, Heidelberg (2009)
11. Perelman, C., Olbrechts-Tyteca, L.: The New Rhetoric: A Treatise on Argumentation. University of Notre Dame Press (June 1969)
12. Sowa, J.: The role of logic and ontology in language and reasoning. In: Poli, R., Seibt, J. (eds.) Theory and Applications of Ontology: Philosophical Perspectives, pp. 231–263. Springer, Netherlands (2010)
13. Uschold, M.: Making the case for ontology. Applied Ontology (2011)
14. Verhagen, T., Meents, S.: A framework for developing semantic differentials in is research: Assessing the meaning of electronic marketplace quality (emq). Serie Research Memoranda 0016, VU University Amsterdam, Faculty of Economics, Business Administration and Econometrics (2007)
15. Wilke, C., Demuth, B.: Uml is still inconsistent! how to improve ocl constraints in the uml 2.3 superstructure. Electronic Communications of the EASST 44 (2011)

Lexical Knowledge Acquisition
Using Spontaneous Descriptions in Texts

Augusta Mela[1], Mathieu Roche[2], and Mohamed El Amine Bekhtaoui[3]

[1] Univ. Montpellier 3, France
[2] LIRMM – CNRS, Univ. Montpellier 2, France
[3] Univ. Montpellier 2, France

Abstract. This paper focuses on the extraction of lexical knowledge
from text by exploiting the "glosses" of words, i.e. spontaneous descriptions identifiable by lexical markers and specific morpho-syntactic patterns. This information offers interesting knowledge in order to build
dictionaries. In this study based on the RESENS project, we compare
two methods to extract this linguistic information using local grammars
and/or web-mining approaches. Experiments have been conducted on
real data in French.

1 Introduction

Automatic acquisition of lexical knowledge from texts aims to detect various
types of lexical units (e.g. terms, named entities, phrases, new words, and words
with a new meaning) as well as their syntactic and semantic properties. In a
multilingual context, using bilingual texts, automatic acquisition consists in detecting the translations of these units. Automatic acquisition constitutes a pre-
cious help in order to build dictionaries, thesaurus, and terminologies for general
and/or specialized domains [1]. This task is also useful for documentary research
thanks to query expansion.

Our work concerns lexical structuration. To situate it, we will restrict ourselves to work on lexical structuration. We adopt the distinction proposed by [2]
concerning the different levels of structuration:

1. Semantic lexical relationships such as synonymy or hypernymy concern
   microstructuration,
2. The classification of units in topics concerns macrostructuration: At this
   level the nature of the link between units is not identified.

In macrostructuration, the existing approaches are mainly based on Harris distributional semantics [3]. According to this theory, the set of the contexts in
which the word appears allows to draw its portrait and to determine the meaning. In practice, given a word, we look for which words we can substitute for it
(paradigmatic properties), with which words it can be combined (syntagmatic
properties), or more loosely in what immediate context it appears (co-occurrence
properties). These questions are not new and lexicographers have always used

G. Bouma et al. (Eds.): NLDB 2012, LNCS 7337, pp. 366–371, 2012.

corpora to answer them. What has changed today is the size of the corpora and the computer tools capable of systematizing the distributional approach.

In microstructuration, the approaches are various:

- Some approaches use the structure of the unit itself (e.g., *le coussin de sécurité* (security cushion) is a hypernym of *coussin de sécurité arrière* (rear security cushion) and *anticapitaliste* is an antonym of *capitaliste*).
- Following the distributional approach, other methods consider that the related units appear in similar context. This one can be limited to the words in the close environment of the target word or can be extended to refer to the entire text. Represented in a word-based vector space, both words are similar if their vector representations are close [4].
- Another type of approach which is close to our work, starts from the principle that semantic relationships are expressed by linguistic markers (lexical or grammatical elements), or paralinguistic elements (punctuation marks, inverted commas). In this approach, morpho-syntactic patterns which combine these marks are manually built and projected into the texts to seek out the items supposedly in the semantic relationships under study. These patterns can also be acquired semi-automatically [5]. The quality of the results of this approach depends on the precision with the patterns have been defined or the quality of the corpus used to infer them.

Section 2 details our approach which combines linguistic and statistic information. Section 3 presents the obtained results, and Section 4 outlines future work.

2   The Word and Its Gloss

2.1   Definition

In the RESENS project (PEPS-CNRS project), linguists and computer scientists work together to study the phenomena of the gloss and to propose methods to access the meanings of words. Glosses are commentaries in a parenthetic situation, often introduced by markers such as *appelé, c'est-à-dire, ou* (called, that is to say, or) which signal the lexical semantic relationship involved: Equivalence, with *c'est-à-dire, ou* ; specification of the meaning, with *au sens* (in the sense); nomination, with *dit, appelé* (called); hyponomy, with *en particulier, comme* (in particular, like); hypernymy, with *et/ou autre(s)* (and/or other), etc. They are in apposition to the glossed word, usually of the nominal category. Their syntactic and semantic topologies have been established by linguists [6]. It has also been noted that their frequency depends on the genre of the texts: They are more often found in didactic texts or work of popularization than in poetry.

A gloss is spontaneous. It shares this characteristic with a so-called natural definition because neither is the fruit of the reflective work of a lexicographer. However, a gloss is parenthetical whereas a natural definition is the main object of the proposition. Thus the gloss can roughly be described by the configuration X marker Y as described in the following section.

2.2   Gloss Extraction

In this section we detail our method, considering the case of glosses with *appelé*. We start from the principle that X and Y are noun phrases (NP) extracted with the approach of [7]. In addition, we take into account that Y can be a coordination of NPs, thus the variant of the abstract pattern of the gloss becomes: X marker $Y_1, Y_2...Y_n$.

- **Local Pattern 1:** The first pattern detects Y_1, the first NP to the right of the marker. For example, the sentence *Un disque microsillon, appelé disque vinyle* (a grammaphone record, called vinyl record) allows to extract $X = $ *un disque microsillon* and $Y_1 = $ *disque vinyle*.

- **Local Pattern 2:** A second pattern takes into account the possibility of coordination in position Y, to extract a sequence such as *un disque microsillon appelé disque vinyle ou Maxi*. Two NPs are extracted: $Y_1 = $ *disque vinyle* and $Y_2 = $ *Maxi*.

- **Global Extraction:** Once the gloss with *appelé* had been detected in a corpus, we look for the NPs situated between the marker and a right hand boundary. This right boundary is either a conjugated verb or a strong punctuation mark.

The following section describes a ranking function which orders the extracted phrases Y_i.

2.3   Gloss Ranking

Our web mining approach is based on the Dice measure. This one calculates the "dependency" between both Nominal Phrases (NP), i.e. dependency in terms of co-occurrence more or less close. This dependency is calculated by querying the Web.

Applied to X (i.e. glossed NP) and Y_i (glossing NP), the measure is defined with the formula (1):

$$Dice1(X, Y_i) = \frac{2|X \cap Y_i|}{|X| + |Y_i|} \tag{1}$$

where $|X \cap Y_i|$ is the number of web pages containing the words X and Y_i one beside the other, $|X|$ is the number of pages containing the term X, and $|Y_i|$ is the number of pages containing Y_i. In order to calculate $|X \cap Y_i|$ by querying the Web, we use quotations (").

Subsequently, we release the proximity constraints between phrases. So the number of web pages where X and Y_i are present together is calculated. In this case, we use a measure called $Dice2$. The numerator of this measure (formula (2)), represents the number of times where X and Y_i are in the same pages.

$$Dice2(X, Y_i) = \frac{2|X \ AND \ Y_i|}{|X|+|Y_i|} \tag{2}$$

From both measurements (i.e. $Dice1$, $Dice2$), two types of combinations are proposed.

1. The first one, $Diexact$, is $Dice1$ if $Dice1$ returns a result. Otherwise $Dice2$ is calculated. So this approach gives priority to $Dice1$.

2. The second one, $Dibary$ (formula (3)), calculates the barycenter of $Dice1$ and $Dice2$.

$$Dibary_k(X, Y_i) = k.Dice1(X, Y_i) + (1 - k).Dice2(X, Y_i) \tag{3}$$
where $k \in [0, 1]$

Our web mining methods can use different search engines (e.g. Google, Yahoo, and Exalead) to calculate Dice measures. Experiments based on 22 texts containing glosses have shown that the API provided by Yahoo have a good behavior. We chose this search engine in our experiments.

3   Experiments

Our corpus is composed of 219 candidate NPs (i.e Y_i) identifyied with the word *"appelé"* (i.e. "called"). These candidates are extracted with a software implemented in Java (see Figure 1).

A linguist expert assignes the mark 1 for the relevant pairs (X, Y), and the mark 2 for the very relevant pairs. When the NP is partially extracted, the couple (X, Y) is evaluated with a score at 1. Irrelevant NP are evaluated at 0.

Evaluation of Extraction Methods

The quality of element X are: 6 NP evaluated as relevant (mark 1), 68 NP evaluated as very relevant (mark 2). Now we can compare the methods in order to extract the 219 glossing NP (i.e. Y_i).

Table 1 presents the results obtained (evaluations at 0, 1, or 2). The results show that the majority of phrases are evaluated as very relevant. The global extraction of Y_i produces a quarter of irrelevant phrases. We can explain this situation because the NP following *"appelé"* are not always semantically related to X. Nevertheless, the global extraction can identify a quantity larger of relevant NPs than the use of simple patterns.

With the marks 1 and 2 considered as relevant, precision, recall, and F-measure are presented in Table 1. This table shows that the best F-measure is obtained with the global extraction of phrases. Note that the use of coordination rules (i.e. Local Pattern 2) is effective (excellent precision and good recall).

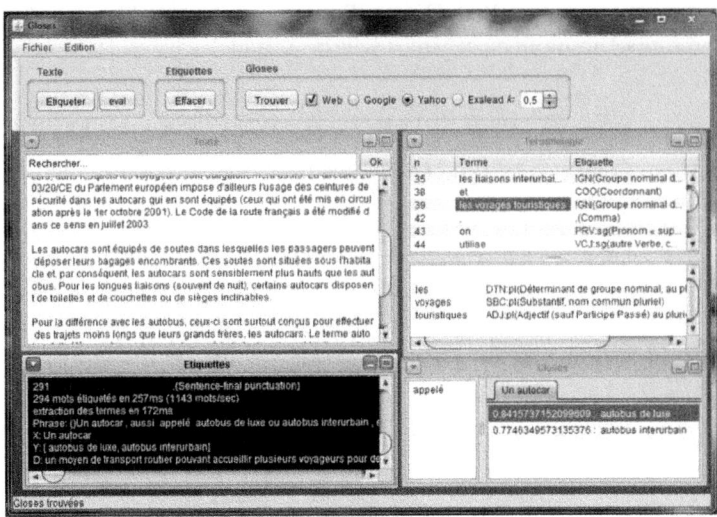

Fig. 1. Extraction of glosses – Software

Table 1. Evaluation of the different methods

Method / Evaluation	Nb of extracted Y_i		
	Evaluation at 1	Evaluation at 2	Evaluation at 0
Local Pattern 1	3 (3.75 %)	**74 (92.50 %)**	3 (3.75 %)
Local Pattern 2	4 (3.67 %)	**101 (92.66 %)**	4 (3.67 %)
Global Extraction	16 (7.30 %)	**150 (68.49 %)**	53 (24.21 %)

Method / Evaluation	Precision	Recall	F-measure
Local Pattern 1	**0.96**	0.46	0.62
Local Pattern 2	**0.96**	0.63	0.76
Global Extraction	0.75	1	**0.86**

Evaluation of Ranking Functions

Now we can evaluate the ranking quality of NP obtained by web-mining approaches. To compare our algorithms, we calculate the sum of the ranks of Y_i evaluated as relevant by the expert. Actually the minimization of this sum is equivalent to maximize the Area Under the ROC Curve [8]. This principle is often used in data-mining field to assess the quality of ranking functions. The method giving the best results returns the lower value.

The sum of the ranks of relevant Y_i obtained with our evaluation corpus is shown in Table 2. This one shows that *Diexact* method gives best results. The influence of k for *Dibary* is low.

Table 2. Sum of relevant elements

Method	Diexact	Dibary					
		$k = 0$	$k = 0.2$	$k = 0.4$	$k = 0.6$	$k = 0.8$	$k = 1$
Sum	**323**	329	329	329	329	329	364

4   Conclusion and Future Work

In this paper, we have presented a method to extract NPs in relationship by a gloss phenomenon. Our methods combine local grammars and statistical associations of units on the web. In the context of the RESENS project, these methods have been manually evaluated on real data.

In our future work, we would like to perform a contrastive analysis of English/French corpora in order to give a new point of view of the phenomenon of spontaneous descriptions. A first study on aligned English/French texts reveals frequent regularities of glosses in a multilingual context. The alignment enables to improve the multilingual lexical acquisition of new words and their translations.

Moreover we plan to test other web mining measures, these ones will be able to take into account other kinds of operators for querying the web (e.g. *Near* operator).

Finally we plan to focus our work on the study of the markers between NPs in order to automatically extract the type of relationships (synonymy, hyponomy, hypernymy, and so forth).

Acknowledgment. We thank Vivienne Mela who improved the readability of this paper.

References

1. Muresan, S., Klavans, J.: A method for automatically building and evaluating dictionary resources. In: Proceedings of LREC (2002)
2. Nazarenko, A., Hamon, T.: Structuration de terminologie: quels outils pour quelles pratiques? TAL 43-1, 7–18 (2002)
3. Daladier, A.: Les grammaires de harris et leurs questions. Languages (99) (1990)
4. Salton, G., Wong, A., Yang, C.S.: A vector space model for automatic indexing. In: Proc. of ACM, vol. 18, pp. 613–620 (1975)
5. Aussenac-Gilles, N., Jacques, M.P.: Designing and Evaluating Patterns for Relation Acquisition from Texts with CAMELEON. Terminology, Pattern-Based Approaches to Semantic Relations 14(1), 45–73 (2008)
6. Steuckardt, A.: Du discours au lexique: la glose. Séminaire ATILF (2006)
7. Daille, B.: Study and implementation of combined techniques for automatic extraction of terminology. In: The Balancing Act: Combining Symbolic and Statistical Approaches to Language, pp. 49–66 (1996)
8. Ferri, C., Flach, P., Hernandez-Orallo, J.: Learning decision trees using the area under the ROC curve. In: Proceedings of ICML, pp. 139–146 (2002)

Interacting with Data Warehouse
by Using a Natural Language Interface

M. Asif Naeem[1], Saif Ullah[2], and Imran Sarwar Bajwa[3]

[1] School of Computing and Mathematical Sciences, Auckland University of Technology,
Private Bag 92006, Auckland, New Zealand
[2] Department of Computer Science & IT, Islamia University of Bahawalpur, Pakistan
[3] School of Computer Science, University of Birmingham, UK
mnae006@aucklanduni.ac.nz, saif_itp@yahoo.com,
i.s.bajwa@cs.bham.ac.uk

Abstract. Writing Online Analytical Processing (OLAP) queries for data warehouses is a complex and skill requiring task especially for the novel users. The situation becomes more critical when a low skilled person wants to access or analyze his business data from a data warehouse. These scenarios require more expertise and skills in terms of understanding and writing the accurate and functional queries. However, these complex tasks can be simplified by providing an easy interface to the users. In order to resolve all such issues, automated software tool is needed, which facilitates both users and software engineers. In this paper we present a novel approach with name QueGen (Query Generator) that generates OLAP queries based on the specification provided in natural English language. Users need to write the requirements in simple English in a few statements. After a semantic analysis and mapping of associated information, QueGen generates the intended OLAP queries that can be executed directly on data warehouses. An experimental study has been conducted to analyze the performance and accuracy of proposed tool.

Keywords: Automatic query generation, Natural language processing, SBVR.

1   Introduction

Business organizations are growing rapidly and they want to be prepared for this rapid growth and volatility. The access of accurate data at the right time to the right place in the right format has become more significant to achieve business success. Customers want the latest information about products in order to purchase, pay and ship. Morning sales in the east corner of the world affect the stock management on the west.

The conventional way of communicating with a data warehouse [1] is to first build a connection stream and then accesses the data contents from the data warehouse using a standardized interfacing mechanism [2]. Simple command shells are typically used and they are often incorporated within every distinct data warehouse product. These command shells are typically simple filters which helps a user to log on to the data warehouse, execute particular commands and receive output. These command

G. Bouma et al. (Eds.): NLDB 2012, LNCS 7337, pp. 372–377, 2012.

shells provide access to the data warehouse from the machine on which the RDBMS (Relational Database Management System) is actually running [3]. After hooking to a particular data warehouse a user or a programmer requires an interface and typically that interface is provided by some technical languages. These languages are called query languages and are constituted of the OLAP commands typically used for retrieving information from data warehouse. OLAP is set of particular commands under SQL [4] which are specifically used to interact with data warehouse repositories.

From an application programmer's point of view, the major step in the data warehouse is to write OLAP queries which inherit from a declarative query language, SQL. The requirements and questions are typically specified in SQL for large web-based and stand-alone applications. However, writing OLAP based queries for a data warehouse is a complex task specifically for the users having little or no knowledge of SQL queries. Moreover, it is quite difficult to remember the SQL commands and use them accurately. An ad-hoc solution was presented in [5] to provide a Q&A interface for data warehouse. However, there are a few short-comings in the presented approach. Firstly, the architecture does not provide the taxonomy used to perform the semantic analysis for the given user's specification of OLAP queries. Secondly, the given approach only considers one type of OLAP queries named aggregation while the other two types drill-down and slicing-dicing are also equally important for query generation. Thirdly, the authors evaluated their approach based on the "number of entities found" only and they did not provide any data about the accuracy and the execution time taken to generate one query.

In this paper, we present a novel approach to provide a natural language (NL) that not only provides a simple and easy way of interacting [14] with a data warehouse but also address the short-comings of the approach presented in [5]. The presented approach is implemented to a prototype tool QueGen that can facilitate both users and software engineers. QueGen has the ability to generate all common OLAP queries from NL (such as English) specification of queries. The details about the architecture and workflow of QueGen are presented in later sections.

2   A NL Interface for OLAP Queries

The presented approach works as a user provides the English specification of a query language and our approach lexically and syntactically analyzes (see Figure 1) the English specification of query using the Stanford POS tagger [12] and the Stanford parser [13]. We have used the off-shelf components for lexical and syntactic analysis as these tools provide high accuracy such as the Stanford POS tagger is 97.0% accurate [12], while the Stanford parser is 84.1% accurate [13]. However, for the semantic analysis of the English queries we have written a semantic analyzer that generates a logical representation based on SBVR vocabulary. Once the logical representation is extracted from the input text, the SBVR based logical representation is mapped to OLAP terms. In such SBVR to OLAP mapping, the appropriate OLAP terms such as such as keyword, table names, field names, field value, etc are assigned to SBVR vocabulary items on the basis of identified semantics. Finally, using these OLAP terms query generation module generates the OLAP query which can be applied on a data warehouse directly.

The English language statements are effortlessly converted into OLAP queries by using QueGen tool. Select query is the common query in OLAP that is used to choose a set of values from a data warehouse table [7].

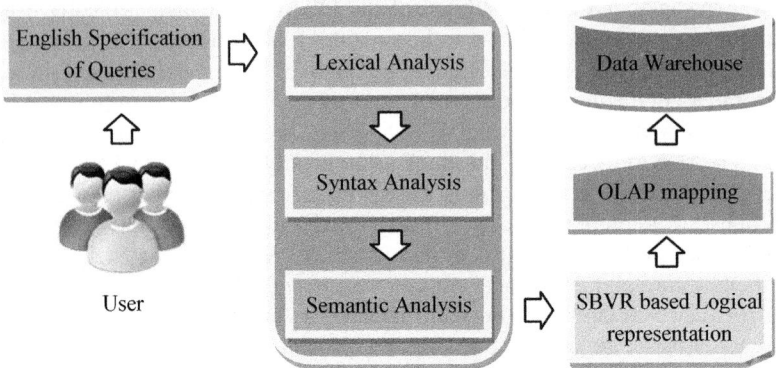

Fig. 1. An abstract level architecture of QueGen

To evaluate the working and performance of QueGen, we implemented a light weight data warehouse for a "Soft Drinks Company" as an example. The "Soft Drinks Company" produced variety of drinks like Coca-Cola, Sprite, Pepsi etc, and supplies these products worldwide. The sketch of start schema we used to implement the data warehouse for the company is shown in Figure 2.

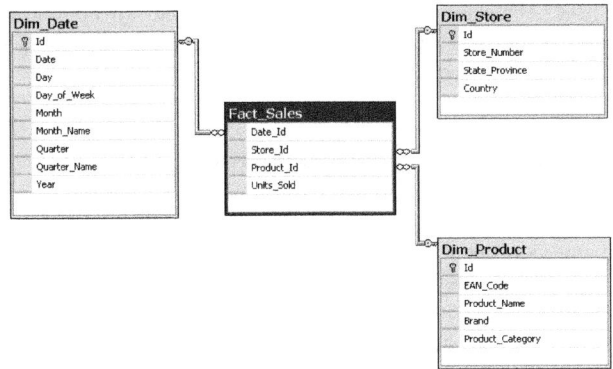

Fig. 2. A sketch of star schema for "Soft Drink Company's" data warehouse

According to the Figure 2, there are three dimension tables namely Dim_Date, Dim_Product, and Dim_Store which are connected with a central fact table, Fact_Sales under one-to-many relationship. Normally the three dimension tables contain master and reference data while the fact table contains transactional data that changes frequently [1].

In the incoming sections, we will show step-by-step that how our approach generates the OLAP query from the example discussed in Section 2.

2.1   Processing NL Queries

The first phase in NL to OLAP queries is detailed parsing of NL text. The NL parsing phase is divided into lexical, syntactic and semantic processing.

Lexical Processing. In lexical processing of English specification of an OLAP query, two steps are involved [15]: Identification of the token is the first step in lexical processing and then afterwards the tokenized text is further passed to Stanford parts-of- speech (POS) tagger v3.0 to identify the basic POS tags. We have used the Stanford POS tagger v3.0 to identify tokens and POS tags from NL text due to its high accuracy that is 97% [12].

Syntactic Processing. We have used the Stanford parser to parse the lexically analyzed text and generate parse tree representation and syntactic dependencies [16].

Semantic Analysis. In this semantic analysis phase, semantic role labeling [12] is performed. As we aim to generate a SBVR based logical representation, we have used the SBVR vocabulary types as the possible semantic roles. We have chosen SBVR for semantic role labeling because SBVR is an adopted standard and easy to map to other formal languages as SBVR is based on higher order logic. The desired role labels are Object Type, Individual Concept, Verb concept, Characteristics, etc. Mapping of English text to SBVR based semantic roles is explained in the remaining part of this Section.

1. *Object Types*: All common nouns are represented as object types.
2. *Individual Concepts:* All proper nouns are mapped to the individual concepts.
3. *Verb Concepts*: The action verbs are represented as verb concepts.
4. *Characteristics*: The attributes are adjectives and possessive nouns (i.e. ending in *'s* or coming after *of*).
5. *Quantifications:* NL quantifiers are mapped to the SBVR quantifiers.

Once the semantic role labeling is performed, the NL text is mapped to a logical representation as shown in the Figure 3. We have written a rule based analyzer that maps semantic roles to a SBVR based logical representation.

(sold
   (characteristic = (sum ? Z)) ∧ (object_type = (∀Y ~ (units ? Y))) ∧
   (object_type = (∀X ~ (fact sales ? X))) ∧ (individual_concept = (Coca-Cola? W)) ∧
   (object_type = (∀V ~ (country ? V)))) ∧ (characteristic = (01-Jan-11 ? U)) ∧
   (characteristic = (31-Dec-11 ? T))))

Fig. 3. SBVR based Logical representation

2.2   Extracting OLAP Query Elements

In this phase, finally the SBVR rule is further processed to extract the OO information. The extraction of each OO element from SBVR representation is described below:

1. *Extracting Tables:* Each *Object Type* or *Individual Concept* is mapped to the table names in the target data warehouse and if matches with a table name then that *Object Type or Individual Concept* is tagged as 'Table Name'.
2. *Extracting Fields*: An *Object Type* or an *Individual Concept* that does not match to any table name, it is mapped to the field names of each table in the target data warehouse and if matches with any field name then that *Object Type* or *Individual Concept* is tagged as 'Field Name'.
3. *Extracting Functions*: An *Object Type, Individual Concept* or a characteristic that does not match to any table name or field name is looked in a list of functions names and if matches with any function name then that *Object Type* or *Individual Concept* is tagged as 'Function Name'.
4. *Extracting Field Values*: A *Characteristic* that is not a function is mapped to field value. Moreover, any *Individual Concept* that does not match to a table name or field name are considered as field name.
5. *Extracting Keywords*: The tokens in English text such as 'show', 'list', 'select', and 'display' are mapped to "select" keyword.

2.3   OLAP Query Generation

This is the final phase in generation of OLAP query from English specification of queries. In this phase the logical representation generated in semantic analysis phase and the keywords extracted in Section 3.2 are combined to generate a particular query. Finally, the SQL (OLAP) query is generated by embedding the extracted information in Section 3.2 in the following template:

```
SELECT <field-name>, [<function-name>(<field-name>)]
FROM <field-name> WHERE <field-name>= <field-value> [AND
<field-name> and …] [GROUP BY <field-name>];
```

Fig. 4. One of the templates used for OLAP query generation

3   Conclusions and Future Work

The task of writing OLAP queries is difficult and error prone for non-technical users or even medium level skilled persons. This task can be simplified by providing an easy interface that is more familiar and well known to the users. In this paper we proposed a novel automated query generator tool with name QueGen. The proposed tool has the capability to generate all three categories of OLAP queries. User needs to write the requirements in simple English language. The proposed tool analyzes the given script. After semantic analysis and mapping of associated information, it generates the intended OLAP queries that can be applied directly on data warehouse. The proposed tool provides a quick and reliable way to generate OLAP queries to save the time and cost of users and business organizations. We also evaluated the performance of the proposed tool with respect to both execution time and accuracy.

Currently, QueGen does not have the ability to generate complex queries which are normally used in the area of data mining. In future we have a plan to extend the feature of QueGen so that it can generate these kinds of queries.

References

1. Inmon, W.H.: Building the Data Warehouse. John Wiley (1992)
2. Allen, J.: Natural Language Understanding. Benjamin- Cummings Publishing Company, New York (1994)
3. Biber, D., Conrad, S., Reppen, R.: Corpus Linguistics: Investigating Language Structure and Use. Cambridge Univ. Press, Cambridge (1998)
4. DeHaan, D., Toman, D., Consens, M.P., Ozsu, T.: A Comprehensive XQuery to SQL Translation using Dynamic Interval Encoding. In: SIGMOD (2003)
5. Kuchmann-Beauger, N., Aufaure, M.-A.: A Natural Language Interface for Data Warehouse Question Answering. In: Muñoz, R., Montoyo, A., Métais, E. (eds.) NLDB 2011. LNCS, vol. 6716, pp. 201–208. Springer, Heidelberg (2011)
6. Object Management Group. Semantics of Business vocabulary and Rules (SBVR) Standard v.1.0. Object Management Group (2008), http://www.omg.org/spec/SBVR/1.0/
7. Thompson, C.A., Mooney, R.J., Tang, L.R.: Learning to parse natural language database queries into logical form. In: Workshop on Automata Induction, Grammatical Inference and Language Acquisition (1997)
8. Zelle, J.M., Mooney, R.J.: Learning semantic grammars with constructive inductive logic programming. In: Proceedings of the 11th National Conference on Artificial Intelligence, pp. 817–822. AAAI Press/MIT Press, Washington, D.C (1993)
9. Losee, R.M.: Learning syntactic rules and tags with genetic algorithms for information retrieval and filtering: An empirical basis for grammatical rules. Information Processing and Management 32(2), 185–197 (1996a)
10. Manning, C.D., Schutze, H.: Foundations of Statistical Natural Language Processing. MIT Press, Cambridge (1999)
11. Partee, B.H., ter Meulen, A., Wall, R.E.: Mathematical Methods in Linguistics. Kluwer, Dordrecht (1990)
12. Manning, C.D.: Part-of-Speech Tagging from 97% to 100%: Is It Time for Some Linguistics? In: Gelbukh, A.F. (ed.) CICLing 2011, Part I. LNCS, vol. 6608, pp. 171–189. Springer, Heidelberg (2011)
13. Cer, D., Marneffe, M.C., Jurafsky, D., Manning, C.D.: Parsing to Stanford Dependencies: Trade-offs between speed and accuracy. In: Proceedings of LREC 2010 (2010)
14. Bajwa, I.S., Mumtaz, S., Naveed, M.S.: Database Interfacing using Natural Language Processing. European Journal of Scientific Research 20(4), 844–851 (2008)
15. Bajwa, I.S., Mumtaz, S., Samad, A.: Object Oriented Software Modeling using NLP Based Knowledge Extraction. European Journal of Scientific Research 32(3), 613–619 (2009)
16. Bajwa, I.S., Lee, M.G., Bordbar, B. Translating Natural Language Constraints to OCL. Journal of King Saud University - Computer and Information Sciences 24(2) (June 2012)

Processing Semantic Keyword Queries for Scientific Literature*

Ibrahim Burak Ozyurt[1], Christopher Condit[2], and Amarnath Gupta[2]

[1] Department of Psychiatry
iozyurt@ucsd.edu
[2] San Diego Supercomputer Center
{condit,gupta}@sdsc.edu
University of California, San Diego, La Jolla, CA 92093

Abstract. In this short paper, we present early results from an ongoing research on creating a new graph-based representation from NLP analysis of scientific documents so that the graph can be utilized for answering structured queries on NL-processed data. We present a sketch of the data model and the query language to show how scientifically meaningful queries can be posed against this graph structure.

1 Introduction

Searching for information in scientific literature is one of the most common activities carried out by scientists of all disciplines. In biomedical sciences, the standard practice is to search through the PubMed collection of paper abstracts and the PubMed Central collection of full-text documents. In PubMed, a user can search by both metadata of documents (like author and title) and provide a set of keywords that appear in the abstract (or the body of full text), and retrieve documents that satisfy the metadata conditions and contain the user-specified keywords in the document. It is well recognized that this basic keyword-based document retrieval model is prone to produce semantic false positive, semantic false negative, and semantically ambiguous results. As an illustration, consider the following redacted abstract.

> Synucleinopathies are a group of neurodegenerative disorders, including Parkinson disease, associated with neuronal amyloid inclusions comprised of the presynaptic protein α-synuclein (α-syn); however the biological events that initiate and lead to the formation of these inclusions are still poorly understood. To assess the effects of Aβ peptides and extracellular Aβ deposits on β-syn aggregate formation, transgenic mice (line M83) expressing A53T human α-syn that are sensitive to developing α-syn pathological inclusions were cross bred to Tg2576 transgenic mice that generated elevated levels of Aβ peptides and develop abundant Aβ plaques. In addition these mice were bred

---

* This work is partly supported by the ontology grant NIH/NINDS R01NS058296 and NIH/Neuroscience Blueprint contract HHSN271200800035C for the Neuroscience Information Framework (NIF).

G. Bouma et al. (Eds.): NLDB 2012, LNCS 7337, pp. 378–384, 2012.

to mice with the P264L presenilin-1 knock-in mutation that further promotes $A\beta$ plaque formation. These mice demonstrated the expected formation of $A\beta$ plaques; however despite the accumulation of hyperphosphorylated β-syn dystrophic neurites within or surrounding $A\beta$ plaques, no additional α-syn pathologies were observed.

If a user query is `Lewy body`, this document would not be selected by traditional IR methods - however, not seleceing it would be a semantic false negative since Lewy body is a neuronal inclusion which is a central theme in the documet. On the other hand, if the user query is `transgenic mice`, the document will be selected but this is a semantic false positive (or, at least less relevant) result because while this document does contain the string "transgenic mice", this paper is not *about* transgenic mice, but the central theme of the paper is about the accumulation of amyloid inclusions. Finally, if the the user query is `amyloidogenic proteins`, which occurs in the document only once, this document has little relevance on traditional techniques, but its rank would improve dramatically if it is recognized that $A\beta$ and α-syn are central themes in the document and they are both subcategories of amyloidogenic proteins.

The basic proposition of this short paper is to present a method where NLP-based analysis of documents combined with an ontological analysis, improves the quality of search. In this regard, the paper makes the following contributions: 1) it presents an information representation that captures the outcome of an NLP analysis; 2) it introduces a variant of a keyword query language where a user has the option to specify the semantics of the intention of the retrieval request – this language uses the properties of the structure in (1); and, 3) it presents through examples, a few cases where the impact of using the NLP analysis is analyzed.

2   A Representation Scheme for Analyzed Documents

In this paper, we ignore the metadata of the document including the title, author, keywords, etc. that are part of a scientific publication. We also ignore the structured content of a documents such as its sections and subsection, and focus solely on the text of the document, which we will call the *body*. The body is represented as a *sequence of parse graphs* such that each parse graph corresponds to a sentence. A parse graph is a parse tree that is additionally adorned with a number of cross links across its leaf elements. The parse tree is generated by any sentence parser, in our case, Penn-Treebank style parse tree produced by the McClosky-Charniak reranking parser adapted to biomedical domain [2,6]. Consider the sentence "Several lines of evidence suggest that Abeta peptides and/or extracellular Abeta deposits may directly or indirectly promote intracellular alpha-syn aggregation." The parse graph for a fragment of this sentence is shown in Figure 1. The top part of the graph is the parse tree. The adornments on the tree are produced from different forms of text analysis.

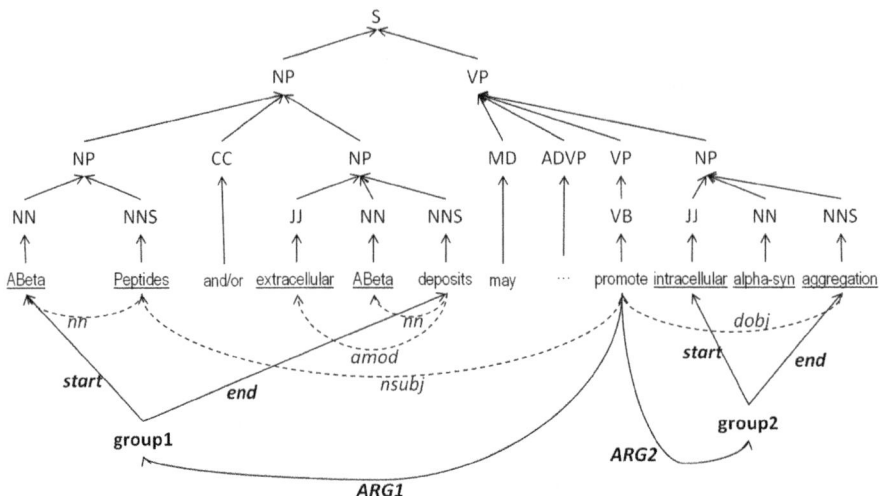

Fig. 1. Parse graphs are tree-structured on top and the tokens are connected by typed, labeled edges. The edges are produced from different NLP analyses.

1. We use the Stanford dependency scheme [4] on the Penn-Treebank style parse trees. This produces a graph over the leaf nodes of the prase trees as partly shown by the dashed edges in Figure 1.

2. We use a *named entity recognition* procedure based on conditional random field (CRF) to detect named entities. The figure shows the recognized entities as underlined text. A fraction of the named entities exist in an ontology. In our specific example, they are are connected to the NIFSTD ontology from the Neuroscience Information Framework [1]. For all such terms, an edge labeled *occurence-of* is created from the term in the parse tree to the ontological term.

3. For certain verbs in the sentence, we perform *semantic role labeling*. The lower part of the figure shows a portion of the semantic role labels associated with the verb "promote". The groups specify the starting and ending tokens of the labeled phrases, and the relationship between the *anchor verb* and the groups are presented as "argument types". In this case, the two arguments determine the promoting entity and the promoted process respectively. We know that the promoting item is composite entity consisting of two entities " Abeta peptides" "extracellular ABeta deposits" because these entities are ontological (inferencing is beyond the scope of this paper); similarly we know from the ontology that the promoted item, viz. aggregation is a nominalized process object.

It is important to recognize that the parse graph may not be semantically minimal, i.e., it may have redundancies. For example, the fact that "extracellular" is an adjective modifier of "deposits" could be inferred from the parse tree itself. But at this point we do not take any steps to normalize the parse graph into a more optimal structure. Further, it is possible to add other analysis to enrich the parsegraphs

Table 1. A portion of the graph-based semantic model is shown. Relationships are binary, and are shown as mappings from the domain to the range.

createdBy:	Doc → Person	isAbout:	Doc → setof(OntoTerm)
createdOn:	Doc → Date	reagentsUsed:	Doc → setof(Reagent)
hasTitle:	Doc → String	reagentUseLoc	Reagent: → setOf(sentenceSequence)
documentType:	Doc → {researchArticle, ...}	problem:	Doc → Statement
hasClaim:	Doc → Claim	statementLocation:	Statement → setOf(sentenceSequence)
claimLoc:	Claim → setOf(sentenceSequence)	Motivation:	Doc → setOf(sentenceSequence)
usesTechnique:	Doc → Technique	Result:	Doc → setOf(sentenceSequence)
techniqueLoc:	Technique → setOf(sentenceSequence)	explanation:	Argument → Argument

and connect parsegraphs of the same document through inter-sentence analyses like coreference resolution. At present however, the parsegraphs are independent and are connected only through common ontological mappings.

3   Structured Semantic Keyword Queries

We base our keyword query language on a Google like query syntax using *words* and *quoted strings* and a set of operators. Among operators of interest, the * operator is a placeholder for zero or more words between two terms; ẽ stands for an approximate query. A modifier is a word from a special vocabulary - one restricts a search to a domain by using the modifier `site:`. While one can add more complexity to this language by extending it with positional predicates and term weights, we would like to focus on a different category of expansions. In general a query expressed in this simple language utilizes neither the semantics of the corpus being queried nor the semantics of the query terms. For example, if the corpus is a scientific document the query does not say whether any query terms should be theme of the document or just appear in the content of the entire document. Similarly, it does not specify if the query words must have a specific semantic sense i.e., if the query is: `cerebellum ''gene expression''`, it is not clear if the query refers to the expression of genes in the *brain region* called "cerebellum" or whether it is about the expression of the *gene* called "cerebellum".

Semantic Model. To capture the semantics of scientific publications, we use a model inspired by prior work on semantic annotation of literature [5,3]. A fraction of the semantic model of a scientific publication can be described through the ontology shown in Table 1. The semantic model uses a combination of classes and properties for generic documents (e.g., createdBy), generic scientific publications (e.g., claim, technique), and domain-specific scientific publications (e.g., reagent). Many class instances map to ontologies. Properties can be both object-valued or literal-valued. All classes and properties are associated with occurrences, which are sets of sentence groups where the respective classes or properties occur in a document. Not shown in Table 1 are the internal representations of arguments and counter-arguments. An argument is a sequence of sentences that establish or refute a viewpoint. An explanation is an argument identified through a causative expression. Finally, a relationship is a 3-tuple where the first and last argument are nominative while the middle argument is typically a verb phrase that bears significant domain semantics. We must point out that unlike an annotation model, where users annotate text regions to terms in a vocabulary or other referenceable entities (or even free text), we use semantic model as a rough guideline for query formulation. The actual mapping of a semantic concept (like motivation) to a text region is done through a set of retrieval rules. For example, the `isAbout` function can be written using various rules, such as creating the union of all ontological terms in the title of the document with all high tf-idf terms in the document; alternately, it can be union of all ontological terms in the title with the highest-probability topic model output of an LDA algorithm. Similar rules can be created for determining the claim, technique or motivation of a document. Many of these rules need to extract sentences and therefore use the parse graph sequences computed from the document. Finding text regions for "motivation", for instance, may use the (simplistic and not-very-accurate) rule to locate the sentence that has a "however" in the first one third of an abstract, such that the sentence has the form `however, <noun-phrase> <negated-verb-phrase>` such that the verb-phrase uses the verbs "studied", "investigated" or "examined". Currently, these rules are being implemented and their effectiveness is being tested.

Query Model. We present the query syntax through a few increasingly complex examples. For these examples, a valid result is a set of relevant documents, with each document presenting sentences that support the selection of the document.

Example 1. *Find all papers on synucleinopathy.* If we interpret this to refer to publications that are on the topic of synucleinopathies and their synonyms, we express it as:

```
document where isAbout(document)= synonyms(subclassOf*
(synucleinopathy)
```

using the `isAbout` function discussed earlier, and the `subclassOf*` operation computes the transitive closure of the `subclassOf` relationship in the ontology.

Example 2. *Which papers have results on membrane proteins related to Parkinson's disease?* This query first needs to find membrane proteins related to Parkinson's disease (PD), and then find publications whose result sections mention these proteins as the central theme. The first subtask may be accomplished by (a) compiling a list L of membrane proteins from a protein ontology, and (b) finding members of L that are associated with PD. One can specify (b) as:

```
V.Arg1 where has-verb(sentence,V), mapsTo(V, Verbnet-concept
(amalgamate-22.2-2)), memberOf(V.Arg1, L), isequivalent(V.Arg2,
synonyms('PD'))
```

where the Arg1 and Arg2 refer to the PropBank style semantic roles of all verbs (like "associate") mapped to a specific word sense corresponding to the Verbnet concept `amalgamate-22.2-2`. The second subtask can now be specified as:

```
neighbors(sentence,2) where sentenceOf(sentence, doc),memberOf
(doc, corpus), isAbout(extract-section('Results',document)) = q1
```

where `q1` is the expression for (b), and the `isAbout` function operates on the "Results" section of the documents in the corpus. The result of the query outputs the matching sentence and its two neighboring sentences.

Example 3. *Find text regions from publications that suggest that oxidized deriva-tives of dopamines play some role in alpha synuclein aggregation.* The following query expression illustrates one way of formulating the query using both the parse tree as well as the semantic role labeling performed on the sentence.

```
tuple(doc.ID, doc.title, neighbors(sentence,1)) where sentenceOf(sentence,
doc), memberOf(doc, corpus), has-structure(sentence, (np1 v1 np2)),
mapsTo(v1, propbank(suggest.01)), has-structure(np2, (np3 v2 np4)),
mapsTo(v2, propbank-frameset.roleset(play.02)), mapsTo(v2.Arg0,
span(synonyms(''alpha synuclein''), synonyms(''aggregation''),4)),
mapsTo(v2.Arg0, synonyms(''dopamine derivative'')),
adj(v2.Arg0, ''oxydized'')
```

Here the `has-structure` predicate recognizes a part of the parse tree struc-ture of the sentence. The first `mapsTo` predicate states that verb v1 maps to the `suggest.01` verb sense as defined in the Propbank structure, and second `mapsTo` states that verb v2 maps to the verb sense `play.02` as defined in the role sets of the frameset defined in Propbank. Similarly the `Arg0` role of v2 maps to the complex concept consisting of some synonym of "alpha-synuclein" and some synomym of "aggregation" within 4 words of each other. Thus a frag-ment like *formation of SNCA aggregates* with match the predicate. Finally, the `adj` predicate states that the `Arg1` role of v2 must have the adjective "oxidized".

4   Conclusion

In this paper, we showed how the current forms of NLP analysis can represented as a graph. We also presented examples of a query language that takes advantage of this structure, and using standard built-in predicates (e.g., `has-structure`, `mapsTo`), can formulate fairly complex semantic questions. We recognize that certain processes like NLP parsing is inherently slow, but in the current paper we only focus on querying the graph structure. Graph construction will greatly improve as better NLP methods are devised. Our work is restricted to creating an information system that can utilize the fruits of NLP research. Future work includes ranking results by a semantic distance of a selected sentence from the user's query. A user interface for the query language is a work in progres.

References

1. Bug, W., Ascoli, G., Grethe, J., Gupta, A., Fennema-Notestine, C., Laird, A., Larson, S., Rubin, D., Shepherd, G., Turner, J., Martone, M.: The NIFSTD and BIRNLex vocabularies: Building comprehensive ontologies for neuroscience. Neuroinformatics 6(3), 175–194 (2008)
2. Charniak, E.: A maximum-entropy-inspired parser. In: ANLP, pp. 132–139 (2000)
3. Ciccarese, P., Ocana, M., Castro, L.G., Das, S., Clark, T.: An open annotation ontology for science on web 3.0. J. of Biomedical Semantics 2(suppl. 2), S4+ (2011)
4. de Marneffe, M.-C., Manning, C.D.: The stanford typed dependencies representation. In: Proc. of the Workshop on Cross-Framework and Cross-Domain Parser Evaluation, pp. 1–8 (2008)
5. Groza, T., Handschuh, S., Möller, K., Decker, S.: SALT - Semantically Annotated LaTeX for Scientific Publications. In: Franconi, E., Kifer, M., May, W. (eds.) ESWC 2007. LNCS, vol. 4519, pp. 518–532. Springer, Heidelberg (2007)
6. McClosky, D., Charniak, E.: Self-training for biomedical parsing. In: Proc. of the 46th Ann. Meeting of the Assoc. for Comput.l Linguistics (Short Papers), pp. 101–104 (2008)

Arabic Rhetorical Relations Extraction for Answering "Why" and "How to" Questions

Jawad Sadek[1], Fairouz Chakkour[2], and Farid Meziane[1]

[1] School of Computing Science & Engineering, University of Salford, Manchester, England
j.sadek@hotmail.com, f.meziane@salford.ac.uk
[2] Computer Engineering Departement, Aleppo University, Aleppo, Syria
feirouzch@yahoo.fr

Abstract. In the current study we aim at exploiting discourse structure of Arabic text to automatically finding answers to non-factoid questions ("Why" and "How to"). Our method is based on Rhetorical Structure Theory (RST) that many studies have shown to be a very effective approach for many computational linguistics applications such as (text generation, text summarization and machine translation). For both types of questions we assign one or more rhetorical relations that help discovering the corresponding answers. This is the first Arabic Question Answering system that attempts to answer the "Why" and "How to" questions.

Keywords: Information Retrieval, Text Mining, Question Answering for Arabic, non-factoid questions, Discourse analysis.

1 Introduction

During recent years the internet has witnessed an explosive growth in the amount of text available. This has motivated researchers in the field of natural language processing to pay special attention to develop systems and tools that are capable of generating direct answers to questions containing specific information the user is looking for instead of a list of relevant documents. These systems are known as question answering (QA) systems.

Researchers in the field of QA have developed many systems for different natural languages. Most of these systems focused on factoid questions like who, what, where and when [1][2][3]. However in [4] a study that was presented for answering why questions in the English language; However, no attempt was made to create a system that can handle why and how to questions for the Arabic language.

In this paper we developed an Arabic text analysis tool which aims at extracting proper Arabic rhetorical relations that can be used within RST structure for automatically discovering answers to why and how to questions in the Arabic language.

2 Methodology for RST-Based Question Answering

Finding answers to *why* and *how to* questions involves searching for arguments in texts. The distinction that RST makes between the part of a text that realizes the

G. Bouma et al. (Eds.): NLDB 2012, LNCS 7337, pp. 385–390, 2012.

primary goal of the writer, termed nucleus, and the part that provides supplementary material, termed satellite, makes it an appropriate tool for analyzing argumentative paragraphs.

Consider the following example, taken from an Arabic website, which explains the method used to extract answers. The text is broken into seven elementary units delimited by square brackets that produced the schema shown in Fig.1.

[حذر بحث علمي حديث من أن قناديل بحر عملاقة قد تهيمن على محيطات العالم]¹ [جراء الصيد الجائر والتغيرات المناخية وأنشطة بشرية أخرى قد تؤدي لفناء الثروة السمكية.]² [وتحذر دراسة أجراها "مركز CSIRO للأبحاث البحرية والجوية" الأسترالي، من نوع ضخم من قناديل البحر، يدعى "نورمورا Normura" وله قابلية النمو ليصل حجمه إلى حجم مصارع سومو ياباني، وقد يزن 200 كيلوغرام، بقطر يبلغ المترين.]³ [ويعمل باحثون على تجربة تقنيات مختلفة للسيطرة على انتشار قناديل البحر،]⁴ [منها استخدام الموجات الصوتية لتفجير تلك المخلوقات التي تتميز بجسم شفاف، وتطوير شبكات خاصة للقضاء عليها.]⁵ [ويعزو الباحثون التزايد الهائل في أعداد قناديل البحر]⁶ [للصيد الجائر للأسماك التي تقتات على قناديل البحر وتتنافس معها على موارد الغذاء.]⁷

[A new research warns that giant jellyfish may dominate world's oceans]¹ [due to overfishing, climate change and other human activities, which could lead to destroy fisheries.]² [A study led by "CSIRO marine and atmospheric research" in Australia warns of giant jellyfish called "Normura" that can grow as big as a sumo wrestler, they weigh up to 200 kilograms and can reach 2 meters in diameter.]³ [Researchers are experimenting with different methods to control jellyfish,]⁴ [some of these methods involve the use of sound waves to explode these creatures that have transparent body and develop special nets to cut them up.]⁵ [Researchers (scientists) said that the cause of this explosion number of jellyfish]⁶ [is the overfishing that feed on small jellyfish and compete with them for their food.]⁷

Given the following question, of *why* type, relating to the above text, we need to extract answer according to the derived schema.

{What is the cause of the increasing number of jellyfish?}   {ما سبب تزايد اعداد قناديل البحر؟}

We notice that question words match the unit6. Furthermore, unit7 provides the cause of the problem stated in unit6, this means that an interpretation relation holds between

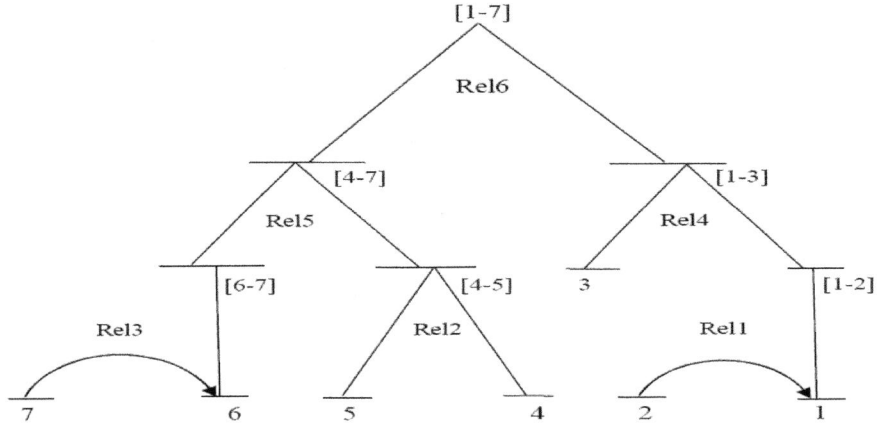

Fig. 1. A scheme representation of the text

unit**7** and unit**6** which is labeled as ***Rel3*** in the schema (Fig.1). Because of the relevance between the question and the unit**6**, we can select the other part of the relation, unit**7**, as a candidate answer.

Now in the case of the following question, belonging to the how-to question type

{How do we control jellyfish blooms?}   { كيف يمكن لنا الحد من انتشار قناديل البحر ؟}

One can observe that unit5 gives some methods for solving the problem mentioned in unit4, so it is concluded that an explanation relation holds between the two units ***(Rel2)***. Since our question corresponds to unit4 we can select the other part of the relation as the answer which is unit5.

3   Text Structure Derivation

3.1   Rhetorical Relations Selection

A set of Arabic rhetorical relations should be complied to use in the application of why and how to question answering systems. We performed an Arabic text analysis with the aim of extracting Arabic relations that lead to answer these types of Arabic questions.

We came up with new four different rhetorical relations (Causal – Evidence – Explanation – Purpose); Table1 shows the definition of the extracted relations according to the constraint stated in [5]. We also selected four rhetorical relations (Interpretation – Base – Result – Antithesis) which were identified by Al-Sanie [6] who used RST for Arabic text summarization.

3.2   Determining the Elementary Units and Rhetorical Relations

As a first step towards automatically deriving the structure of a text, we first need to determine the elementary units of a text and then find the rhetorical relations that hold between these units. Punctuation and cue phrase can play an important role in solving a variety of natural language processing tasks [7]. Cue phrases are the connective (words, phrases, letters…) that are used by writer as cohesive ties between adjacent clauses and sentences which help the reader understanding the text. Our work is based on cue phrases and punctuation as indicators of the boundaries between elementary textual units and to hypothesize the rhetorical relations that hold between them.

In this context, we analyzed Arabic texts and studied Arabic style of linking linguistic units at all levels [8][9] to generate a set of cue phrases. For example the relation Explanation can be identified on the basis of the occurrence of the cue phrases ("وأشار"، "أكد" ، "وقال"....). Also (....."بواسطة"،"عن طريق"،"من خلال") can signal an Evidence relation.

Table 1. Definitions of the extracted Arabic rhetorical relations

Rhetorical Relation	Definitional Element	Definition
Casual	Constraints on the nucleus, N:	None.
	Constraints on the satellite, S:	S presents a cause for the situation presented in N.
	Constraints on the $N + S$:	Without the presentation of S, reader might not know the particular cause of the situation in N.
	The effect:	Reader recognizes the situation presented in S as a cause of the situation presented in N.
Evidence	Constraints on the nucleus, N:	Writer state information may need to be believed by reader (R).
	Constraints on the satellite, S:	Writer presents what supports his claim in N.
	Constraints on the $N + S$:	R's comprehension of S increases his belief of N.
	The effect:	Reader's belief of N is increased.
Explanation	Constraints on the combination of nuclei:	Reader (R) won't comprehend information before reading both nuclei.
	The effect:	R's completely comprehension of writer's notion.
Purpose	Constraints on the nucleus, N:	Presents an activity or event that needs justification to be convinced by reader (R).
	Constraints on the satellite, S:	Writer provides a justification.
	Constraints on the $N + S$:	S presents a purpose for the event stated in N.
	The effect:	R recognizes the aim of the activity presented in N.

Each cue phrase associated with the features mentioned in [7] so that rhetorical relations can be identified based on their values. We assigned some of these features to the extracted cue phrases (Relation, Position, Status, Linking, Regular Expression, and Action) and added the feature question position which specifies the part of a relation that is relative to the question.

The break action feature value specifies where to create an elementary unit boundary in the input text; it takes one of the following values:

● **Normal:** instructs to insert a unit boundary immediately before the occurrence of the cue phrase.

● **Normal_then_comma:** instructs to insert a unit boundary immediately before the occurrence of the cue phrase and another unit boundary immediately after the occurrence of the first comma. If no comma is found before the end of the sentence, a unit boundary is created at the end of the sentence.

● **Normal_then_to:** instructs to insert a unit boundary immediately before the occurrence of the cue phrase and another unit boundary immediately after the occurrence of the first preposition "إلى" (to).

● **Nothing:** no unit boundary is inserted, but assigns an action value to the next cue.

● **End:** instructs to insert a unit boundary immediately after the occurrence of the cue.

The algorithm presented in Fig .2 identifies elementary units of a text and derive rhetorical relations that relate them, where rhetorical relation has the form: *rhet_rel* (relation name, left span, right span). In step 7 the algorithm hypothesizes all possible relations signaled by the cue phrase under scrutiny.

```
Input: A text T.
Output: A list RR of relations that hold among units.
1. RR:= null;
2. Determine the set C of all cue phrases in T;
3. Use the position and action properties for each cue phrases in C
   in order to insert textual boundaries;
4. for each cue ∈ c
5.    rr:= null;
6.        while there is a relation that cue can relate
7.            rr:= rr ⊕ rhet_rel(name(cue), l(cue), r(cue));
8.        end while
9.    RR: = RR ∪ {rr};
10. end for
```

Fig. 2. Algorithm that extracts relations for a given text

4   Evaluation

We developed a system using the Java programming language and performed an experiment similar to the one described in [4]. We selected a number of texts (150-350 words each) taken from Arabic news websites. No corrections have been made to the content of text in case of any mistake (grammatically or orthographic). We distributed the texts to 15 people from different disciplines and were asked to read and extract why and how to-questions which answers could be found in the text. They were also asked to answer the extracted questions. As a result we collected a total of 98 why and how to questions and answers pair.

We posed the 98 questions we collected to our system and compared the answers retrieved by the system to the subject-formulated answers and considered the answer as correct if it matches the answer selected by the subject. The system was able to answer 54 questions correctly (55% of all questions). Table 2 presents the frequency distribution of the rhetorical relations extracted in this work.

If we consider the number of referred questions for Result and Base relations as shown in Table 2, we notice that they lead to three answers only. This is because of the nature of the texts used in our experiment which are news texts. However, these two types of relations can play a much more important role if other types of texts such as organizations' reports are used.

Table 2. Shows relations frequency that led to correct answers

Relation	#correct answers	%correct answers	Relation	#correct answers	%correct answers
Interpretation	12	22.2	Result	1	1.9
Explanation	11	20.3	Purpose	10	18.5
Antithesis	3	5.6	Casual	9	16.7
Evidence	6	11.1	Base	2	3.7

5   Conclusion

Deriving the discourse structure is very important for extracting answers to argument questions. We focused on doing manually Arabic texts analysis with the aim of extracting a set of Arabic rhetorical relations associated with a set of cue phrases that lead to identify the correct answers for why and how to questions. As a first Arabic question answering system that attempts to answer the "Why" and "How to" questions, the evaluation of our experiment gave good preliminary results. In future we plan to study different types of texts which may increase the number of rhetorical relations extracted in this work. Furthermore, we expect that the overall performance of our system will be reduced if longer and more specialized texts such as scientific and economic documents are used. Hence, the use of simple cue phrases need to be expanded to include more complex patterns based on the text structure and the syntax of the Arabic language.

References

1. Kannan, G., Hammoui, A., Al-Shalabi, R., Swalha, M.: A new Question Answering System for Arabic Language. American Journal of Applied Science, 797–805 (2009)
2. Benajiba, Y., Rosso, P., Lyhyaoui, A.: Implementation of the Arabic QA Question Answering System's Computers. In: ICTC (2007)
3. Hammou, B., Abu-salem, H., Lytinen, S., Evens, M.: QARAB: A question answering system to support the Arabic language. In: Workshop on Computational Approaches to Semitic Languages. ACL (2002)
4. Suzan, V., Lou, B., Nelleke, O.: Discourse-based answering of why-questions. Treatment Automatic des languages. Special issue on computational Approaches to Discourse and Document Processing 47(2), 21–41 (2007)
5. Mann, W.C., Thompson, S.: A Rhetorical Structure Theory: Toward a functional theory of text organization (1988)
6. Mathkour, H., Touir, A., Al-Sanea, W.: Parsing Arabic Texts Using Rhetorical Structure Theory. Journal of Computer Science 4(9), 713–720 (2008)
7. Daniel, M.: The Theory and Practice of Discourse Parsing and Summarization. The MIT press, London (2000)
8. Jattal, M.: Nezam al-Jumlah, pp. 127–140. Aleppo University (1979)
9. Haskour, N.: Al-Sababieh fe Tarkeb Al-Jumlah Al-Arabih. Aleppo University (1990)

Facets of a Discourse Analysis of Safety Requirements

Patrick Saint-Dizier

IRIT - CNRS, 118 route de Narbonne, 31062 Toulouse Cedex, France
stdizier@irit.fr

Abstract. In this paper, we provide a linguistic analysis and an implementation of the discourse structure of safety requirements within the framework of the <TextCoop> discourse semantics platform. The main structures are introduced and a detailed evaluation is carried out.

Understanding the contents of requirements is a major objective to be able to automatically organize them, develop traceability, explore overlaps or various types of contradictions, extract their scope and themes, explore semantic and dependency relations between requirements, etc. Even if requirements follow in some context precise authoring recommendations, they often have a complex language structure, with a large diversity of expression modes and nuances. This is in particular the case for a subclass of requirements: safety requirements that include public safety regulations as well as business rules.

Discourse analysis allows the recognition of the backbone of a requirement: how it is organized (in the case of safety requirements: context, conditions, warnings and hints, actions to realize, expectations, goals, etc.). Then, a semantic analysis of the contents of each structure allows to get a more precise information. This article explores the discourse analysis of requirements within the context of the LELIE project which is briefly presented below.

1 The LELIE Project

The main goal of the ANR LELIE project is to produce an analysis, a model and a piece of software based on language processing and artificial intelligence that detects and analyses potential risky situations in technical documents. Risks (health, ecology) Risks emerge in particular when these documents are not correctly written, not updated or contain gaps in safety recommendations. This is a very frequent situation in almost any kind of technical document. We concentrate on procedural documents e.g. [2] and on requirements, e.g. [3], which are, by large, the main types of technical documents. Given a set of procedures over a certain domain produced by a company, and possibly given some domain knowledge (ontology or terminology and lexical), the challenge is to annotate procedures wherever potential risks are identified. Procedure authors are then invited to revise these documents.

Requirements, in particular those related to safety, often exhibit complex language structures which make their understanding and concrete use quite challenging especially in emergency situations. For example, it is quite frequent to observe in an instruction several negations, pronouns, complex cross-references and embedded conditions.

G. Bouma et al. (Eds.): NLDB 2012, LNCS 7337, pp. 391–396, 2012.

Safety requirements are oriented towards action: little space should be left to personal interpretations resulting from gaps or misunderstandings. It is therefore crucial to analyse their structure, get an adequate understanding of their contents and possibly suggest some writing revisions.

To overcome this situation, from original safety requirements, a number of organizations and industries reformulated and customized them for specific activities and users (e.g. the French INRS). Authoring norms have also emerged, e.g. EEC norms for chemical product storage and uses. Besides public regulations, safety requirements also include a number of business requirements associated with certain types of products (e.g. chemicals) or activities. In spite of several levels of reformulation and validation, safety requirements remain quite complex to use.

Our goals in the long term are:

- to detect inappropriate ways of authoring requirements: complex expressions, implicit elements (e.g. verb arguments), references, scoping difficulties, inappropriate terminology or granularity levels, inappropriate style w.r.t. domain authoring norms,
- to make the semantic contents of requiremements more accessible and explicit to users, via contents elicitation,
- to propose a more accurate level of theme extraction and indexing, e.g. to possibly help to reformulate of re-structure sets of requirements, or to improve traceability,
- to develop techniques for (partial) overlap and inconsistency checking, via lexical and textual inference,
- to develop a model for the confrontation of safety requirements with procedures (of maintenance, production, installation) in order, on simple cases, to detect whether safety precautions as stated in requirements, are indeed clearly specified in procedures, at the correct place.

2 Safety Requirements

To manage the difficulty of such a task, we first focus on the discourse analysis of safety requirements from a number of use cases we have defined, characterized by various types of authoring constraints, functional levels, application domains and target users with various levels of expertise. Most requirements make heavy use of domain knowledge: it is here restricted to equipment and product ontologies.

We constructed a development corpus for requirements in French (English is ongoing and is quite similar) from the following types of sources:

- Large public or professionals: *paint scouring, wood lathe uses, electric welding, charging batteries, cooking for groups, working on high positions (roofs), maintenance of agricultural engines,* etc.
- Professionals: *isolated worker, working in cold areas, working in noisy environments, food processing tasks and precautions, oil spill and evacuation plans, X-ray equipments uses, fire in oil refineries, security in nuclear plants, radiation biosafety, various chemical product manipulations, plastic fiber uses,* etc.

Documents come from public administrations (French, EEC, Canadian) and companies (transportation, energy, food processing, etc.). Finally, the corpus concerns several types of activity: standard uses and production, emergency situations, installation, maintenance, storage, etc.

Our corpus is composed at the moment of 460 requirements, i.e. a total of about 550 pages of plain text. Pictures and graphics are not taken into account at this stage, however, they are not so frequent and are essentially meant to improve understanding.

3   Discourse Structures of Requirements

The standard form of a safety requirement is composed of three fields, outlining its business or functional structure. These fields are linguistically distinct but they are often interleaved throughout the whole document:

- **The Context of application** field roughly indicates when the actions that follow are relevant and must be carried out. It may have several forms, which are often quite elliptic: (1) title-subtitles hierarchy, which can have several forms, e.g. general statements (*Isolated worker, working in cold areas*), or (2) conditional expressions (*if you use nitric acid, when using* ...). Besides these central elements, this field may also contains definitions, various types of restrictions of application, general purpose information, reformulations, comments, references to legal or business information, and a number of general purpose warnings and advice.
- **The Actor** field is in general very simple. For complex requirements, several types of actors may be required, in that case, each instruction indicates who is doing what.
- **The Action** field is basically composed of instructions and specific warnings and advice. These may be associated with 'low level' discourse structures such as definitions, illustrations, reformulations, concessions, etc. For requirements covering complex activities, this field may be decomposed into several functional fields (identified via subtitles) which are treated sequentially, in particular: (1) actions per equipment or product used, or (2) actions for a given equipment per type of use, e.g.: direct use, maintenance, cleaning, storage, training and supervision. For more casual users, a 'you should do / you must never do' classification is sometimes adopted. Some Action fields end by a 'Remember' or 'Advice Plus' section.

In the remainder of this paper, we investigate the linguistic structure of the main discourse structures we have identified in safety requirements. These structures are implemented as rules in the <TextCoop> environment, our platform for discourse semantic analysis [9,10]. It is used to process the discourse structure of various types of documents [1]. <TextCoop> is based on a linguistic and logic-based approach to discourse analysis: it provides a logic-based language (Dislog) and an environment for authoring rules and developing lexical resources [8]. It also allows the integration of knowledge and reasoning within the rules, which is an important feature to resolve structure ambiguities, accurate content analysis and scoping problems [11]. Source documents are in general in Word, Visio, or XML, sometimes in pdf.

4   The Discourse Structure of Safety Requirements

In a theory of discourse such as the RST [4, 5, 6, 7], relations are binary: a satellite is bound to a kernel (e.g. an illustration with what it illustrates). Identifying and precisely characterizing these structures is in general quite challenging. Much better results can be expected on specific textual genres with limited complexity, which is the case for procedures and requirements. At the moment the following discourse structures have been identified and characterized in detail for requirements, these have been adapted from our work on explanation analysis [1]:

- **title** (9 rules), basically typographic considerations are used, titles are often very elliptic with missing verb or object. If not made explicit, the title hierarchy is difficult to identify. It gives the general purpose of the requirement at stake.
- **instructions** (15 rules) are based on the fact that they contain an action verb often in the infinitive or imperative form; they express the 'what to do' in requirements,
- **advice** (27 rules) and **warnings** (23 rules) are complex structures from argumentation theory, they are formed of a statement and one or more supports [10,11], a warning indicates a danger or the need to care very much about an action, the support(s) indicates the risks if not carefully realized. An advice is more optional in nature: it indicates ways to improve the result of an action or ways to realize it e.g. more comfortably; its support(s) indicates the level and nature of the gain. Advice therefore introduce a specific subclass of requirements, with a different orientation.
- **illustration** (20 rules) is often associated with an element within an instruction, to better characterize it,
- **restatement** (12 rules) is a way to shed a different light on a statement,
- **purpose** (9 rules), often embedded into instructions, develops low-level motivations,
- **condition** (13 rules) ranges over one or a few instructions, it defines restrictions of application or use,
- **circumstance** (15 rules) indicates the environment in which an action must be carried out,
- **frame** (17 rules) is more general than condition or circumstance: it scopes over the entire document or a whole section and specifies the context of application,
- **concession** (8 rules) is not very frequent, it is used to indicate an alternative to an action, it is often very constrained,
- **elaboration** (19 rules) is a high-level discourse relation, it is also not very frequent in requirements (but more frequent in procedures), it is mainly used to focus on an action and to develop it when its realization may be difficult for some users,
- **definition** (9 rules) appears mainly in the Context field of the requirement,
- **goal** (14 rules) may be high level and plays the role of a title, it may also be associated with a group of instructions where it indicates their objectives, from that point of view it is stronger than the purpose relation,
- some forms of **causes** (11 rules) which develops forms close to argument supports but in structures others that warnings and advice.

Here is a short example that illustrates the kind of analysis which is realized:

<requirement> <title> Monitoring safe operation of industrial trucks </title>
<warning> Working practices should be monitored by a responsible supervisor to ensure that safe systems of work are followed. </warning>
<purpose> This list is a basic guide - <elaboration> it is not exhaustive and is not intended to be a substitute for the guidance and training. </elaboration> </purpose>
<subtitle1> Operators should always: </subtitle1>

\<instruction\> observe floor loading limits - \<restatement\> find out the weight of the laden truck. \</restatement\> \</instruction\>
\<instruction\> plan their way first. \</instruction\>
\<warning\> ensure the load is not wider than the width of the gangways.\</warning\>
\<instruction\> watch out for pedestrians and bystanders. \</instruction\> ....
\<illustration\>(see paragraph 390-394)\</illustration\>. \</instruction\> ...
\<subtitle1\> Operators should never: \</subtitle1\>
\<warning\> lift loads that exceed the truck's rated capacity. \</warning\>
\<warning\> travel with a bulky load obscuring vision. \</warning\>
\<warning\> travel on soft ground \<concession\> unless the industrial truck is suitable for this purpose \</concession\>. \</warning\> ....
\<subtitle1\> Remember: \</subtitle1\>
\<warning\> never allow unauthorised people to operate the industrial truck. \</warning\>
\</requirement\>

5   Results and Performances

An important feature of our description is that it mainly requires domain indepen-dent linguistic resources. Discourse rules, over various domains, require the following resources (due to space limitations a subset of relations are mentioned below):

structure	discourse marker	connector	negation	pronouns	prepo-sition	punctuation, typography
instruction						X
advice concl.				X		
advice support		X		X		
warning concl.			X	X	X	
warning support		X	X	X	X	
illustration	X					X
restatement	X					X
purpose		X				X
condition		X				X
circumstance		X				X

structure	modals, auxiliaries	action verbs and verb classes (7100)	adverbs (75)	expr. with polarity (290)	adhoc (360)	know-ledge
instruction	X	action verbs	X			
advice concl.	X	communication	X		X	
advice support	X	change verbs		positive	X	
warning concl.	X	communication	X		X	
warning support	X	change verbs		negative	X	
illustration	X	X			X	X
restatement	X	epistemic			X	X
purpose					X	X
condition						X
circumstance	X				X	

From a test corpus (62 requirements, about 31 500 words), with the same distribution as for the development corpus, we have the following coverage and accuracy rates, expressed in terms of recall and precision. Our strategy was to favor precision over recall. The following figures are based on a comparison of the system performances w.r.t. manual annotations. A structure is correct if it is correctly identified and well delimited.

structure	number manually annotated	precision (%)	recall (%)
instruction	554	98	96
advice concl.	49	87	76
advice support	42	91	82
warning concl.	112	91	88
warning support	88	93	90
illustration	38	92	87
restatement	47	86	79
purpose	101	89	86
condition	168	93	82
circumstance	121	95	92

6   Perspectives

We have here developed the analysis of the discourse structure of safety requirements with the purpose of improving risk analysis in the industrial procedures they apply to. This is the first step before realizing contents and coherence controls. This analysis should improve requirement understanding, and to explore overlaps or contradictions, relations between requirements and how to enhance traceability.

Acknowledgements. This project is supported by the French ANR.

References

1. Bourse, S., Saint-Dizier, P.: The language of explanation edicated to technical documents. Syntagma 27 (2011)
2. Delin, J., Hartley, A., Paris, C., Scott, D., Vander Linden, K.: Expressing Procedural Relationships in Multilingual Instructions. In: Proceedings of the IWNLG7, USA, (1994)
3. Hull, E., Jackson, K., Dick, J.: Requirements Enginnering. Springer (2011)
4. Mann, W., Thompson, S.: Rhetorical Structure Theory: Towards a Functional Theory of Text Organisation. TEXT 8(3), 243–281 (1988)
5. Marcu, D.: The Rhetorical Parsing of Natural Language Texts. ACL (1997)
6. Rösner, D., Stede, M.: Customizing RST for the Automatic Production of Technical Manuals. In: Dale, R., Hovy, E., Rösner, D., Stock, O. (eds.) IWNLG 1992. LNCS, vol. 587, pp. 199–214. Springer, Heidelberg (1992)
7. Saito, M., Yamamoto, K., Sekine, S.: Using Phrasal Patterns to Identify Discourse Relations. ACL (2006)
8. Takechi, M., Tokunaga, T., Matsumoto, Y., Tanaka, H.: Feature Selection in Categorizing Procedural Expressions. In: IRA 2003 (2003)
9. Saint-Dizier, P.: Programming in DISLOG: some foundational elements. In: LTC 2011, Poznan (2011)
10. Saint-Dizier, P.: Processing Natural Language Arguments with the <TextCoop> Platform. Journal of Argumentation and Computation 3-1 (2012)
11. Walton, D., Reed, C., Macagno, F.: Argumentation Schemes. Cambridge University Press (2008)

Author Index